Numerical Methods in Mechanics of Materials, Third Edition

With Applications from Nano to Macro Scales

Numerical Methods in Mechanics of Materials, Third Edition

With Applications from Nano to Macro Scales

Ken P. Chong
Arthur P. Boresi
Sunil Saigal
James D. Lee

CRC Press
Taylor & Francis Group
Boca Raton London New York

CRC Press is an imprint of the
Taylor & Francis Group, an **informa** business

CRC Press
Taylor & Francis Group
6000 Broken Sound Parkway NW, Suite 300
Boca Raton, FL 33487-2742

First issued in paperback 2019

ISBN-13: 978-1-138-71916-3 (hbk)
ISBN-13: 978-0-367-88625-7 (pbk)

Library of Congress Cataloging-in-Publication Data

Names: Chong, K. P. (Ken Pin), 1942- author.
Title: Numerical methods in mechanics of materials : with applications from nano to macro scales / Ken P. Chong, Arthur P. Boresi, Sunil Saigal, James D. Lee.
Description: Third edition. | Boca Raton : Taylor & Francis, CRC Press, 2017. | Includes bibliographical references and indexes.
Identifiers: LCCN 2017051559| ISBN 9781138719163 (hardback) | ISBN 9781315146010 (ebook)
Subjects: LCSH: Materials--Mechanical properties--Mathematics. | Strength of materials--Mathematical models. | Numerical analysis.
Classification: LCC TA404.8 .C535 2017 | DDC 620.1/12301518--dc23
LC record available at https://lccn.loc.gov/2017051559

Visit the Taylor & Francis Web site at
http://www.taylorandfrancis.com

and the CRC Press Web site at
http://www.crcpress.com

Contents

3 Finite difference methods 37

4 The finite element method 73

Preface

The widespread use of digital computers and simulation has had a profound effect in engineering and science. A realistic and successful solution of an engineering problem usually begins with an accurate physical model of the problem and a proper understanding of the assumptions employed. With advances in big data, cloud computing, simulation-based engineering and sciences, computer hardware, and appropriate software, we can model and analyze complex physical systems and problems. However, efficient and accurate use of numerical results obtained from computer programs requires considerable background and advanced working knowledge to avoid blunders and the blind acceptance of computer results. In this book, we attempt to provide the essential background and knowledge necessary to avoid these pitfalls. In particular, we consider the most commonly used numerical methods employed in the solution of physical problems. Except for relatively simple cases in which exact closed form solutions exist, the successful use of numerical methods depends heavily upon the use of digital computers, ranging in power from microcomputers to supercomputers, depending on the complexity of the problem.

This is a unique and useful book, offering in-depth coverage of the numerical methods for scales from nano to macro, in nine self-contained chapters with extensive applications/problems and up-to-date references and bibliographies. The topics covered include the following:

- Trends and new developments in simulation and computation
- Weighted residuals methods
- Finite difference methods
- Finite element methods
- Finite strip/layer/prism methods
- Boundary element methods
- Meshless methods
- Multiphysics in molecular dynamics simulation
- Multiscale modeling from atomistic theory to continuum theory

The mathematical formulation of these methods is perfectly general. However, in this book, the applications deal mainly with the field of engineering mechanics and materials. An extensive list of references is provided.

Practicing engineers and scientists should find this book very readable and useful. Students may use the book both as a reference and as a text. This book is a sequel to an earlier book by three of the authors (*Elasticity in Engineering Mechanics*, published by Wiley, New York, 2011).

Ken P. Chong
Arthur P. Boresi
Sunil Saigal
James D. Lee

Authors

Ken P. Chong, PE was the former interim division director, engineering advisor, and director of Structural Systems, Mechanics and Materials at the National Science Foundation (NSF), 1989–2009. Currently, he is a research professor at George Washington University. He earned a PhD in engineering mechanics from Princeton University. He specializes in solid mechanics and materials, computational mechanics, nanomechanics, smart structures, and structural mechanics. He has been the principal investigator of over 20 federally funded research projects (from NSF, DOD, DOE, DOI, etc.). He was a senior research engineer with the National Steel Corp. for 5 years after graduation from Princeton. After that, he was a professor for 15 years at a state university. He has published 200 refereed papers; authored several textbooks on engineering mechanics, including *Elasticity in Engineering Mechanics* and *Approximate Solution Methods in Engineering Mechanics*; and edited 10 books, including *University Programs in Computer-Aided Engineering, Design, and Manufacturing*, *Materials for the New Millennium*, and *Modeling and Simulation-Based Life-Cycle Engineering*. He was the editor of the Elsevier *Journal of Thin-Walled Structures* from 1987 to 2013. He is coeditor of the UK *Journal of Smart and Nano Materials*, a CRC/Spon book series on structures, and serves on several editorial boards. He has given over 50 keynote lectures at major conferences, the Mindlin, Sadowsky, and Raouf Lectures and has received awards, including the fellow of ASME, American Academy of Mechanics, SEM, and USACM; Edmund Friedman Professional Recognition Award; Honorary Doctorate, Shanghai University; Honorary Professor, Harbin Institute of Technology; NCKU Distinguished Alumnus, Distinguished Member, ASCE; NSF highest Distinguished Service Award; and the ASME Belytschko Mechanics Award. He was the ASME Thurston Lecturer for 2014.

Arthur P. Boresi was professor emeritus in the Mechanical Science and Engineering Department at the University of Illinois at Urbana–Champaign, where he taught for more than 20 years and later was the head of the Department of Civil and Architectural Engineering at the University of Wyoming in Laramie. He is currently professor emeritus in the Civil and Architectural Engineering Department of the University of Wyoming. He has published over 200 refereed papers and several books, including *Elasticity in Engineering Mechanics*, *Advanced Mechanics of Materials*, *Engineering Statics*, and *Engineering Dynamics*. He had chaired and organized national conferences for ASCE and other societies. He is a fellow of AAM, ASME and ASCE.

Sunil Saigal is a distinguished professor in the Department of Civil and Environmental Engineering and former dean of engineering at New Jersey Institute of Technology as well as a former NSF program director. His research in the area of computational mechanics has spanned numerous subdisciplines and industrial applications. Much of his research has been focused on interactions with the industry, and these contributions have included development of boundary element shape optimization, in collaboration with United Technologies; formulations for powder packing, in collaboration with Alcoa and DuPont; development of computational models for nonlinear soil behavior, in collaboration with ANSYS; cohesive element formulations for postcrack behavior of glass–polymer composites, in collaboration with DuPont; explicit algorithms for high-velocity impact, in collaboration with Naval Surface Warfare Center; and computational simulations of acetabular hip component, in collaboration with the University of Pittsburgh Medical Center. These interdisciplinary efforts involved diverse fields ranging from anatomy to computer science to materials and solid mechanics. Recognition of his work has occurred through awards/honors, including Leighton and Margaret Orr Award for Best Paper, ASME, 2004; George Tallman Ladd Research Award, Carnegie Mellon University, 1990; Presidential Young Investigator Award, NSF, 1990; Ralph R. Teetor Award, Society of Automotive Engineers, 1988. He is the coauthor of five books on engineering mechanics and holds a patent for a method of manufacturing hot rolled I-beams. He is the author of over 100 peer-reviewed articles in archival journals. He has served on the editorial boards of several journals, including *International Journal for Numerical Methods in Engineering*, *Engineering with Computers*, and *International Journal for Computational Civil and Structural Engineering*. He is a fellow of numerous societies including, ASCE, ASME, and AAAS.

James D. Lee is a professor of engineering and applied science in the Department of Mechanical and Aerospace Engineering, George Washington University, Washington, DC. He was an associate professor at West Virginia University and University of Minnesota. He worked at General Tire and Rubber Company for one year, National Institute of Standard and Technology for four years, and NASA/Goddard Space Flight Center for one year. He earned a PhD degree from the Department of Mechanical and Aerospace Engineering, Princeton University, in 1971. He has been doing research in many fields, including nanoscience, multiscale modeling, mechanobiology, microcontinuum physics, continuum mechanics, fracture mechanics, finite element method, meshless method, optimal control theory, robotics, etc. He has been the principal investigator of federally funded research projects from NASA, NSF, and DOT. He has published 120 journal papers, 20 book chapters, two textbooks (*Meshless Methods in Solid Mechanics and Elasticity in Engineering Mechanics*), and numerous conference papers and presentations. He has received the Distinguished Researcher Award from the School of Engineering and Applied Science, The George Washington University. He is a fellow of ASME and honorary fellow of Australian Institute of High Energetic Materials.

Chapter 1

The role of numerical methods in engineering

1.1 INTRODUCTION

The widespread use of digital computers and simulation has had a profound effect on engineering and science in the current 4th Industrial Revolution mainly due to digital transformation (Kemp, 2016). With advances in big data (Zikopoulos and Eaton, 2011), cloud computing (Armbrust et al., 2009; Mell and Grance, 2011; Zhang et al., 2010), simulation-based engineering and sciences (Jain, 1990; Oden et al., 2006), computer hardware, and appropriate software, we can model and analyze complex physical systems and problems. However, efficient and accurate use of numerical results obtained from computer programs requires considerable background and advanced working knowledge to avoid blunders and the blind acceptance of computer results.

The solution of an engineering problem generally requires the description of the response of a material body (computer chips, machine part, structural element, or mechanical system) to a given excitation (such as force). In an engineering sense, this description is usually required in numerical form, the objective being to assure the designer or engineer that the response of the system will not violate design requirements. These requirements may include the consideration of deterministic and probabilistic concepts (Thoft-Christensen and Baker, 2012; Wen, 1984; Yao, 1985) and system engineering (Hazelrigg, 1996). In a broad sense, the numerical results are predictions as to whether the system will perform as desired. For example, the solution to the elasticity problem may be obtained by a direct numerical process (numerical stress analysis) or in the form of a general solution (which ordinarily requires further numerical evaluation).

The successful solution of a complex engineering problem begins with an accurate physical model of the problem. In turn, this physical model is transformed into a mathematical model. The solution of the mathematical model is usually obtained by numerical methods that are by definition approximate. Thus, the accuracy, convergence, data set such as material properties, software, model, and other aspects need to be verified and validated (Kleijnen, 1995; Oden et al., 2006; Pardo, 2016). The success of these numerical techniques rests in turn upon high-speed digital computers and quality of the software. Other engineering aspects such as durability and sustainability also need to be considered (Chong et al., 2013; Monteiro et al., 2001).

The finite difference method (FDM) (Mitchell and Griffiths, 1980; Pletcher et al., 2012) and the finite element method (FEM) (Bathe, 2006; Cook, 2007; Hughes, 2012; Zienkiewicz and Taylor, 2005) are widely used numerical techniques. These methods are classified as *domain methods*, in that the engineering system is analyzed either in terms of discretized finite grids (FDMs) or finite elements (FEMs) throughout the entire region of the system (body). Another method that has emerged as a powerful tool is the boundary element method (BEM) (Brebbia, 1984; Brebbia and Connor, 1989; Cruse, 1988; Rizzo, 1967).

In certain problems, this method has some distinct advantages over FDM and FEM, for several reasons. In particular, a discretization of *only* the boundary of the domain of interest is necessary for BEM—hence the name *boundary element method.*

All three of the methods noted previously, and a variety of other specialized techniques, provide powerful means of treating complex boundary value problems of engineering. In a particular case, depending on requirements, one of these methods may be more efficient than the others in generating a solution. For example, it may be more advantageous to use BEM for certain classes of linear problems characterized by infinite or semi-infinite domains, stress concentrations, three-dimensional structural effects, and so on (Beskos, 1989; Brebbia and Connor, 1989).

A fourth method, the finite strip method (FSM) (and the associated finite layer method [FLM] and finite prism method [FPM]) (Cheung, 2013), falls somewhere between domain methods (FDM and FEM) and boundary methods (BEM) in that it reduces the dimensions of the problem before discretization of the domain. Other hybrid methods that combine the advantages of several formulations have been proposed. For example, Golley and Grice (1990), Golley et al. (1987), and Petrolito et al. (1990) combined FEM and FSM to study plate bending problems.

In this book, we consider FDMs (Chapter 3), FEMs (Chapter 4), finite strip (finite layer and finite prism) methods (Chapter 5), boundary elements (Chapter 6), and meshless methods (Chapter 7). In Chapter 2, we consider the fundamentals of approximate methods of analysis, including boundary solutions in terms of weighted residual methods (WRM). The various approximation techniques (FDM, FEM, FSM, BEM, etc.) may be represented as special cases of weighted residual formulations. They can be studied using the concepts of approximation and weighting that are fundamental to WRM and that are popular with engineers and mathematicians.

The rest of the chapters cover molecular dynamics, multiphysics problems, and multiscale methods (Boresi et al., 2011; Liu et al., 2010). In Chapter 8, molecular dynamic simulation is discussed, which has been accepted as a widely employed simulation technique for the study of material behaviors at nanoscale. There are certain physical phenomena for which the experiments just cannot possibly be executed, yet by allowing us to obtain numerical results, it provides us with an in-depth knowledge in the study of materials. The multiphysics approaches reach a new height for modeling and simulation. It opens up a new opportunity to connect engineering applications with basic science. In Chapter 8, the general governing equation of nonequilibrium molecular dynamics, covering thermomechanical–electromagnetic coupling effects, is derived. It includes an introduction of Maxwell's equations and Lorentz force at nanoscale, reformulation of Nosé–Hoover thermostat and proof of the objectivity of Nosé–Hoover thermal velocity and virial stress tensor. A sample problem designed to study the coupling effect of temperature and magnetic field has been solved by the Heun method.

In Chapter 9, it is an established fact that multiscale modeling is an effective way of studying material behaviors over a realistic length/time scale. Molecular dynamics provides a solid foundation for the bottom–up sequential multiscale modeling, from which one may calculate material parameters, including the elastic constants, thermal conductivity, specific heat, and thermal expansion coefficients, for thermoelasticity. With these preparations, we further present the newly formulated concurrent multiscale theory. The key challenge in constructing a concurrent multiscale theory hinges at the formulation of the interfacial conditions, which determine the communication between the atomic region and genuine continuum region. The interfacial conditions are constructed naturally by anchoring finite element nodes at centroids of clusters of atoms simulated by molecular dynamics. Also, the amount of heat flowing out from a node must be equal to that coming into the corresponding cluster of atoms and vice versa. In this way, a concurrent multiscale modeling theory from atoms to genuine continuum is constructed. To test the capability of this multiscale theory, we conducted crack propagation simulations and observed crack branching and closure. A crack preexisting in the continuum region can propagate into the critical atomic region. None of these simulations needs any fracture criterion.

1.2 FIELDS OF APPLICATION

Approximation and numerical methods (Moin, 2010) are employed generally in all fields of engineering, mathematics, and science. For example, textbooks have been written dealing with applications in specialized subjects such as elasticity, plasticity, porous media flow, structural mechanics, fluid mechanics, aerodynamics, and so on (Boresi et al., 2011; Holzer, 1985; Lubliner, 2008; Nelson, 1989; Pletcher et al., 2012). In this book we are concerned mainly with formulations of common and new numerical methods and with recent developments in computer implementations of these methods. Selected references are given, with an emphasis on engineering applications in the solid mechanics and material area.

1.3 FUTURE PROGRESS AND TRENDS

Papers on the applications of the theory of elasticity to engineering problems form a significant part of the technical literature in solid mechanics (Boresi et al., 2011; Chong and Davis, 1999; Dvorak, 1999). Many of the solutions presented in current papers employ numerical methods and require the use of high-speed digital computers. This trend is expected to continue into the foreseeable future, particularly with the widespread use of microcomputers and workstations as well as the increased availability of supercomputers (Fosdick, 1996; Louder, 1985). For example, FEMs have been applied to a wide range of problems, such as plane problems, problems of plates and shells, and general three-dimensional problems, including linear and nonlinear behavior, and isotropic and anisotropic materials. Furthermore, through the use of computers, engineers have been able to consider the optimization of large engineering systems (Atrek et al., 1984; Kirsch, 2012; Zienkiewicz and Taylor, 2005), such as the Space Shuttle. In addition, computers have played a powerful role in the fields of computer-aided design and computer-aided manufacturing (Ellis and Semenkov, 1983), as well as in *virtual testing* and model-based simulation (Chong et al., 2002; Fosdick, 1996). Recent advances in computation, in conjunction with synergistic combination with data-based models (Garboczi et al., 2000), make virtual testing a valuable tool in designer materials with optimized properties.

The FEM has limitations in solving problems with large deformations, crashes, fracture propagation, penetration, and other moving boundary problems. Mesh-free and particle methods, including smoothed particle hydrodynamics methods, mesh-free Galerkin methods, and molecular dynamics methods (Li and Liu, 2002), are especially useful for these classes of problems. Recently, hypersingular residuals in the mesh-based BEM have been extended to the meshless boundary node method with good potential for solving a wide range of problems efficiently (Chati et al., 2001).

In the past, engineers and material scientists have been involved extensively with the characterization of given materials. With the availability of advanced computing, along with new developments in material sciences, researchers can now characterize processes, design, and manufacture materials with desirable performance and properties. Using nanotechnology (Bhushan, 2010; Chong, 2004; Reed and Kirk, 1989; Roco, 2011; Siegel et al., 1999; Timp, 1999; Wolf, 2015), engineers and scientists can build *designer materials* molecule by molecule.

Tremendous progress is also being made in the measurement and application of microforces (see, e.g., Bowen et al., 1998; Saif and MacDonald, 1996). Advances have been made in instrumentation, such as the atomic force microscope, scanning electron microscope, high-resolution transmission electron microscope, and surface roughness evolution spectroscope. Figure 1.1 summarizes the gauge length and strain resolution of various instruments, spanning scales from 10^{-9} to 10^{0} m. One of the challenges is to model short-term microscale material behavior through mesoscale and macroscale behavior into long-term structural systems performance (Figure 1.2). Accelerated tests to simulate various environmental forces and impacts are needed. Supercomputers and/or workstations used in parallel are useful tools (1) to solve this multiscale and size-effect problem by taking into account the large number of variables and unknowns to project microbehavior into infrastructure systems performance and (2) to model or extrapolate short-term test results into long-term life-cycle behavior.

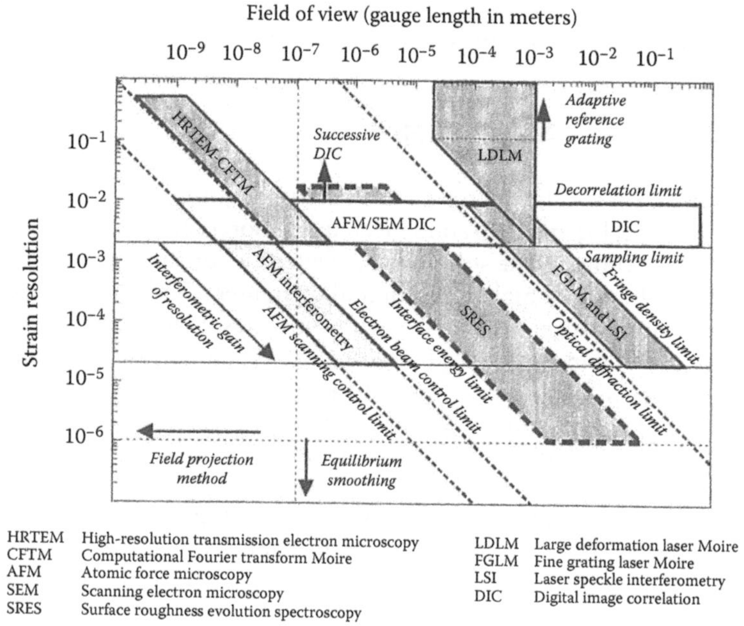

Figure 1.1 Map of deformation-measurement techniques. (Courtesy of K.-S. Kim, Nano & Micromechanics Laboratory, Brown University.)

Materials		**Structures**	**Infrastructure**	
Nano level (10^{-9}) m	Micro level (10^{-6})	Meso level (10^{-3})	Macro level (10^{+0})	Systems level (10^{+3}) m
Molecular scale	*Micrometers*		*Meters*	*Up to kilometer scale*
Nano mechanics Self-assembly Nanofabrication	Micro mechanics Micro structures Smart materials	Meso mechanics Interfacial structures Composites	Beams Columns Plates	Bridge systems Lifelines Airplanes

Figure 1.2 Scales in materials and structures. (From Boresi, A.P., Chong, K.P., Lee, J.D., *Elasticity in Engineering Mechanics*, 3rd ed., Wiley, New York, 2011.)

Currently, the *transcendent technologies* (Chong and Davis, 1999) of our time are as follows:

- *Microelectronics:* Moore's law, the doubling of computer capabilities every 18 months since 1970; unlimited scalability
- *Information technology:* confluence of computing and communications; Internet of Things (IoT)
- *Nanotechnology:* engineering at the molecular scale (Roco, 2011)
- *Biotechnology:* molecular secrets of life.

Engineering mechanics and the associated numerical solution form the backbone of these transcendent technologies.

We stand at the threshold of an explosive expansion of computer applications in numerical solutions of problems in all areas of engineering (Beskos, 1989; Nelson, 1989; Pletcher et al., 2012). Major advances in finite element model technology have been outlined by Belytschko (1989) and Belytschko et al. (2013). Numerical methods for solution of finite element systems are described by Wilson (1989). In addition, a broad range of topics on the application of modern computers to the solution of structural engineering problems is discussed in the proceedings of the 1989 Structures Congress (Nelson, 1989). These topics include artificial intelligence, parallel processing, optimization, knowledge-based systems, and computer-aided analysis and design. The present status and possible future developments in BEMs in structural analysis are discussed by Banerjee and Wilson (2005), Cruse (1988), and Beskos (1989). It is anticipated that future work in meshless and hybrid methods will play a greater role in approximation methods (Chati et al., 2001; Chong et al., 2002; Costabel and Stephan, 1988; Golley and Grice, 1990).

Another major development is the Simulation-based Engineering Science (SBES), which plays a key role in the current or 4th Industrial Revolution (Oden et al., 2006). An example is the Material Genome Initiative the US President announced in 2011 using vast databases and advanced computer modeling/simulation (cf. SBES) to design new material systems *in half the time* required (see https://www.whitehouse.gov/mgi). Other major trends include viable multiscale modeling, nanotechnology, biotechnology, advanced manufacturing, 4D printing, automation and robotics, Internet of Things, big data, artificial intelligence, cloud computing, and many others that affect the productivity, efficiency, and quality of life.

REFERENCES

Armbrust, M., Fox, A., Griffith, R., Joseph, A. D., Katz, R. H., Konwinski, A., Lee, G., Patterson, D. A., Rabkin, A., Stoica, I., and Zaharia, M. 2009. *Above the Clouds: A Berkeley View of Cloud Computing*, Electrical Engineering and Computer Sciences, University of California at Berkeley, Berkeley, CA. Technical Report No. UCB/EECS-2009-28.

Atrek, E., Gallagher, R. H., Ragsdell, K. M., and Zienkiewicz, O. C. (eds.) 1984. *New Directions in Optimal Structural Design*, Wiley, New York.

Banerjee, P. K., and Wilson, R. B. 2005. *Developments in Boundary Element Methods*, Elsevier, London.

Bathe, K. J. 2006. *Finite Element Procedures*, Prentice Hall, Upper Saddle River, NJ.

Belytschko, T. 1989. Major advances in finite element model technology, in *Proc. Sessions Related to Computer Utilization at Structures Congress '89*, Nelson, J. K., Jr. (ed.), American Society of Civil Engineers, Reston, VA, pp. 11–20.

Belytschko, T., Liu, W. K., Moran, B., and Ekhodary, K. 2013. *Nonlinear Finite Elements for Continua and Structures*, Wiley, New York.

Beskos, D. E. 1989. *Boundary Element Methods in Structural Analysis*, American Society of Civil Engineers, Reston, VA.

Bhushan, B. (ed.) 2010. *Springer Handbook of Nanotechnology*, Springer Science & Business Media, New York.

Boresi, A. P., Chong, K. P., and Lee, J. D. 2011. *Elasticity in Engineering Mechanics*, 3rd ed., Wiley, New York.

Bowen, W. R., Hilal, N., Lovitt, R. W., and Wright, C. J. 1998. Direct measurement of the force of adhesion of a single biological cell using an atomic force microscope, *Colloids Surf. A Physicochem. Eng. Aspects*, **136**, 231–234.

Brebbia, C. A. 1984. *The Boundary Element Method for Engineers*, Wiley, New York.

Brebbia, C. A., and Connor, J. J. (eds.) 1989. *Advances in Boundary Elements*, Springer-Verlag, New York.

Chati, M. K., Paulino, G. H., and Mukherjee, S. 2001. The meshless standard and hypersingular boundary node methods, *Int. J. Numer. Methods Eng.*, **50**, 2233–2269.

Cheung, Y. K. 2013. *Finite Strip Method*, Elsevier, Amsterdam, the Netherlands.

Chong, K. P. 2004. Nanoscience and engineering in mechanics and materials, *J. Phys. Chem. Solids*, **65**, 1501–1506.

Chong, K. P., and Davis, D. C. 1999. Engineering mechanics and materials research in the information technology age, *Mechanics*, **28**(7/8), 1–3.

Chong, K. P., Saigal, S., Thynell, S., and Morgan, H. S. (eds.) 2002. *Modeling and Simulation-Based Life-Cycle Engineering*, Spon, New York.

Chong, K. P., Wang, X., and Chong, S.-L. 2013. Translational research in sustainability and mechanics, *Int. J. Sustain. Mater. Struct. Syst.*, **1**(2), 105–122.

Cook, R. D. 2007. *Concepts and Applications of Finite Element Analysis*, Wiley, New York.

Costabel, M., and Stephan, E. P. 1988. Coupling of finite elements and boundary elements for transmission problems of elastic waves in R^3, in *Advanced Boundary Element Methods*, Cruse, T. A. (ed.), Springer-Verlag, New York, pp. 117–124.

Cruse, T. A. (ed.) 1988. *Advanced Boundary Element Methods*, Springer-Verlag, New York.

Dvorak, G. J. (ed.) 1999. Research trends in solid mechanics, *Int. J. Solids Struct*, **37**(1/2): special issues.

Ellis, T. M. R., and Semenkov, O. I. (eds.) 1983. *Advances in CAD/CAM*, North Holland, Amsterdam, the Netherlands.

Fosdick, L. D. (ed.) 1996. *An Introduction to High-Performance Scientific Computing*, MIT Press, Cambridge, MA.

Garboczi, E. J., Bentz, D. P., and Frohnsdorff, G. F. 2000. Knowledge-based systems and computational tools for concrete, *Concrete Int.*, **22**(12), 24–27.

Golley, B. W., and Grice, W. A. 1990. *Prismatic Folded Plate Analysis Using Finite Strip-Elements.* University College, Australian Defence Force Academy, University of New South Wales, Canberra.

Golley, B. W., Grice, W. A., and Petrolito, J. 1987. Plate-bending analysis using finite strip-elements, *J. Struct. Eng.*, **113**(6), 1282–1296.

Hazelrigg, G. A. 1996. *Systems Engineering: An Approach to Information-Based Design*, Pearson College Division, London.

Holzer, S. M. 1985. *Computer Analysis of Structures*, Elsevier Science, New York.

Hughes, T. J. 2012. *The Finite Element Method: Linear Static and Dynamic Finite Element Analysis*, Courier Corporation, North Chelmsford, MA.

Jain, R. 1990. *The Art of Computer Systems Performance Analysis: Techniques for Experimental Design, Measurement, Simulation, and Modeling*, John Wiley & Sons, Hoboken, NJ.

Kemp, R. 2016. Fourth industrial revolution, *The Lawyer*, **31**(21), 12.

Kirsch, U. 2012. *Structural Optimization*, Springer Science & Business Media, New York.

Kleijnen, J. P. 1995. Verification and validation of simulation models, *Eur. J. Oper. Res.*, **82**(1), 145–162.

Li, S. F., and Liu, W. K. 2002. Meshfree and particle methods and their applications, *Appl. Mech. Rev.*, Sept. 2002, 1–34.
Liu, W. K., Qian, D., Gonella, S., Li, S., Chen, W., and Chirputkar, S. 2010. Multiscale methods for mechanical science of complex materials: Bridging from quantum to stochastic multiresolution continuum, *Int. J. Numer. Methods Eng.*, **83**(8–9), 1039–1080.
Louder, R. 1985. Access to supercomputers, *Mosaic*, **16**(3), 26–32.
Lubliner, J. 2008. *Plasticity Theory*, Courier Corp., North Chelmsford, MA.
Mell, P., and Grance, T. 2011. *The NIST Definition of Cloud Computing*, NIST, Gaithersburg, MD.
Mitchell, A. R., and Griffiths, D. F. 1980. *The Finite Difference Method in Partial Differential Equations*. John Wiley, Hoboken, NJ.
Moin, P. 2010. *Fundamentals of Engineering Numerical Analysis*, Cambridge University Press, Cambridge.
Monteiro, P. J. M., Chong, K. P., Larsen-Basse, J., and Komvopoulos, K. (eds.) 2001. *Long-Term Durability of Structural Materials*, Elsevier, Oxford, UK.
Nelson, J. K., Jr. (ed.) 1989. *Computer Utilization in Structural Engineering*, Proc. Structures Congress '89, American Society of Civil Engineers, Reston, VA.
Oden, J. T. et al. 2006. *Simulation-Based Engineering Sciences*, NSF Blue Ribbon Panel Report, http://www.nsf.gov/pubs/reports/sbes_final_report.pdf.
Pardo, S. A. 2016. Validation and verification, in *Empirical Modeling and Data Analysis for Engineers and Applied Scientists*, Springer International Publishing, pp. 197–201.
Petrolito, J., Grice, W. A., and Golley, B. W. 1989. Finite strip-elements for thick plate analysis, *J. Struct. Eng.*, **115**(6), 1163–1182.
Pletcher, R. H. Tannehill, J. C., and Anderson, D. A. 2012. *Computational Fluid Mechanics and Heat Transfer*, CRC Press, Boca Raton, FL.
Reed, M. A., and Kirk, W. P. (eds.) 1989. *Nanostructure Physics and Fabrication*, Academic Press, San Diego, CA.
Rizzo, F. J. 1967. An integral equation approach to boundary value problems of classical elastostatics, *Q. J. Appl. Math.*, **25**, 83–95.
Roco, M. C. 2011. The long view of nanotechnology development: The National Nanotechnology Initiative at 10 years, *J. Nanopart. Res.*, **13**(2), 427–445.
Saif, M. T. A., and MacDonald, N. C. 1996. A milli-Newton micro loading device. *Sens. Actuators A Phys.*, **52**, 65–75.
Siegel, R. W., Hu, E., and Roco, M. C. (eds.) 1999. *WTEC Panel Report on Nanostructure Science and Technology*, Kluwer Academic Publishers, Norwell, MA.
Thoft-Christensen, P., and Baker, M. J. 2012. *Structural Reliability Theory and Its Applications*, Springer Science & Business Media, New York.
Timp, G. (ed.) 1999. *Nanotechnology*, Springer-Verlag, New York.
Wen, Y.-K. (ed.) 1984. *Probabilistic Mechanics and Structural Reliability*, American Society of Civil Engineers, Reston, VA.
Wilson, E. L. 1989. Numerical methods for solution of finite element systems, in *Proc. Sessions Related to Computer Utilization at Structures Congress '89*, Nelson, J. K., Jr. (ed.), American Society of Civil Engineers, Reston, VA, pp. 21–30.
Wolf, E. L. 2015. *Nanophysics and Nanotechnology: An Introduction to Modern Concepts in Nanoscience*, John Wiley & Sons, Hoboken, NJ.
Yao, J. T. P. 1985. *Safety and Reliability of Existing Structures*, Pitman Advanced Publishing Program, Boston, MA.
Zhang, Q., Cheng, L., and Boutaba, R. 2010. Cloud computing: State-of-the-art and research challenges. *J. Internet Serv. Appl.*, **1**(1), 7–18.
Zienkiewicz, O. C, and Taylor, R. L. 2005. *The Finite Element Method*, Butterworth-Heinemann, Oxford, UK.
Zikopoulos, P., and Eaton, C. 2011. *Understanding Big Data: Analytics for Enterprise Class Hadoop and Streaming Data*, McGraw-Hill Osborne Media, Columbus, OH.

BIBLIOGRAPHY

Banichuk, N. V. 2013. *Problems and Methods of Optimal Structural Design* (Vol. 26), Springer Science & Business Media, New York.

Belytschko, T., Krongauz, Y., Organ, D., Fleming, M., and Krysl, P. 1996. Meshless methods: An overview and recent developments, *Comput. Methods Appl. Mech. Eng.*, **139**(1), 3–47.

Bourne, D. E., and Kendall, P. C. 2014. *Vector Analysis and Cartesian Tensors*, Academic Press, San Diego, CA.

Brand, L. 1957. *Vector and Tensor Analysis*, Wiley, New York.

Brebbia, C. A. 1982. *Boundary Element Methods in Engineering*, Springer, New York, pp. 1–649.

Chapra, S. C., and Canale, R. P. 2012. *Numerical Methods for Engineers* (Vol. 2), McGraw-Hill, New York.

Chau, K. T. 2018. *Theory of Differential Equations in Engineering and Mechanics*, CRC Press, Boca Raton, FL.

Chen, J., and Lee, J. D. 2011. The Buckingham catastrophe in multiscale modelling of fracture, *Int. J. Theor. Appl. Multiscale Mech.*, **2**(1), 3–11.

Chen, Y., Lee, J. D., and Eskandarian, A. 2006. *Meshless Methods in Solid Mechanics*, Springer Science & Business Media, New York.

Chong, K. P., Dewey, B. R., and Pell, K. M. 1989. *University Programs in Computer-Aided Engineering, Design, and Manufacturing*, American Society of Civil Engineers, Reston, VA.

Danielson, D. A. 1997. *Vectors and Tensors in Engineering and Physics*, Addison-Wesley, Reading, MA.

Dym, C. L., and Shames, I. H. 1973. *Solid Mechanics: A Variation Approach*, McGraw-Hill, New York.

Edelen, D. G. B., and Kydoniefs, A. D. 1980. *An Introduction to Linear Algebra for Science and Engineering*, 2nd ed., Elsevier Science, New York.

Eisberg, R., and Resnick, R. 1974. *Quantum Physics*, 2nd ed., Wiley, Hoboken, New Jersey.

Eisele, J. A., and Mason, R. M. 1970. *Applied Matrix and Tensor Analysis*, Wiley-Interscience, New York.

Gere, J. M., and Weaver, W. 1984. *Matrix Algebra for Engineers*, 2nd ed., Prindle, Weber, and Schmidt, Boston.

Ghanem, R. G., and Spanos, P. D. 2003. *Stochastic Finite Elements: A Spectral Approach*, Courier Corporation, North Chelmsford, MA.

Haile, J. M. 1992. *Molecular Dynamics Simulation* (Vol. 18), Wiley, New York.

Hay, G. E. 2012. *Vector and Tensor Analysis*, Courier Corporation, North Chelmsford, MA.

Hoffman, J. D., and Frankel, S. 2001. *Numerical Methods for Engineers and Scientists*, CRC Press, Boca Raton, FL.

Isaacson, E., and Keller, H. B. 1994. *Analysis of Numerical Methods*, Courier Corporation, North Chelmsford, MA.

Jeffreys, H. 1987. *Cartesian Tensors*, Cambridge University Press, New York.

Kemmer, N. 1977. *Vector Analysis*, Cambridge University Press, Cambridge.

Li, J., Lee, J. D., and Chong, K. P. 2011. Multiscale analysis of composite material reinforced by randomly-dispersed particles, *Int. J. Smart Nano Mater.*, **3**(1), 2–13.

Li, J., Wang X., and Lee J. D. 2012. Multiple time scale algorithm for multiscale modeling, *Comput. Model. Eng. Sci.*, **85**, 463–480.

Liu, W. K., Karpov, E. G., and Park, H. S. 2006. *Nano Mechanics and Materials: Theory, Multiscale Methods and Applications*, Wiley, Hoboken, NJ.

Luré, A. I. 1964. *Three-Dimensional Problems of the Theory of Elasticity*, Wiley-Interscience, New York.

Nash, W. A. 1993. *The Mathematics of Nonlinear Mechanics*, CRC Press, Boca Raton, FL.

Nemat-Nasser, S., and Hori, M. 1999. *Micromechanics*, Elsevier, Amsterdam, the Netherlands.

Oden, J. T. 1986. *Qualitative Methods in Nonlinear Mechanics*, Prentice Hall, Upper Saddle River, NJ.

Rappe, A. K., Casewit, C. J., Colwell, K. S., Goddard, W. A., and Skiff, W. M. 1992. UFF, a full period table force field for molecular mechanics and molecular dynamics simulations, *J. Am. Chem. Soc.*, **114**, 10024–10035.
Roco, M. C. 2005. Environmentally responsible development of nanotechnology, *Environ. Sci. Technol.*, **39**(5), 106A–112A.
Roco, M. C., and Bainbridge, W. S. 2003. *Converging Technologies for Improving Human Performance: Nanotechnology, Biotechnology, Information Technology and Cognitive Science*, Kluwer Academic Publishers, Norwell, MA.
Sakurai, J. J., and Napolitano, J. 2011. *Modern Quantum Mechanics*, Addison-Wesley, Reading, MA.
Spanos, P. D. (ed.) 1999. *Computational Stochastic Mechanics*, A. A. Balkema, Rotterdam, the Netherlands.
Stillinger, F. H., and Weber, T. A. 1985. Computer simulation of local order in condensed phases of silicon, *Phys. Rev. B*, **31**, 5262.
Temple, G. 2012. *Cartesian Tensors: An Introduction,* Courier Corporation, North Chelmsford, MA.
Ting, T. C. T. 1996. *Anisotropic Elasticity*, Oxford University Press, New York.
Ting, T. C. T. 2000. Recent developments in anisotropic elasticity, *Int. J. Solids Struct.*, **37**(1), 401–409.
Weaver, W., and Gere, J. M. 2012. *Matrix Analysis Framed Structures,* Springer Science & Business Media, New York.
Zhang, Y., Pei, Q., and Wang, C. 2012. Mechanical properties of graphenes under tension: A molecular dynamics study, *Appl. Phys. Lett.*, **101**(8), 081909.

Chapter 2

Numerical analysis and weighted residuals

2.1 INTRODUCTION

Exact analytical solutions to certain engineering boundary value problems exist (Boresi et al., 2011). However, in many cases, the boundary value problems of engineering cannot be solved exactly by currently available analytical methods. In such cases, numerical solutions are sought.

In some situations, we may be able to simplify the physical model suitably and obtain exact solutions to the modified problem. In other circumstances, it may be advantageous to employ a direct attack on the equations through finite difference methods (Chapter 3) or by piecewise polynomial methods (such as finite element methods; Chapter 4). Alternatively, in some problems, we may employ the classical method of separation of variables in conjunction with series methods (Chapter 5).

There exists a broad area of mathematics known as *approximation theory* (Shisha, 1968). The term is usually reserved for that branch of mathematics devoted to the approximation of general functions by means of simple functions. For example, in practice, we may wish to approximate a real arbitrary function $f(x)$ by means of a polynomial $p(x)$ in some finite interval of space, say, $a \leq x \leq b$. The motivation for such an approximation is often one of simplification, particularly when the function $f(x)$ is too complicated to manipulate. Because the approximation, say $F(x)$, ordinarily differs from $f(x)$, questions arise immediately as to the manner in which we should proceed. As noted by Shisha, approximation theory considers such problems as the following:

1. What kinds of functions $F(x)$ should we consider for approximating $f(x)$? In other words, the question of function form or trial function for the approximation is considered.
2. How do we measure the accuracy or goodness of the approximation? To answer this question, we must establish some measure or norm of accuracy or goodness of approximation.
3. Of all possible forms or trial functions, is there one that approximates $f(x)$ well? How well? Is there, by some standard of measure (norm), one approximation that fits $f(x)$ best? That is, for a given norm, is there some best approximation of $f(x)$? If there is a best approximation, is it unique?
4. How can we get approximations to $f(x)$ in practice?

The problem of obtaining approximate or numerical solutions to initial value or boundary value problems differs from that of obtaining the best approximation of known functions in that the exact solution—call it $y(x)$—is unknown. In addition, the solution is required to satisfy a differential equation as well as initial values and/or boundary values. Hence, although we wish to approximate $y(x)$ by some trial function $Y(x)$ in the best possible sense, our problem is greatly complicated. Nevertheless, the ideas and concepts of general approximation theory can be employed in approximation solutions of initial value and/or boundary value problems of engineering.

In this chapter, we discuss briefly concepts common to a broad class of approximation methods. In particular, we treat the *method of weighted residuals*, the most general of trial function methods (Collatz, 2012; Crandall, 1983; Finlayson, 2013). Depending on the norm employed, we may show that the method of weighted residuals leads to well-known approximation methods (e.g., the Galerkin method, collocation, least squares). In other words, the method of weighted residuals *unifies* many approximation methods that are currently in use (Cook, 2007; Finlayson and Scriven, 1966; Hughes, 2012; Reddy, 1991). For example, we will see that the Rayleigh–Ritz method (Langhaar, 1989) is a weighted residual method with a particular choice of weighting function.

In Chapter 3, we treat in detail numerical solutions of engineering problems by finite difference methods. In Chapter 4, we develop finite element methods and apply them to two- and three-dimensional problems. In Chapter 5, we formulate finite strip, finite layer, and finite prism methods for special applications. In Chapters 6 and 7, we present boundary element and meshless methods. In Chapters 8 and 9, we present multiphysics, molecular dynamics, and multiscale modeling.

2.2 APPROXIMATION PROBLEM (TRIAL FUNCTIONS; NORMS OR MEASURES OF ERROR)

Here, we are concerned primarily with numerical methods of solving boundary value problems in engineering. In boundary value problems in engineering, we are generally faced with determining a solution to a differential equation (or a system of differential equations) in a region R. The solution is required to meet certain conditions on the boundary B of the region. In many cases, the given equation or equations do not possess a known exact solution. Accordingly, unless we are able to obtain the unknown exact solution, we are forced to find an approximation of the exact solution. The form of this approximation is often cast in terms of a *trial solution* or *trial function* $F(x; A)$, which is assumed to be compatible with the exact solution $E(x)$, where x is defined in a compact subset (region) of space. For example, x may denote n real variables, or it may denote three spatial coordinates in the domain R; in one-dimensional problems, x denotes a single real variable. The symbol A stands for a collection of parameters $a_1, a_2, a_3,\ldots, a_n$. Thus, $A = A{:}(a_1, a_2, a_3,\ldots, a_n)$. Unfortunately, there is no general scientific method of determining which of the unlimited number of approximating functions (trial functions or forms) ordinarily available will lead to the most efficient approximation of $E(x)$. In practice, the choice of trial functions is often made on the basis of experience or intuition. For example, one may sense intuitively that a certain form of trial function (say, a polynomial or a Fourier series) may be suitable but not have any method at hand to actually determine the required approximation.

The choice of a method of estimating the accuracy of the approximation (choice of norm) ordinarily is less important than that of trial function (choice of form). Generally speaking, if $F(x; A)$ is compatible with $E(x)$, almost any reasonable norm will lead to an efficient approximation of $E(x)$. However, if $F(x; A)$ is not compatible with $E(x)$, an efficient approximation will not ordinarily be attained, regardless of the norm employed. Of course, the choice of the norm may affect the complexity of estimating the accuracy of the approximation. In some cases, this factor may dictate the choice of norm. In other words, once the form or family of trial functions for the approximation is selected, we wish to select the best approximation possible within the family. Then the norms on which this best approximation rests are essentially unlimited in number. To simplify the choice somewhat, we restrict the discussion that follows to two rather broadly employed methods of error measurement: the method of weighted residuals and the variational method. In either case, for linear boundary value problems, the method leads to consideration of the solution of a set of simultaneous linear algebraic equations.

In Section 2.3, we consider the method of weighted residuals as applied to ordinary differential equations, and in Section 2.4, to partial differential equations. In Section 2.5, we outline briefly the variation method and show its relation to the method of weighted residuals.

2.3 METHOD OF WEIGHTED RESIDUALS (ORDINARY DIFFERENTIAL EQUATIONS)

2.3.1 Preliminary remarks

In certain simplified situations, the boundary value problem may be reduced to one of ordinary differential equations in a single dependent variable. (See, for example the axially symmetric and spherically symmetric problems of elasticity; Boresi et al., 2011. See also the finite strip method of Section 5.2.) Furthermore, the method of separation of variables leads to ordinary differential equations (see Sections 6.9, 7.10, and 7.15; Boresi et al., 2011). Consequently, in this section, we consider a method of evaluation of ordinary differential equations based on the method of weighted residuals. First, we outline the general approximation method for ordinary differential equations.

As noted in Section 2.2, an approximate solution to differential equations is often sought by assuming that the exact solution $E(x)$ may be approximated by an expression of the form $F(x; a_1, a_2, a_3,\ldots, a_n)$, where $a_1, a_2, a_3,\ldots, a_n$ are arbitrary parameters to be chosen to best fit the exact solution $E(x)$. This best fit is related to a particular measure (norm) of the approximation. In selecting the form $F(x; a_1, a_2, a_3,\ldots, a_n)$, we have certain options available. For example, let $R + B = D$ be the domain of the boundary value problem, where R is the interior of D bounded by surface B. Then we may choose F in one of the following ways (Collatz, 2012):

1. The differential equation is satisfied exactly in R, and the a_i is selected to make F fit the boundary conditions on B in some best sense (norm). This method of selecting the a_i is called *the boundary method*.
2. The boundary conditions are satisfied exactly on B, and the a_i is selected so that F satisfies the differential equation in the interior R in some best sense (norm). This method is called *the interior method*.
3. The differential equation is not satisfied in R, nor are the boundary conditions satisfied on B. The a_i is chosen to satisfy the differential equation in R and the boundary conditions on B in some best sense. This method of determining the a_i is called a *boundary–interior method* or simply a *mixed method*.

In ordinary differential equations, interior methods are most often used, as even if we know the general solution to the differential equation (boundary method), we still must solve a set of n algebraic (linear or possibly nonlinear) equations in the a_i to satisfy the boundary conditions. In the boundary value problem of ordinary differential equations, the boundary conditions are usually specified at two points of the region, say, $x = 0$ and $x = L$. Thus, we speak of the two-point boundary value problem.

In boundary value problems of partial differential equations, both boundary and interior methods are used. However, in many cases, boundary methods are preferred, as satisfaction of boundary conditions, as far as integration is concerned, requires the evaluation of integrals over the boundary B rather than evaluation of integrals through the interior region R, as do interior methods. When the differential equations and boundary conditions are very complicated, some simplification may be possible with the use of mixed methods.

2.3.2 Method of weighted residuals

The method of weighted residuals seeks to produce a best approximate solution to a differential equation (subject to boundary conditions) through the use of trial functions. This use is also a feature of variational methods (see Section 2.5 and Chapter 4, Langhaar, 1989). Special widely used cases of the method of weighted residuals include the Galerkin method, the method of collocation, and the method of least squares (Allen and Isaacson, 2011; Botha and Pinder, 1983; Collatz, 2012). The general approach may be outlined as follows: Consider the ordinary differential equation

$$G[y]-f(x)=0 \qquad \text{for} \qquad x_0 \le x \le x_1 \tag{2.1a}$$

with boundary condition

$$B[y]=0 \qquad \text{for} \qquad x=x_0,\ x=x_1 \tag{2.1b}$$

where $f(x)$ is a known function of x, and G and B denote differential operators of x. In general, G and B may be nonlinear operators, for example,

$$G=C_1\frac{d^2}{dx^2}+C_2\left(\frac{d}{dx}\right)^2+C_3$$

In linear boundary value problems, G and B are linear differential operators, for example,

$$G=C_1\frac{d^2}{dx^2}+C_2\frac{d}{dx}+C_3$$

The Cs may be functions of x. The general solution of Equation 2.1a is a function $y = y(x)$ that satisfies the differential equation.

In the method of weighted residuals, we may assume an approximate solution $\bar{y}$ of Equation 2.1a of the form

$$y \approx \bar{y}(x:a_1,a_2,a_3,\ldots,a_n)=\sum_{i=1}^{n} a_i\phi_i(x) \tag{2.2a}$$

where the a_i are undetermined parameters and the $\phi_i(x)$ are *trial functions* chosen so that $\bar{y}$ satisfies boundary conditions (the interior method). For example, if the boundary conditions are $y = 0$ for $x = a$ and for $x = b$, a possible choice of $\phi_i(x)$ is

$$\phi_i(x)=(x-a)(x-b)x^{i-1} \tag{2.2b}$$

In some problems, a boundary condition may be nonlinear (in y). Then it may not be possible to select $\phi_i(x)$ such that $\bar{y}$ satisfies this boundary condition, and we are led to the mixed method of determining the a_i. The residual $r(x) = B(\bar{y})$ formed by substituting $\bar{y}$ and its derivatives for y and its derivatives into the boundary condition (see Equation 2.1b) is then not identically zero. We may obtain an equation in the a_i by liquidating the residual $r(x)$ in some way, for example, by arbitrarily setting it equal to zero. This equation must then be solved with the other equations by the method of weighted residuals.

For the *interior method*, the method of weighted residuals (Collatz, 2012, calls it the orthogonality method) requires that the integral of the residual, $R(x) = G[\bar{y}] - f(x)$, appropriately weighted by functions $w_i(x)$, $i = 1, 2, 3,\ldots, n$, vanish over the interior region $x = [0, L]$. (In the terminology of Collatz, 2012, $R(x)$ is said to be orthogonal to $w_i(x)$ over the interior region $x = [0, L]$.) Thus,

$$\int_0^L w_i(x)R(x)\,dx = 0 \qquad i = 1,2,3,\ldots,n \tag{2.3}$$

where $w_i(x)$ is n linearly independent functions. Theoretically, the $w_i(x)$ should be members of a complete set of functions. In practice, the $w_i(x)$ is often chosen to be the first n functions of a complete set, such as $\sin(2\pi x/L)$, $\sin(4\pi x/L),\ldots, \sin(2\pi n x/L)$. The various trial function techniques are characterized by the choices for w_i (Finlayson, 2013), as discussed in the following.

The *collocation method* requires that the residual $R(x)$ vanish exactly at n points. Thus,

$$R(x_i) = 0 \qquad i = 1,2,3,\ldots,n \tag{2.4}$$

Because Equation 2.4 may be obtained from Equation 2.3 with $w_i(x) = \delta(x - x_i)$, where $\delta(x - x_i)$ is the Dirac delta function defined by the conditions,

$$\delta(x - x_i) = 0,\ x \neq x_i \qquad \int_{-\infty}^{+\infty} \delta(x - x_i)\,dx = 1 \tag{2.5}$$

the weighting function for the method of collocation is the Dirac delta function. Thus,

$$\int_0^L \delta(x - x_i)R(x)\ dx = R(x_i) = 0 \qquad i = 1,2,3,\ldots,n \tag{2.6}$$

and this weighting is equivalent to making the residual vanish at n chosen points x_i. The collocation method has the mathematical advantage that the integration is trivial. However, because the forced solution is made to agree exactly at n selected points, it may fluctuate widely between points. For multivariable problems, Yang and Zhao (1985) addressed this difficulty by requiring residuals to vanish on selected *location lines*. Seemingly, this method has certain advantages over the point-location (collocation) method in multivariable problems.

The *method of least squares* requires that the integral of the residual squared be minimal. Thus, it requires that the integral

$$I = \int_0^L p(x)R^2(x)\,dx \tag{2.7}$$

be a minimum with respect to the a_i, where $p(x)$ is an arbitrary positive function. [Often, $p(x)$ is taken equal to 1.] Hence, the trial parameters a_i are determined from the condition

$$\frac{\partial I}{\partial a_i} = 0 \qquad i = 1,2,3,\ldots,n \tag{2.8}$$

Accordingly, in general, the weighting functions in the method of least squares are

$$w_i(x) = p(x)\frac{\partial R(x)}{\partial a_i} \tag{2.9}$$

Unfortunately, in the method of least squares, the integrations involved in Equation 2.3 are often very complicated.

In the *Galerkin method*, the most popular weighted method, the weighting functions $w_i(x)$ are taken to be the trial functions ϕ_i themselves. Hence, the residual is forced to be orthogonal to the trial functions $\phi_i(x)$. Thus, in the Galerkin method, the trial parameters a_i are determined by the n equations

$$\int_0^L \phi_i(x)R(x)\,dx = 0 \qquad i = 1,2,3,\ldots,n \tag{2.10}$$

Accordingly, by the requirements of the method of weighted residuals, the trial functions $\phi_i(x)$ should be terms of a complete set of functions. However, this condition, required for mathematical purposes, is often ignored in practice. Indeed, in practice, at best, the residual is made orthogonal to only the first few terms in a complete set.

In the *method of moments*, the residual is made orthogonal to the terms of a system of functions that need not be the same as the trial functions. Often, the method of moments is taken to be defined by the choice $w_i(x) = x^{i-1}$ regardless of the choice of $\phi_i(x)$. However, broadly speaking, the method of moments may be considered equivalent to the method of weighted residuals (or the method of orthogonalization). More narrowly, the method of moments is defined by the conditions

$$\int_0^L x^{i-1}R(x)\,dx = 0 \qquad i = 1,2,3,\ldots,n \tag{2.11}$$

The *partition method* (or the *subdomain collocation method*) requires that the integral of the residual equals zero over m subintervals $s_i = [x_i, x_{i-1}]$, $i = 1, 2, 3,\ldots, m$ of $x = [0, L]$. Thus,

$$\int_{x_{i-1}}^{x_i} R(x)\,dx = 0 \qquad i = 1,2,3,\ldots,m \tag{2.12}$$

Hence, the weighting functions $w_i(x)$ of Equation 2.3 may be considered to be the unit step functions (see Chapter 4, where trial functions for finite element methods are discussed; see also Zienkiewicz and Morgan, 2006).

$$w_i = \begin{cases} 1 & x \text{ in } s_i \\ 0 & x \text{ not in } s_i \end{cases} \tag{2.13}$$

2.3.3 Boundary methods

In the preceding discussion, we considered primarily interior methods in which the trial functions are chosen such that $\bar{y}$ satisfies the boundary conditions but not the differential equation. For boundary methods, we select trial functions so that the differential equation will be satisfied but not the boundary conditions. Then the procedure follows through as for interior methods, except that the averages over the interior (Equation 2.3) are replaced by appropriate equations on the boundary.

For discussions of mixed methods, refer to the comprehensive review of the method of weighted residuals given by Finlayson and Scriven (1966); see also Collatz (2012), Finlayson (2013), and Cook (2007).

2.3.4 Convergence theorems

A few theorems have been proved concerning the convergence of methods of weighted residuals. These theorems pertain primarily to linear problems (Finlayson and Scriven, 1966; see also Bathe, 1995). When dealing with nonlinear problems, we must consider the possibility of the existence of more than one solution. The results obtained by successive approximations (with $n = 1, 2, 3, \ldots, N$) are often compared to justify convergence. Essential in this process is the choice of convergence tolerances (Bathe, 1995). If the tolerances are too coarse, inaccurate results may result. If the tolerances are too fine, excessive computational effort may be required, with little gain in accuracy. Also, a physical knowledge of the problem assists in judging whether the solution appears to be reasonable.

Example 2.1: Suspended Heavy Chain; Collocation Method, $n = 1$

A heavy chain is suspended from its endpoints located at equal heights at $x = 0$ and at $x = 1$. The deflection of the chain is approximated by solution of the differential equation $y'' + k(y')^2 + 1 = 0$, where k is the weight per unit length of chain divided by the tension in the chain at midspan. Primes denote derivatives relative to x. Assume an approximation $\bar{y}$ (Equation 2.2a), with ϕ_i given by Equation 2.2b.

(a) For $n = 1$, with $a = 0$, $b = 1$, $k = \frac{1}{2}$, $\bar{y}(0) = \bar{y}(1) = 0$, let $x_1 = \frac{1}{2}$ and determine a_1 by the collocation method (Equation 2.4).
(b) Compare the results with the exact solution.

$$y = 2\ln\frac{\cos\left(x/\sqrt{2} - 1/2\sqrt{2}\right)}{\cos\left(1/2\sqrt{2}\right)} \tag{2.14}$$

Solution: Let

$$\bar{y} = \sum_{n=1}^{N} a_n\phi_n(x) \qquad \text{with} \qquad \phi_n = (a-x)(b-x)x^{n-1}$$

(a) For $N = 1$, $a = 0$, $b = 1$, $k = \frac{1}{2}$, $\bar{y}(0) = \bar{y}(1) = 0$, we get $\phi_1(x) = x(x - 1)$, Hence, $\bar{y}(x) = a_1x(x - 1)$, $\bar{y}'(x) = a_1(2x - 1)$, $\bar{y}'' = 2a_1$, and $R(x) = \bar{y}'' + k(\bar{y}')^2 + 1 = 2a_1 + \frac{1}{2}a_1^2(2x - 1)^2 + 1$. For $x_1 = \frac{1}{2}$, $R(x_1) = 2a_1 + 1 = 0$. Therefore, $a_1 = -\frac{1}{2}$ and

$\bar{y} = -(x^2 - x)/2$. Consider the values $x_1 = \frac{1}{2}$ and $x_2 = \frac{3}{4}$. For these values of x, $\bar{y}\left(\frac{1}{2}\right) = 0.125$ and $\bar{y}\left(\frac{3}{4}\right) = 0.09375$.

b) By Equation 2.14, $y(0) = y(1) = 0$, $y\left(\frac{1}{2}\right) = 0.1277$, and $y\left(\frac{3}{4}\right) = 0.09628$.

Example 2.2: Suspended Heavy Chain; Collocation Method, $n = 2$

Repeat Example 2.1 for $n = 2$ with $x_1 = \frac{1}{4}$, $x_2 = \frac{1}{2}$. Compare $\bar{y}\left(\frac{1}{2}\right)$ with the results of Example 2.1. Compare $\bar{y}\left(\frac{1}{4}\right)$ with $\bar{y}\left(\frac{3}{4}\right)$.

Solution: Let $n = 2$ in Example 2.1. Then, by Equation 2.2b, $\phi_2(x) = (x - a)(x - b)x$. Therefore, for $a = 0$, $b = 1$, $\phi_2(x) = x^3 - x^2$. Hence, by Equation 2.2a, with $\phi_1 = (x^2 - x)$, from Example 2.1,

$$\bar{y} = a_1\phi_1(x) + a_2\phi_2(x) = a_1(x^2 - x) + a_2(x^3 - x^2) \tag{2.15}$$

Therefore,

$$\bar{y}' = a_1(2x - 1) + a_2(3x^2 - 2x)$$

$$\bar{y}'' = 2a_1 + a_2(6x - 2)$$

and

$$R(x) = \bar{y}'' + \frac{1}{2}(\bar{y}')^2 + 1 = 2a_1 + a_2(6x - 2) + \frac{1}{2}[(2x-1)]^2 a_1^2 + 2a_1a_2(2x-1)(3x^2 - 2x)$$
$$+ a_2^2(3x^2 - 2x)^2] + 1$$

Hence,

$$\begin{aligned} R\left(\frac{1}{4}\right) &= 2a_1 - \frac{1}{2}a_2 + \frac{1}{8}a_1^2 + \frac{5}{32}a_1a_2 + \frac{25}{512}a_2^2 + 1 = 0 \\ R\left(\frac{1}{2}\right) &= 2a_1 + a_2 + \frac{1}{32}a_2^2 + 1 = 0 \end{aligned} \tag{2.16}$$

The solution of Equation 2.16 is

$$a_1 = -0.5103 \qquad a_2 = 0.0206 \tag{2.17}$$

Equations 2.15 and 2.17 yield

$$\bar{y}\left(\frac{1}{2}\right) = 0.125 \qquad \text{(see Example 2.1b)}$$

and

$$\bar{y}\left(\frac{1}{4}\right) = 0.09472 \quad \bar{y}\left(\frac{3}{4}\right) = 0.09278$$

Example 2.3: Suspended Heavy Chain; Galerkin Method

Repeat Example 2.1 using the Galerkin method (Equation 2.10). Compare the results with Example 2.1 for $\bar{y}\left(\frac{1}{2}\right)$.

Solution: From Example 2.1, with $n = 1$, $x = 0$, $x = 1$, $\phi_1(x) = x(x - 1)$, $\bar{y}(x) = a_1(x^2 - x)$, and $R(x) = 2a_1 + \frac{1}{2}a_1^2\ (4x^2 - 4x + 1) + 1$, we have, by Equation 2.3,

$$\int_0^1 (x^2 - x)\left[2a_1 + \frac{1}{2}a_1^2(4x^2 - 4x + 1) + 1\right] dx = 0$$

Integration yields $20a_1 + a_1^2 = -10$. The solution of this equation is $a_1 = -0.513$. Hence, $\bar{y}(x) = -0.513(x^2 - x)$. Thus, we obtain $\bar{y}\left(\frac{1}{2}\right)_{\text{Galerkin}} = 0.128$. This result agrees well with that obtained by the exact solution, Example 2.1b.

Example 2.4: Suspended Heavy Chain; Method of Moments

Repeat Example 2.3 by the method of moments (Equation 2.11).

Solution: As in Example 2.2, we have $R(x) = 2a_1 + \frac{1}{2}a_1^2\ (4x^2 - 4x + 1) + 1$. Hence, by Equation 2.11, with $i = 1$, we have

$$\int_0^1 \left[2a_1 + \frac{1}{2}a_1^2(4x^2 - 4x + 1) + 1\right] dx = 0$$

Integration yields $12a_1 + a_1^2 = -6$. The solution to this equation is $a_1 = -0.5228$. Hence, $\bar{y}(x) = -0.5228(x^2 - x)$ and $\bar{y}\left(\frac{1}{2}\right) = 0.1307$, compared to $y\left(\frac{1}{2}\right) = 0.128$.

2.4 METHOD OF WEIGHTED RESIDUALS (PARTIAL DIFFERENTIAL EQUATIONS)

Many of the methods discussed in Section 2.3 for ordinary differential equations carry over to partial differential equations in a straightforward manner. However, the difficulties inherent in the approximate solutions of ordinary differential equations are magnified for partial differential equations. Indeed, even more than for ordinary differential equations, there is a need for further study of the existence and uniqueness of exact solutions. Perhaps, more important to the stress analyst is the fact that the question of convergence of approximating trial functions is answered only for limited classes of problems (Allen and Isaacson, 2011; Zienkiewicz and Morgan, 2006). Unfortunately, calculations applied to partial differential equations may well lead to incorrect results. Finally, again, most frustrating to the stress analyst, there is the fact that approximate methods may converge very nicely *but to values that are unrelated to the correct solution*. In early work on finite element methods, the latter phenomenon was observed frequently (see Chapter 4). It remains an important consideration in finite difference methods as well (see Chapter 3). Nevertheless, because the finite difference method is applicable to boundary value problems in general, it finds great favor. In particular, the approximate equations are easy to set up and even with a coarse mesh may give an approximate solution that is sufficiently accurate for the purposes of numerical stress analysis. One of the greatest disadvantages of the method is its slow rate of convergence, a difficulty when fine accuracy is required. (Collatz, 2012, presents many examples of this problem in his Chapter V; see also our Chapter 3.)

In the following, we briefly outline the method of weighted residuals as applied to boundary value problems of partial differential equations. The equations of elasticity are treated by difference methods in Chapter 3 and by finite element methods in Chapter 4. Because a number of important problems of elasticity reduce to the treatment of partial differential equations of second order, we restrict our attention to these types of equations. Treatment of higher-order equations follows by analogy.

For partial differential equations we seek to approximate the exact solution, say, $u(x_1, x_2,..., x_n)$ by a trial function $\phi(x_1, x_2,..., x_n; a_1, a_2,..., a_p)$, where $x_1, x_2,..., x_n$ denote independent variables and $a_1, a_2,..., a_p$ are p arbitrary parameters. The trial function ϕ satisfies either the differential equation (boundary method) or the boundary conditions (interior method), whichever is more convenient. Substitution of ϕ into either the boundary conditions or the differential equations yields the residual error function $R(x_i, a_p)$. By choosing the parameters a_p so that the residual R is liquidated, say, in the sense of Equation 2.3, we determine $\phi(x_i, a_p)$ to approximate $u(x_i, a_p)$ in the weighted residual sense.

Example 2.5: Torsion Problem by the Interior Method

Consider the torsion problem defined by the equation (see Boresi et al., 2011, Section 7.3).

$$\begin{aligned} \nabla^2 u &= -2 \quad \text{in the interior of } D \\ u &= 0 \quad \text{on the boundary of } D \end{aligned} \tag{2.18}$$

where for simplicity, region D is taken as the square $|x| \le 1$, $|y| \le 1$.

For the approximating function $\phi(x, y; a_0, a_1..., a_p)$, we may assume an expression that satisfies the differential equation (boundary method) or the boundary conditions (interior method). For example, in the interior method, to satisfy the boundary conditions, we may choose a polynomial that meets any symmetry conditions that exist but whose coefficients are otherwise arbitrary. The coefficients are then selected to meet the boundary condition requirements. Thus, we take

$$\phi(x, y; a_0, a_1, \ldots, a_p) = a_0 + a_1(x^2 + y^2) + a_2x^2y^2 + a_3(x^4 + y^4) + a_4(x^4y^2 + x^2y^4) + \ldots \tag{2.19}$$

Retaining only three terms in Equation 2.19, as $u = 0$ on the boundary B, we find the one-parameter trial function

$$\phi(x,y;a) = a(1-x^2-y^2+x^2y^2) = av_1(x,y) \tag{2.20}$$

where $v_1(x, y) = 1 - x^2 - y^2 + x^2y^2$. Retaining five terms in Equation 2.19, we obtain the two-parameter trial function

$$\phi(x,y;a,b) = av_1(x,y) + bv_2(x,y) \tag{2.21}$$

where

$$v_1(x,y) = 1 - x^2y^2 - x^4 - y^4 + x^2y^2(x^2+y^2)$$
$$v_2(x,y) = x^2 + y^2 - 2x^2y^2 - x^4 - y^4 + x^2y^2(x^2+y^2)$$

Proceeding in this manner, we may include more parameters in the trial function ϕ for the interior method.

Alternatively, the differential equation (Equation 2.18) may be satisfied by the superposition of a particular integral (solution) and a series of harmonic functions that satisfy the homogeneous equation $\nabla^2 u = 0$ and also any symmetries of the problem (see Boresi et al., 2011, Section 7.5). Thus, for a particular integral, we take $-\frac{1}{2}(x^2 + y^2)$, and for the homogeneous equation, we take the real parts of the complex variable $z = x + iy$ raised to the $4n$ power, where $n = 1, 2, 3,\ldots$. Then we have the $n + 1$ parameter trial function for the boundary method:

$$\begin{aligned}\phi(x,y;a_0,a_1,\ldots,a_n) = &-\frac{1}{2}(x^2+y^2) + a_0 + a_1(x^4 - 6x^2y^2 + y^4)\\ &+ a_2(x^8 - 28x^6y^2 + 70x^4y^4 - 28x^2y^6 + y^8)\\ &+ \ldots + a_n \operatorname{Re}(x+iy)^{4n}\end{aligned} \tag{2.22}$$

where Re denotes the real value. The method of weighted residuals may now be used to determine the parameters a_n. To determine the parameters a_n, we consider the collocation method and the Galerkin method in the following examples.

2.4.1 Collocation method

First, consider Equation 2.20 to illustrate collocation as an *interior* method. Substitution into the differential Equation 2.18 yields the residual

$$R(x,y;a) = \nabla^2\phi + 2 = -2a(2-x^2-y^2) + 2 \tag{2.23}$$

The coefficient a must be determined by setting the residual $R(x, y; a)$ equal to zero at some value of (x, y). The choice of the point (x, y) is rather arbitrary, although in the case of polynomials involving several parameters, $a_1, a_2,\ldots$, an attempt is usually made to select the collocation points $(x_1, y_1), (x_2, y_2),\ldots$, more or less uniformly spaced in the region D. Generally, there is no reliable procedure to assess the effect of the choice of collocation points. For the one-parameter example considered here, the parameter a is chosen so that $R(x, y; a)$ vanishes at one point. Thus, Equation 2.23 yields the result

$$a = \frac{1}{2 - x^2 - y^2} \tag{2.24}$$

for an arbitrary position of the single collocation point. Because for the interior of region D, $-1 < x < 1$ and $-1 < y < 1$, a may vary from the value $\frac{1}{2}$ (for $x = y = 0$) to a value as large as desired as $|x| \to 1$, $|y| \to 1$. Hence, the value of $\phi(0, 0; a)$ changes drastically with the choice of collocation point, a rather unsatisfactory result.

For the two-term approximation $\phi(x, y; a_1, a_2)$ of $u(x, y)$, we must consider two collocation points. Because of the symmetries that exist in the problem, this results in collocation at 16 points in the interior of region D. For example, if we take the points $x_1 = \frac{1}{2}$, $y_1 = \frac{1}{4}$, and $x_2 = \frac{3}{4}$, $y = \frac{1}{2}$ for collocation, by symmetry the points $\left(x = \pm\frac{1}{2}, y = \pm\frac{1}{4}\right)$, $\left(x = \pm\frac{3}{4}, y = \pm\frac{1}{2}\right)$, $\left(x = \pm\frac{1}{2}, y = \pm\frac{3}{4}\right)$, and $\left(x = \pm\frac{1}{4}, y = \pm\frac{1}{2}\right)$ are collocation points. Again, the value of $\phi(0, 0; a_1, a_2)$ changes with collocation points but not widely as for the one-parameter approximation.

2.4.2 Galerkin method

Again consider Equation 2.20 to illustrate the Galerkin method (Cook, 2007) as an *interior* method. As in the collocation method, the residual is given by Equation 2.23. Hence, because of the symmetry of the problem, the coefficient a is determined by the equation

$$\int_0^1 \int_0^1 v_1(\nabla^2\phi + 2)\ dx\ dy = 0 \tag{2.25}$$

where v_1 is given by Equation 2.20 and $(\nabla^2\phi + 2)$ is given by Equation 2.23. Evaluation of the integral of Equation 2.25 yields $a = 0.625$ and $\phi(0, 0; a) = 0.625$.

For the two-term approximation with (a, b), the Galerkin method requires that

$$\int_0^1 \int_0^1 v_i(\nabla^2\phi + 2)\ dx\ dy = 0 \qquad i = 1,2 \tag{2.26}$$

where $\phi = av_1 + bv_2$, and (v_1, v_2) are given by Equation 2.21. Integration of Equation 2.26 yields

$$\begin{aligned} 1768a + 640b &= 735 \\ 320a + 176b &= 105 \end{aligned} \tag{2.27}$$

Hence, $a = 0.584386$, $b = -0.465930$, and $\phi(0, 0) = 0.584386$. Further refinement is possible by retaining more terms in the trial function representation (Equation 2.19).

2.5 VARIATION METHOD (RAYLEIGH–RITZ METHOD)

Generally speaking, the calculus of variations is a mathematical technique concerned with the determination of a function (say, y) that makes stationary a certain integral (say, J) of the function (Reddy, 1991). The integral or functional J takes on a particular numerical value for each function y. Hence, the functional J is said to be a function of the function y. For example, we may write for the one-dimensional case

$$J(y) = \int y(x)\, dx \tag{2.28}$$

Then for each function $y(x)$, $J(y)$ takes on a particular numerical value.

The fundamental problem of the calculus of variation is to determine a function y such that increments δy (called variations) of y result in second-order increments (or higher) in the functional J. Then J is said to be stationary. Symbolically, this means that when

$$y \rightarrow y + \delta y \tag{2.29}$$

then

$$J(y) \rightarrow J(y) + O(\delta y^2) \tag{2.30}$$

Thus, the term $\delta J(y)$ of first degree in y must vanish identically. The requirement that $J(y)$ be stationary, that is, $\delta J(y) = 0$, for arbitrary variations δy of y leads to an equation for y (the Euler equation; Langhaar, 1989). In physical problems, the solution y of this equation is subject to certain boundary conditions. Then, the variation δy of y must not violate these boundary conditions. For example, in the one-dimensional case, if $y(a) = C_1$ and $y(b) = C_2$, then $\delta y(a)$ and $\delta y(b)$ must be zero.

In the simplest variational problems, the integrand of the functional J does not contain derivatives higher than the first order in y. Thus for the case of one independent variable x,

$$J = \int_{x_0}^{x_1} F(x, y, y')\, dx \tag{2.31}$$

where $F(x, y, y')$ is a given function, y is a function of x, and the prime denotes derivative with respect to x. For J to be stationary for $y \rightarrow y + \delta y$, $\delta J = 0$. Thus, by Equations 2.29, 2.30, and 2.31,

$$\delta J = \int_{x_0}^{x_1} \delta F\, dx = 0 \tag{2.32}$$

where the first variation δF of F is given by

$$\delta F = \frac{\partial F}{\partial y}\delta y + \frac{\partial F}{\partial y'}\delta y' \tag{2.33}$$

where

$$\delta y' = \frac{d}{dx}(\delta y) \tag{2.34}$$

By Equations 2.32, 2.33, and 2.34, integration by parts yields

$$\delta J = \int_{x_0}^{x_1} \delta y \left(\frac{\partial F}{\partial y} - \frac{d}{dx}\frac{\partial F}{\partial y'} \right) dx + \frac{\partial F}{\partial y'}\delta y \Bigg|_{x_0}^{x_1} \tag{2.35}$$

Since δy is arbitrary (except for possible requirements on the boundary $x = x_0$, $x = x_1$), for $\delta J = 0$, the integrand of Equation 2.35 must vanish identically. Thus,

$$\frac{\partial F}{\partial y} - \frac{d}{dx}\frac{\partial F}{\partial y'} = 0 \tag{2.36}$$

Equation 2.36 is called the *Euler equation* of the functional J. For the one-dimensional case, it is an ordinary differential equation.

If the function $y(x)$ is required to satisfy boundary conditions at $x = x_0$, $x = x_1$, then

$$y(x_0) = C_1 \qquad y(x_1) = C_2 \tag{2.37}$$

where C_1 and C_2 are constants. It follows that $\delta y(x_0) = \delta y(x_1) = 0$, and the terms outside the integral on Equation 2.35 vanish identically. Alternatively, if y is free to take on arbitrary variations at $x = x_0$, $x = x_1$, for $\delta J = 0$, it is necessary that

$$\frac{\partial F}{\partial y'}\Bigg|_{x_0} = \frac{\partial F}{\partial y'}\Bigg|^{x_1} = 0 \tag{2.38}$$

These conditions enter naturally from Equation 2.35 when y is not constrained by boundary conditions at x_0, x_1 Accordingly, Equation 2.38 is called the *natural boundary conditions* of the variation problem of seeking a function $y(x)$ that makes J stationary.

In some physical problems, such as stability problems, it is necessary to determine whether the stationary value of J corresponds to a maximum value, a minimum value, or neither (a *saddle-point value*). Then it is necessary to examine the higher-degree terms in Equation 2.30. In other words, the nature of the second variation $\delta^2 J = \delta(\delta J)$ must be examined (Langhaar, 1989, Chapter 6).

More generally, if $F = F(x, y', y'', y''',\ldots)$, the Euler equation for F is

$$\frac{\partial F}{\partial y} - \frac{d}{dx}\frac{\partial F}{\partial y'} + \frac{d^2}{dx^2}\frac{\partial F}{\partial y''} - \frac{d^3}{dx^3}\frac{\partial F}{\partial y'''} + \cdots = 0 \tag{2.39}$$

If $F = F(x, y, z, y', z', y'', z'',\ldots)$, where y and z are functions of x, we obtain two Euler equations for F in the form

$$\begin{aligned} &\frac{\partial F}{\partial y}-\frac{d}{dx}\frac{\partial F}{\partial y'}+\frac{d^2}{dx^2}\frac{\partial F}{\partial y''}-\frac{d^3}{dx^3}\frac{\partial F}{\partial y'''}+\cdots=0 \\ &\frac{\partial F}{\partial z}-\frac{d}{dx}\frac{\partial F}{\partial z'}+\frac{d^2}{dx^2}\frac{\partial F}{\partial z''}-\frac{d^3}{dx^3}\frac{\partial F}{\partial z'''}+\cdots=0 \end{aligned} \tag{2.40}$$

If $F = F(x, y, w, w_x, w_y)$, where x and y are independent variables, $w = w(x, y)$, and subscripts x and y denote partial differentiation, the Euler equation is the partial differential equation

$$\frac{\partial F}{\partial w}-\frac{d}{dx}\frac{\partial F}{\partial w_x}-\frac{d}{dy}\frac{\partial F}{\partial w_y}=0 \tag{2.41}$$

Generalization of Equation 2.41 for several dependent variables and higher derivatives parallels the generalizations to Equations 2.39 and 2.40.

2.5.1 Approximation techniques based on variational methods

The approximation problem outlined in Section 2.2 may be attacked by variation methods. For example, let the Euler equation for the foundational J be of the form

$$H\phi - f = 0 \tag{2.42}$$

where H is a differential operator, ϕ is the function for which an exact solution is unknown, and f is a known function of the independent variables. Accordingly, let ϕ_i be a set of N trial functions for ϕ. Then, by Section 2.2, we seek N parameters a_i such that

$$\bar{\phi} = \sum_{i=1}^{N} a_i\phi_i \tag{2.43}$$

is the best approximation of ϕ obtainable with the trial functions ϕ_i. All trial function methods are concerned with this type of problem. The variational method of seeking an answer proceeds as follows.

Consider the one-dimensional case $\phi = \phi(x)$. The parameters a_i are constants. Assume that we know a functional V where

$$V = \int_{x_0}^{x_1} F(x,\phi)\, dx \tag{2.44}$$

whose Euler equation is Equation 2.42. Then substitution of $\bar{\phi}$ for ϕ into Equation 2.44 yields

$$\bar{V} = \int_{x_0}^{x_1} F(x,a_1,a_2,\ldots,a_N)\, dx \tag{2.45}$$

where $\bar{V}$ is an approximation of V.

Since V is stationary with respect to ϕ, we require $\bar{V}$ to be stationary with respect to $\bar{\phi}$ (i.e., with respect to $a_1, a_2,\ldots, a_N$). Accordingly, we require that

$$\delta\bar{V} = \frac{\partial\bar{V}}{\partial a_1}\delta a_1 + \frac{\partial\bar{V}}{\partial a_2}\delta a_2 + \cdots + \frac{\partial\bar{V}}{\partial a_N}\delta a_N = 0 \tag{2.46}$$

Since a_i values are arbitrary, Equation 2.46 yields the N equations

$$\frac{\partial\bar{V}}{\partial a_i} = 0 \qquad i = 1,2,\ldots,N \tag{2.47}$$

for the N parameters a_i. In this way, we obtain an approximate solution to Equation 2.42 that is the best in that it renders $\bar{V}$ stationary. In addition, we obtain a value of V that is accurate to second order in the error in the approximate solution for ϕ. Hence, if V_0 is the exact value of V, and if $\delta\phi$ is the error in the approximate solution of ϕ (i.e., $\bar{\phi} = \phi + \delta\phi$), then

$$\bar{V} = V_0 + O(\delta\phi^2) \tag{2.48}$$

The concept can be extended directly to multidimensional problems (Langhaar, 1989, Chapter 3). Often, in the multidimensional case, a semidirect approach is employed (Kantorovich and Krylov, 1964). For example, for two-dimensional problems $\phi = \phi(x, y)$, Equation 2.43 is replaced by

$$\bar{\phi} = \sum a_i(y)\phi_i(x) \tag{2.49}$$

where now the parameters a_i are unknown functions of y. Then substitution of Equation 2.49 into a functional

$$V = \int_{x_0}^{x_1} dx \int_{y_0}^{y_1} F(x, y, \phi)\, dy \tag{2.50}$$

yields

$$\bar{V} = \int_{y_0}^{y_1} G(y, a_1, a_2, \ldots, a_N)\, dy \tag{2.51}$$

Then, setting the variation of $\bar{V}$ equal to zero, we obtain a set of ordinary differential equations that must be solved for the $a_i(y)$.

In statical elasticity problems, the functional V is the potential energy of the system, and the Euler equations are the equations of equilibrium. By the approximation method outlined earlier, a mechanical system (an elastic body, say) with infinitely many degrees of freedom is reduced to a system with finite degrees of freedom, since the a_i, $i = 1, 2, 3,\ldots, N$ may be thought of as generalized coordinates (Langhaar, 1989, Section 3.11). Rayleigh (1976) employed this idea in the study of vibrations of elastic bodies. Ritz (1909) refined and generalized Rayleigh's method, Ritz's method being equivalent to the procedure that leads to Equation 2.47. In elasticity problems, the condition $\delta V = 0$ is called the *principle of stationary potential energy*. Ritz's method approximates the condition $\delta V = 0$ by the conditions $\partial \bar{V} / \partial a_i = 0$, $i = 1, 2, \ldots, N$. These equations determine the a_i. Consequently, an approximation of ϕ is obtained by Equation 2.43.

Example 2.6: Rayleigh–Ritz Method Applied to a Simple Beam

The Rayleigh–Ritz method may be employed to approximate a continuous system by a system with a finite number of degrees of freedom. By way of illustration, consider a simply supported beam of length L with lateral point load P applied at its midsection. The Bernoulli–Euler beam theory leads to the result

$$y(x) = \frac{PL^2 x}{16EI} - \frac{Px^3}{12EI} \tag{2.52}$$

for the deflection $y(x)$, where x denotes the axial coordinate from one end of the beam, E denotes the modulus of elasticity, and I is the moment of inertia of the cross-sectional area of the beam.

The internal potential energy (strain energy) of the beam due to bending is

$$U = \frac{1}{2}\int_0^L EI(y'')^2\, dx \tag{2.53}$$

where the primes denote differentiation with respect to x. The potential energy of external loads is

$$\Omega = -Py\left(\frac{L}{2}\right) \tag{2.54}$$

Hence, the total potential energy of the beam is

$$V = U + \Omega = -Py\left(\frac{L}{2}\right) + \frac{1}{2}\int_0^L EI(y'')^2\, dx \tag{2.55}$$

As a one-degree-of-freedom approximation, we let the deflection $y(x)$ be approximated by

$$\bar{y} = a \sin\frac{\pi x}{L} \tag{2.56}$$

where a is an unknown parameter. Substitution of Equation 2.56 into Equation 2.55 yields an approximation for V, namely,

$$\bar{V} = -aP + \frac{\pi^4 a^2 EI}{4L^3} \tag{2.57}$$

Table 2.1 Approximation of y(x)

x/L	EIy/PL^3	$EI\bar{y}/PL^3$	Error (%)
0	0	0	0
0.25	0.01432	0.01422	0.7
0.50	0.0208	0.0202	2.9

Substitution of Equation 2.57 into 2.47 yields

$$a = \frac{2PL^3}{\pi^4 EI} \tag{2.58}$$

Hence,

$$\bar{y} = \frac{2PL^3}{\pi^4 EI} \sin\frac{\pi x}{L} \tag{2.59}$$

Equation 2.59 represents a fairly good approximation of $y(x)$, Equation 2.52, as shown in Table 2.1. However, the derivatives of y are not approximated as accurately by derivatives of $\bar{y}$. For example, the bending moment of the beam is related to the second derivative of y by the relation

$$M = -EIy'' \tag{2.60}$$

or by Equation 2.52,

$$M = \frac{1}{2}Px$$

Alternatively, substitution of $\bar{y}''$ for y'' in Equation 2.60 yields the approximation $\bar{M}$ of M:

$$\bar{M} = \frac{2PL}{\pi^2} \sin\frac{\pi x}{L} \tag{2.61}$$

Hence, $M(L/2) = 0.25PL$, whereas $\bar{M}(L/2) = 0.203PL$, an error of approximately 19%. This result is fairly typical of approximations of the function y by $\bar{y}$ (primary functions), the higher derivatives of y (derived functions) being less accurately approximated by those of $\bar{y}$ than y is approximated by $\bar{y}$.

2.6 RITZ METHOD REVISITED AND TREFFTZ METHOD

As shown in Section 7.10 of Boresi et al. (2011), the torsion problem for a simply connected region is represented by the equations

$$\begin{aligned} \nabla^2\phi &= -2G\beta \qquad \text{over } R \\ \phi &= 0 \qquad \text{on C} \end{aligned} \tag{2.62}$$

In terms of the stress function ϕ, the stress components τ_{xz}, τ_{yz} are given by the equations

$$\tau_{xz} = \frac{\partial \phi}{\partial y} \qquad \tau_{yz} = -\frac{\partial \phi}{\partial x} \tag{2.63}$$

and the twisting moment M, by

$$M = 2\int\int \phi \, dx \, dy \tag{2.64}$$

Equation 2.64 is essentially a boundary condition over the end planes (see Boresi et al., 2011, Section 7.3).

2.6.1 Ritz method applied to a rectangular cross-section

We seek to obtain an approximate solution to the torsion problem of a rectangular cross-section $-a \le x \le a$, $-b \le y \le b$ (Boresi et al., 2011, Section 7.10) by the Ritz method. This method (Section 2.5) is an interior method (Section 2.3) in that trial functions $\bar{\phi}$ are chosen to satisfy the boundary conditions. The arbitrary constants in the trial functions are then selected to provide a minimum for the integral over the cross-section of the square of the error gradient (Ritz, 1909; Timoshenko and Goodier, 1982). Thus, it may be shown that the error (residual) integral is minimized if

$$\int\int_R \text{grad}^2\bar{\phi} \, dx \, dy = \text{minimum} \tag{2.65}$$

provided $\bar{\phi}$ is such that the boundary conditions (the second of Equations 2.62 and 2.64) are satisfied.

The trial function $\bar{\phi}$ is chosen to satisfy the boundary condition on C (i.e., $\bar{\phi} = 0$ on C). The boundary condition on the end planes (Equation 2.64) may be introduced into Equation 2.65 by means of the Lagrange multiplier method (see Boresi et al., 2011, Sections 1.29 and 2.11). Thus, we define the auxiliary function to include the end plane boundary condition

$$F = \int\int_R \left[\text{grad}^2\bar{\phi} - L\bar{\phi}\right] dx \, dy \tag{2.66}$$

where L is the Lagrange multiplier. The problem now is to determine $\bar{\phi}$ such that F is minimal and $\bar{\phi} = 0$ on C.

We may show that the problem of minimizing the function F is equivalent to minimizing the potential energy of the torsional system. We first note that the strain energy U in torsion is given by the equation (see Boresi et al., 2011, Equation 4.6.13)

$$U = \frac{1}{2G}\int\int_R \left(\tau_{xz}^2 + \tau_{yz}^2\right) dx \, dy \tag{2.67}$$

Substitution of Equations 2.63 into 2.67 yields (in terms of the trial function $\bar{\phi}$)

$$U = \frac{1}{2G}\iint_R \left[\left(\frac{\partial\bar{\phi}}{\partial y}\right)^2 + \left(\frac{\partial\bar{\phi}}{\partial x}\right)^2\right] dx\,dy = \frac{1}{2G}\iint_R \text{grad}^2\bar{\phi}\; dx\,dy \tag{2.68}$$

The potential energy Ω of the twisting moment (the external load) is given by

$$\Omega = -M\beta \tag{2.69}$$

By Equations 2.64 and 2.69, we have (in terms of the trial function $\bar{\phi}$)

$$\Omega = -2\iint_R \bar{\phi}\beta\, dx\,dy \tag{2.70}$$

Hence, the total potential energy $V = U + \Omega$ of the system is given by (in terms of the trial function $\bar{\phi}$)

$$2GV = \iint_R [\text{grad}^2\,\bar{\phi} - 4G\beta\bar{\phi}]\, dx\,dy \tag{2.71}$$

Comparison of Equations 2.66 and 2.71 shows that minimizing F is equivalent to minimizing $2GV$ with $L = 4G\beta$.

To satisfy the lateral boundary conditions $\phi = 0$ on C (Figure 2.1; see also Figure 7.10.1 of Boresi et al., 2011), we may take

$$\bar{\phi} = (x^2 - a^2)(y^2 - b^2)\sum_{m=0}^{M}\sum_{n=0}^{N} A_{mn}x^m y^n \tag{2.72}$$

Since ϕ is even in (x, y), only even values of m and n need be considered.

2.6.2 First approximation

To simplify the calculations, we consider a square cross-section ($a = b$). Then, as a first approximation, we may take

$$\bar{\phi}_1 ,= A_{00}(x^2 - a^2)(y^2 - a^2) \tag{2.73}$$

Substitution of Equation 2.73 into 2.66 yields, after integration,

$$F = a^6\left(\frac{256}{45}a^2A_{00}^2 - \frac{16}{9}LA_{00}\right) \tag{2.74}$$

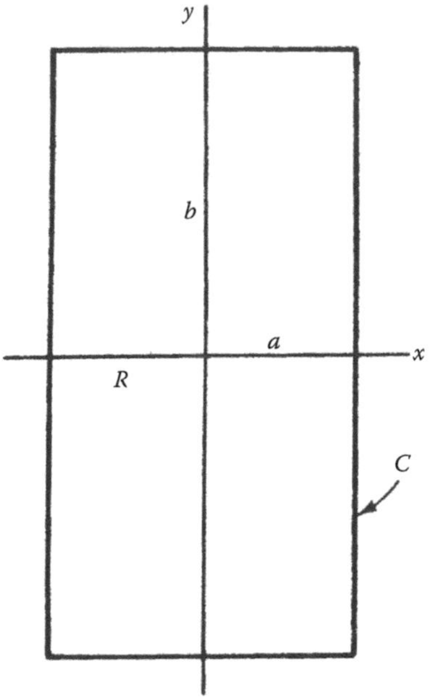

Figure 2.1 Bar with rectangular section. (From Boresi, A. P., and Chong, K. P., *Elasticity in Engineering Mechanics*, Wiley, New York 2000.)

and

$$\frac{\partial F}{\partial A_{00}} = \frac{512}{45}a^8 A_{00} - \frac{16}{9}a^6 L = 0 \tag{2.75}$$

With the condition $L = 4G\beta$, Equation 2.75 yields the result

$$A_{00} = \frac{5G\beta}{8a^2} \tag{2.76}$$

and by Equations 2.63, 2.64, 2.73, and 2.76,

$$\begin{aligned}
(\tau_{xz})_1 &= \frac{5}{4}\frac{G\beta}{a^2}(x^2 - a^2)y \\
(\tau_{yz})_1 &= -\frac{5}{4}\frac{G\beta}{a^2}(y^2 - a^2)x \\
M_1 &= \frac{20}{9}G\beta a^4
\end{aligned} \tag{2.77}$$

2.6.3 Second approximation

We next take (Equation 2.72)

$$\bar{\phi}_2 = \left[A_{00} + A_{22}(x^2 + y^2)\right](x^2 - a^2)(y^2 - a^2) \tag{2.78}$$

Proceeding in the manner outlined earlier, we obtain the following results:

$$\begin{aligned}(\tau_{xz})_2 &= \left[7770 + \frac{1575}{a^2}(x^2 - a^2 + 2y^2)\right]\frac{G\beta(x^2 - a^2)y}{6648a^2} \\ (\tau_{yz})_2 &= -\left[7770 + \frac{1575}{a^2}(y^2 - a^2 + 2x^2)\right]\frac{G\beta(y^2 - a^2)x}{6648a^2} \\ M_2 &= 2.245G\beta a^2\end{aligned} \tag{2.79}$$

Comparison of the results for the first and second approximations to the exact results (Boresi et al., 2011 Section 7.10) shows that the following hold for the Ritz method:

For given β:

$$M_{\text{exact}} > M_{N+1} > M_N > \cdots > M_2 > M_1 \tag{2.80}$$

For given M:

$$\beta_{\text{exact}} < \beta_{N+1} < \beta_N < \cdots < \beta_2 < \beta_1 \tag{2.81}$$

However, the comparisons of the stresses provide no such simple relations. For example, for $x = a$, $y = 0$,

$$\left|(\tau_{yz})_2\right| > \left|(\tau_{yz})_1\right|$$

whereas for $x = \sqrt{2}\, a/2$, $y = 0$,

$$\left|(\tau_{yz})_2\right| < \left|(\tau_{yz})_1\right|$$

The overall nature of the stress pattern is reflected by M and indicates that over a portion of the cross-section, the stresses predicted are low in absolute value for a given value of twist β.

2.6.4 Trefftz method applied to a rectangular cross-section

In contrast to the Ritz method, the Trefftz method (Trefftz, 1927) is based on choosing trial functions $\bar{\phi}$ such that the differential equation (Equation 2.62) is satisfied over R, and the arbitrary constants in the trial functions are chosen to minimize the integral, over the cross-section, of the error gradient. That is, the method is a *boundary method*. With this method, the approximate magnitude of the twisting moment is larger than its exact value, in contrast to the Ritz method (Equation 2.80). Hence, by using the Ritz method in conjunction with the Trefftz method, the exact value of the twisting moment M (hence the twist β) may be bounded from above and below.

Since the Trefftz method is a boundary method, we take the trial solution form

$$\bar{\phi} = -\frac{1}{2}G\beta(x^2 + y^2) + \sum_{i=1}^{N} B_i P_i \tag{2.82}$$

where $-G\beta(x^2 + y^2)/2$ is a particular solution of Equation 2.62, the B_i are constants, and the P_i are potential functions ($\nabla^2 P_i = 0$).

If ϕ is the exact (true) solution, we may take the error gradient to be grad($\bar{\phi} - \phi$). Thus, we require the minimization relative to B_i of

$$E = \iint_R \text{grad}^2(\bar{\phi} - \phi)\, dx\, dy \tag{2.83}$$

that is

$$\frac{\partial E}{\partial B_i} = 2\iint_R \text{grad}(\bar{\phi} - \phi)\,\text{grad}\frac{\partial \bar{\phi}}{\partial B_i}\, dx\, dy = 0 \tag{2.84}$$

where, by Equation 2.82,

$$\frac{\partial \bar{\phi}}{\partial B_i} = P_i \tag{2.85}$$

Thus,

$$\iint_R \text{grad}(\bar{\phi} - \phi)\,\text{grad}(P_i)\, dx\, dy = 0 \tag{2.86}$$

or, with reference to Equation 1.16.6 of Boresi et al. (2011),

$$\int_C (\bar{\phi} - \phi)\frac{\partial P_i}{\partial n} dS - \iint_R (\bar{\phi} - \phi)\nabla^2 P_i\, dx\, dy = 0 \tag{2.87}$$

where n is the outward-directed normal vector on C. Since $\nabla^2 P_i = 0$, Equation 2.87 reduces to

$$\int_C \bar{\phi}\frac{\partial P_i}{\partial n} dS = \int_C \phi\frac{\partial P_i}{\partial n} dS \tag{2.88}$$

Hence, since in this case $\phi = 0$ on C, our problem is further reduced to requiring that

$$\int_C \bar{\phi}\frac{\partial P_i}{\partial n} dS = 0 \tag{2.89}$$

Again, for simplicity, we consider the square cross-section $a = b$, and as a first approximation, we take

$$\bar{\phi} = \bar{\phi}_1 = -\frac{G\beta}{2}(x^2 + y^2) + B_1(x^4 - 6x^2y^2 + y^4) \tag{2.90}$$

Thus, $P_1 = x^4 - 6x^2y^2 + y^4 = P$.

For the square cross-section Equation 2.89 becomes, with Equation 2.90,

$$\int_C \bar{\phi}\frac{\partial P}{\partial n}dS = 2\int_{-a}^{a} \bar{\phi}\frac{\partial P}{\partial x}\bigg|_{x=a} dy + 2\int_{-a}^{a} \bar{\phi}\frac{\partial P}{\partial y}\bigg|_{y=a} dx = 0 \tag{2.91}$$

Substitution of Equation 2.90 into 2.91 yields, after integration and solution for B_1,

$$\bar{\phi}_1 = -\frac{G\beta}{2}\left[x^2 + y^2 + \frac{7}{36a^2}(x^4 - 6x^2y^2 + y^4)\right] \tag{2.92}$$

and

$$\begin{aligned} \tau_{xz} &= \frac{\partial \bar{\phi}_1}{\partial y} = -\frac{G\beta}{2}\left[2y + \frac{28}{36a^2}(y^3 - 3x^2y)\right] \\ \tau_{yz} &= -\frac{\partial \bar{\phi}_1}{\partial x} = \frac{G\beta}{2}\left[2x + \frac{28}{36a^2}(x^3 - 3y^2x)\right] \end{aligned} \tag{2.93}$$

Since Equation 2.92 does not satisfy identically the condition $\phi = 0$ on C, Equation 2.64 gives an inaccurate result for M. Hence, we compute M by means of the expression for M following Equation 7.3.5 of Boresi et al., 2011; that is,

$$M = -\iint_R \left(x\frac{\partial \bar{\phi}}{\partial x} + y\frac{\partial \bar{\phi}}{\partial y}\right)dx\,dy \tag{2.94}$$

Substitution of Equation 2.92 into 2.94 yields

$$M = 2.253G\beta a^4 \tag{2.95}$$

This procedure may be continued by considering additional terms in ϕ (Equation 2.82).

2.6.5 Bounds on torsion solution

The Ritz solution gives an upper bound for β and a lower bound for M, whereas the Trefftz method provides a lower bound for the trial function $\bar{\phi}$ (Trefftz, 1928). Since $\bar{\phi}$ is proportional to β, the Trefftz method therefore gives a lower bound for β and an upper bound for M, that is,

$$\begin{aligned} &\beta_{\text{Trefftz}} < \beta_{\text{exact}} < \beta_{\text{Ritz}} \\ &M_{\text{Trefftz}} > M_{\text{exact}} > M_{\text{Ritz}} \end{aligned}$$

In the example discussed previously, the results are

$$0.4444\frac{M}{Ga^4} < \beta_{exact} < 0.445\frac{M}{Ga^4}$$

$$2.253\,G\beta a^4 > M_{exact} > 2.245\,G\beta a^4$$

giving very close bounds on the exact results.

For stresses, the comparison is more difficult. However, comparisons for maximum stress can be given. For example, for the square cross-section, τ_{max} occurs at the midpoint of a side, say, at $x = a$, $y = 0$.

Then, for the Ritz solution, Equations 2.77 and 2.79 yield $(\tau_{max})_1 = 1.25G\beta a$, $(\tau_{max})_2 = 1.41G\beta a$, and Equation 2.93 yields for the Trefftz method $\tau_{max} = 1.39G\beta a$. The series solution (Boresi et al., 2011, Section 7.10; Timoshenko and Goodier, 1982) yields $\tau_{max} = 1.35G\beta a = (\tau_{max})_{exact}$. Unfortunately, the stresses are not bounded.

REFERENCES

Allen, M. B., III, and Isaacson, E. L. 2011. *Numerical Analysis for Applied Science*, Wiley, New York.

Bathe, K. J. 1995. *Finite Element Procedures*, Prentice Hall, Upper Saddle River, NJ.

Boresi, A. P., and Chong, K. P. 2000. *Elasticity in Engineering Mechanics*, Wiley, New York.

Boresi, A. P., Chong, K. P., and Lee, J. D. 2011. *Elasticity in Engineering Mechanics*, Wiley, New York.

Botha, D. K., and Pinder, G. F. 1983. *Fundamental Concepts in the Numerical Solution of Differential Equations*, Wiley, New York.

Collatz, L. 2012. *The Numerical Treatment of Differential Equations* (Vol. 60), Springer-Verlag, Berlin, Germany.

Cook, R. D. 2007. *Concepts and Applications of Finite Element Analysis*, Wiley, New York.

Crandall, S. H. 1983. *Engineering Analysis*, Krieger, Publ. Co., Inc., Melbourne, FL.

Finlayson, B. A. 2013. *The Method of Weighted Residuals and Variational Principles* (Vol. 73), SIAM, Philadelphia, PA.

Finlayson, B. A., and Scriven, L. E. 1966. The method of weighted residuals: A review, *Appl. Mech. Rev.*, **19**(9).

Hughes, T. J. R. 2012. *The Finite Element Method: Linear Static and Dynamic Analysis*, Courier Corp., North Chelmsford, MA.

Kantorovich, L., and Krylov, V. 1964. *Approximate Methods of Higher Analysis*, Wiley, New York.

Langhaar, H. L. 1989. *Energy Methods in Applied Mechanics*, Krieger Publishing, Melbourne, FL.

Rayleigh, Lord J. S. 1976. *Theory of Sound*, Dover Publications, Mineola, NY.

Reddy, J. N. 1991. *Applied Functional Analysis and Variational Methods in Engineering*, Krieger Publishing, Melbourne, FL.

Ritz, W. 1909. Uber eine neue Methode zur Lösung gewisser Variations probleme der mathematischen Physik, *J. Reine Angew. Math.*, **135**, 1–61.

Shisha, O. 1968. Trends in approximation theory, *Appl. Mech. Rev.*, **21**, 4.

Timoshenko, S. P., and Goodier, J. N. 1982. *Theory of Elasticity*, McGraw-Hill, New York.

Trefftz, E. 1927. Ein Gegenstück zum Ritzschen Verfahren, *Verh. Zweiten Kongress Tech. Mech.* (Zurich), pp. 131–137.

Trefftz, E. 1928. Konvergenz und Fehlerschätzung beim Ritzschen Verfahren, *Math. Ann.*, **100**, 503–521.

Yang, H. Y., and Zhao, Z. G. 1985. The mixed method of line-location with its applications in elasticity, *Comput. Struct.*, **21**, 671–680.

Zienkiewicz, O. C., and Morgan, K. 2006. *Finite Elements and Approximation*, Courier Corp.

BIBLIOGRAPHY

Brebbia, C. A. (ed.) 1983. *Progress in Boundary Element Methods* (Vol. 2), Springer-Verlag, New York.

Brebbia, C. A., Telles, J. C. F. and Wrobel, L. 2012. *Boundary Element Techniques: Theory and Applications in Engineering*, Springer Science & Business Media, New York.

Cakmak, A. S., and Botha, J. F. 1995. *Applied Mathematics for Engineers*, Computational Mechanics, Billerica, MA.

Dhatt, G, and Touzot, G. 1984. *The Finite Element Method Displayed*, Wiley, New York.

Finlayson, B. A. 1992. *Numerical Methods for Problems with Moving Fronts*, Ravenna Park Publishing, Seattle, WA.

Finlayson, B. A. 2013. *The Method of Weighted Residuals and Variational Principles* (Vol. 73), SIAM.

Krasnoselskii, M. A., Vainikko, G. M., Zabreyko, R. P., Rutícki, Y. B. and Stet'senko, V. V. 2012. *Approximate Solution of Operator Equations*, Springer Science & Business Media.

Langhaar, H. L. 1969. Two numerical methods that converge to the method of least squares, *J. Franklin Inst.*, **288**, 165–173.

Lapidus, L., and Pinder, G. F. 2011. *Numerical Solution of Partial Differential Equations in Science and Engineering*, Wiley, New York.

Mikhlin, S. G. 1964. *Variational Methods in Mathematical Physics*, Macmillan, New York.

Mikhlin, S. G, and Smolitskii, K. L. 1967. *Approximate Methods for Solution of Differential and Integral Equations*, Elsevier Science, New York.

Morse, P. M., and Feshbach, H. 1953. *Methods of Theoretical Physics*, McGraw-Hill, New York, pp. 122–123.

Yang, T. Y. 1986. *Finite Element Structural Analysis*, Prentice Hall, Upper Saddle River, NJ.

Chapter 3

Finite difference methods

3.1 PRELIMINARY REMARKS AND CONCEPTS

The basic concept in finite difference methods, as applied to boundary-value problems, is the representation of governing differential equations and associated boundary conditions by appropriate finite difference equations. This replacement is accomplished by approximating derivatives in the differential equations with finite difference quotients that are combinations of dependent (unknown) function values at specified values of the independent variables. By writing the difference equations at specified values of the independent variables, we are led to systems of simultaneous algebraic equations that may be solved by elementary means with the aid of high-speed computers. Accordingly, we interpret a finite difference method as a numerical procedure that approximates known *exact** differential equations and boundary conditions—say, of an elasticity problem. Then we solve the resulting approximate equations exactly or approximately. On the other hand, we shall see in Chapter 4 that finite element methods approximate the elastic continua by assemblages of discrete elastic systems. Then we solve the resulting discrete systems exactly or approximately.

To illustrate the approximation of derivatives by corresponding difference quotients, consider the one-dimensional case. Let $f(x)$ be a continuous function of the single independent variable x, which is defined in the range $a_0 \le x \le a_n$ or $[a_0, a_n]$. We divide $[a_0, a_n]$ into n equal or unequal parts and denote the subdividing points as $a_0, a_1, a_2, \ldots, a_n$. These points are called *pivotal points*. From the theory of interpolation polynomials, we construct an interpolation polynomial P_n of degree n satisfying the conditions $f(a_i) = P_n(a_i)$, $i = 0, 1, 2, \ldots, n$. Such an interpolation polynomial is expressible as a linear combination of function values $f(a_1)$ (Isaacson and Keller, 1994). Accordingly, we write $f(x) = P_n(x) + R_n(x)$ for all x in $[a_0, a_n]$, where $R_n(x)$ is a remainder term or simply the error in representation of $f(x)$ by $P_n(x)$. Since the derivatives $d^r f(x)/dx^r = d^r P_n(x)/dx^r + d^r R_n(x)/dx^r$, we use the derivatives of the interpolation polynomial to replace approximately the derivatives of the interpolated function $f(x)$.

If the derivatives of the error terms can be estimated, they provide a means of judging the accuracy of the approximation. Upon suppressing the error term, we have the approximation $d^r f(x)/dx^r \approx d^r P_n(x)/dx^r$, where now the $P_n(x)$ is no longer expressed in terms of exact pivotal function values $f(a_i)$, but rather in terms of a linear combination of approximation values of $f(a_i)$, which we denote by $F(a_i)$. Proceeding in this manner, we express all derivatives in the governing differential equations and the boundary conditions in terms of finite difference equations. Evaluation of these finite difference equations at the specified pivotal

* The term *exact* is used in a sense that the differential equations are derived exactly under the framework of the basic assumptions of the theory.

points in the domain of a boundary-value problem leads to a set of algebraic equations in $F(a_i)$. For linear boundary-value problems, these algebraic equations form a system of linear simultaneous equations. Otherwise, they form a system of nonlinear algebraic equations.

3.1.1 Finite differences, finite elements, and weighted residual methods

Much effort has been directed toward the unification of various approximation processes used in the numerical solution of physical problems governed by suitable differential equations (Bushnell, 1973; Zienkiewicz and Morgan, 2006). The finite difference process discussed in this chapter appears to present an entirely different type of approach from that of the weighted residual methods of Chapter 2 and also from the finite element methods of Chapter 4. However, Zienkiewicz and Morgan (2006) have shown that in several one- and two-dimensional cases, the use of finite differences has led to equations that are either identical or extremely similar to those obtained by simple finite elements. They have also shown that the common link between these approximation processes is expansion of the unknown function in terms of shape or basis functions and unknown parameters and the determination of such parameters from a set of weighted residual equations. They coined the phrase *generalized finite element method* to include all the approaches mentioned. With such a generalized approach, computer program organization and the theory to encompass all the approximation processes are unified. Furthermore, arguments concerning the superiority of finite difference methods over finite element methods (or vice versa) become meaningless, as each subclass possesses its own merits in special circumstances.

3.2 DIVIDED DIFFERENCES AND INTERPOLATION FORMULAS

In the theory of interpolating polynomials, *divided differences* play an important role. In this section, we present a few useful properties of divided differences. Consider arbitrary pivotal points $a_0, a_1, a_2, \ldots, a_n$ of the independent variable x and the corresponding function values $f(a_0), f(a_1), f(a_2), \ldots, f(a_n)$. The zero-order divided difference is defined as $f[a_0] = f(a_0)$. First-, second-, and higher-order divided differences are defined by

$$
\begin{aligned}
f[a_0,a_1] &= \frac{f(a_1)-f(a_0)}{a_1-a_0} = \frac{f(a_0)}{a_0-a_1} + \frac{f(a_1)}{a_1-a_0} \\
f[a_0,a_1,a_2] &= \frac{f[a_1,a_2]-f[a_0,a_1]}{a_2-a_0} = \frac{f(a_0)}{(a_0-a_1)(a_0-a_2)} \\
&\quad + \frac{f(a_1)}{(a_1-a_0)(a_1-a_2)} + \frac{f(a_2)}{(a_2-a_0)(a_2-a_1)} \\
&\;\;\vdots \\
f[a_0,a_1,a_2,\ldots,a_n] &= \frac{f[a_1,a_2,\ldots,a_n]-f[a_0,a_1,a_2,\ldots,a_{n-1}]}{a_n-a_0} \\
&= \sum_{j=0}^{n} \frac{f(a_j)}{(a_j-a_0)(a_j-a_1)\ldots(a_j-a_{j-1})(a_j-a_{j+1})\ldots(a_j-a_n)}
\end{aligned}
\tag{3.1}
$$

By Equation 3.1, note that the divided differences remain invariant with respect to any permutation of pivotal points and are simply linear combinations of the pivotal function values $f(a_j)$ at the pivotal points $a_0, a_1, \ldots, a_n$.

The basic form of an interpolation formula due to Newton may be derived directly from the definitions of divided differences. Consider the first-order divided difference with arguments a_0 and x:

$$f[a_0,x]=\frac{f(x)-f(a_0)}{x-a_0}$$

Hence,

$$f(x)=f[a_0]+(x-a_0)f[a_0,x] \tag{3.2}$$

By expanding expressions $f[a_1, a_0, x]$, $f[a_2, a_0, a_1, x]$,... and $f[a_n, a_0, \ldots, a_{n-1}, x]$ according to Equation 3.1, we further have

$$\begin{aligned} f[a_0,x]&=f[a_0,a_1]+(x-a_1)f[a_0,a_1,x]\\ &\vdots\\ f[a_0,a_1,\ldots,a_{n-1},x]&=f[a_0,a_1,\ldots,a_n]+(x-a_n)f[a_0,a_1,\ldots,a_n,x] \end{aligned} \tag{3.3}$$

Successive substitution of Equation 3.3 into 3.2 yields

$$\begin{aligned} f(x)=f[a_0]&+(x-a_0)f[a_0,a_1]+(x-a_0)(x-a_1)f[a_0,a_1,a_2]\\ &+\cdots+(x-a_0)(x-a_1)(x-a_2)\cdots(x-a_{n-1})f[a_0,a_1,\ldots,a_n]+R_n(x) \end{aligned} \tag{3.4}$$

where $R_n(x)$, the remainder or error term, is given by

$$R_n(x)=\omega_n(x)f[a_0,a_1,\ldots,a_n,x] \tag{3.5}$$

where

$$\omega_n(x)=(x-a_0)(x-a_1)\cdots(x-a_n) \tag{3.6}$$

Equation 3.4 may be written in the form

$$f(x)=P_n(x)+R_n(x) \tag{3.7}$$

where $P_n(x)$ is the nth degree (Newton) interpolation polynomial for the function $f(x)$ in terms of divided differences. Accordingly, $P_n(x)$ depends on the $(n + 1)$ pivotal function values at the $(n + 1)$ distinct points, 0, 1, 2, ..., n. Suppressing the error term in Equation 3.7, we obtain the approximation $f(x) \approx P_n(x)$. If an approximate representation of the rth $(r \leq n)$ derivative of $f(x)$ is desired, we may differentiate Equation 3.7 r times to obtain

$$\frac{d^r f(x)}{dx^r} \approx \frac{d^r P_n(x)}{dx^r} \quad \text{or simply} \quad f^{(r)}(x) \approx P_n^{(r)}(x) \tag{3.8}$$

with the associated error $R_n^{(r)}(x)$, where the r superscript in parentheses denotes the order of differentiation. An estimate of the error term $R_n^{(r)}(x)$, may reveal the accuracy and the rate of convergence of the approximation.

For assessment of the error, a convenient representation of $R_n^{(r)}(x)$, is required. For this representation, we note two important properties of divided differences. They are stated as follows (with brief proofs).

Property I: Let $a_0, a_1, \ldots, a_{p-1}, x$ be $(p + 1)$ distinct pivotal points, and let $f(x)$ possess p continuous derivatives in the closed interval $I = [a_0, x]$ contained between the smallest and largest pivotal points. Then there exists a point $\xi = \xi(x)$ in the interval I such that

$$f[a_0, a_1, \ldots, a_{p-1}, x] = \frac{f^{(p)}(\xi)}{(p)!} \tag{3.9}$$

Proof: Denote by $P_{p-1}(x)$ the $(p - 1)$th degree interpolation polynomial for $f(x)$ with respect to the p pivotal points $a_0, a_1 \ldots, a_{p-1}$. Then $f(x) - P_{p-1}(x)$ and $\omega_{p-1}(x)$ (see Equation 3.6) vanish at the p points. Define a linear combination of these functions as

$$G(x) = f(x) - P_{p-1}(x) - \alpha\omega_{p-1}(x) \tag{3.10}$$

and note that $G(x_i) = 0$ for $x_i = a_i$ $(i = 0, 1, 2, \ldots, p - 1)$. For any arbitrary value x distinct from a_i, $\omega_{p-1}(x) \neq 0$; hence, we may select the constant α so that $G(x)$ vanishes. Since $G(x_i) = 0$ at $x_i = a_0, a_1, \ldots, a_{p-1}, x$, there are $(p + 1)$ zeroes for $G(x)$ in the interval I. By Rolle's theorem (Thomas and Finney, 1984), $dG(x)/dx$ vanishes at least p times inside I. Repeated application of Rolle's theorem shows that $d^2G(x)/dx^2 = 0$ at least $(p - 1)$ times and $d^rG(x)/dx^r = 0$ at least $(p - r + 1)$ times inside $I(r \leq p)$. Accordingly, there must be a point $\xi = \xi(x)$ inside the interval I that yields $G^p(\xi) = 0$. Equation 3.10 then yields

$$\alpha = \frac{f^{(p)}(\xi) - P_{p-1}^{(p)}(\xi)}{\omega_{p-1}^{(p)}(\xi)}$$

The fact that $P_{p-1}(x)$ is a polynomial of degree $(p - 1)$ at most and the fact that $\omega_{p-1}(x)$ is a polynomial of degree p show, respectively, that $P_{p-1}^{(p)}(\xi) = 0$ and $\omega_{p-1}^{(p)}(\xi) = p!$. Thus the constant α becomes $\alpha = f^{(p)}(\xi)/p!$. Since we have required $G(x) = 0$ at the point x in I, it follows from Equations 3.7 and 3.10 that

$$\omega_{p-1}(x)\frac{f^{(p)}(\xi)}{p!} = f(x) - P_{p-1}(x) \tag{3.11}$$

On the basis of Equations 3.5 and 3.11, Equation 3.9 is proved. Furthermore, note that Equation 3.11 holds for previously excluded values of x corresponding to $a_0, a_1, \ldots, a_{p-1}$. Consequently, Equation 3.9 is valid for all x.

Property II: Let $f^{(p)}(x)$ be continuous in the closed interval I, and let $a_0, a_1 \ldots, a_n, x$ be in I. Then

$$\frac{d^p}{dx^p} f[a_0, a_1, a_2, \ldots, a_n, x] = (p!) f[a_0, a_1, a_2, \ldots, a_n, x, x, \ldots, x] \tag{3.12}$$

where x is repeated $p + 1$ times on the right-hand side.

Proof: Consider a function

$$g(x) = f[a_0, a_1, a_2, \ldots, a_n, x] \tag{3.13}$$

By definition, $g^{(p)}(x)$ is also continuous in I. Letting $b_0, b_1, b_2, \ldots, b_p$ be another set of $p + 1$ points in I, we form (Hartree, 1961)

$$\begin{aligned} g[b_0, b_1] &= \frac{g(b_1) - g(b_0)}{b_1 - b_0} \\ &= \frac{f[a_0, a_1, \ldots, a_n, b_1] - f[a_0, a_1, \ldots, a_n, b_0]}{b_1 - b_0} \\ &= f[a_0, a_1, a_2, \ldots, a_n, b_0, b_1] \\ &\vdots \\ g[b_0, b_1, \ldots, b_p] &= f[a_0, a_1, \ldots, a_n, b_0, b_1, \ldots, b_p] \end{aligned} \tag{3.14}$$

On the basis of Equation 3.9, it follows, by taking b_p as x, that

$$\frac{g^{(p)}(\xi)}{p!} = f[a_0, a_1, a_2 \ldots, a_n, b_0, b_1, \ldots, b_p] \tag{3.15}$$

where ξ is inside the interval I. On account of the continuity of $g^{(p)}(x)$ in the interval I, we let $x = b_0 = b_1 = \cdots = b_p$. Thus, Equation 3.15 reduces to (see Appendix 3A)

$$\frac{g^{(p)}(x)}{p!} = f[a_0, a_1, a_2, \ldots, a_n, x, x, \ldots, x] \tag{3.16}$$

where x is repeated $p + 1$ times on the right-hand side. Finally, by means of Equations 3.13 and 3.16, we deduce Equation 3.12.

Returning to the error term associated with the approximation of Equation 3.8, we note that

$$R_n^{(r)}(x) = \frac{d^r}{dx^r} \omega_n(x) f[a_0, a_1, a_2, \ldots, a_n, x] \tag{3.17}$$

Differentiating the product of two functions according to Leibnitz's rule, we obtain

$$R_n^{(r)}(x) = \sum_{i=0}^{r} \binom{r}{i} \frac{d^i}{dx^i} \omega_n(x) \frac{d^{r-i}}{dx^{r-i}} f[a_0, a_1, a_2, \ldots, a_n, x]$$

where

$$\binom{r}{i} = \frac{r!}{(r-i)!i!}$$

In view of Equation 3.12, we write

$$R_n^{(r)}(x) = \sum_{i=0}^{r} \frac{r!}{i!} \omega_n^{(i)}(x) f[a_0, a_1, a_2, \ldots, a_n, x, x, \ldots, x] \tag{3.18}$$

where x is repeated $(r - i + 1)$ times on the right-hand side. Now, a generalization of Equation 3.9 yields

$$f[a_0, a_1, a_2, \ldots, a_n, x, x, \ldots, x] = \frac{f^{(n+r-i+1)}(\xi_i)}{(n+r-i+1)!} \tag{3.19}$$

where $\min(a_0, a_1, \ldots, a_n, x) < \xi_i(x) < \max(a_0, a_1, \ldots, a_n, x)$ and where x is repeated $r - i + 1$ times on the left-hand side.

Accordingly, substitution of Equation 3.19 into Equation 3.18 leads to

$$R_n^{(r)}(x) = \sum_{i=0}^{r} \frac{r!}{(n+r-i+1)!i!} \omega_n^{(i)}(x) f^{(n+r-i+1)}(\xi_i) \tag{3.20}$$

Since the behavior of higher derivatives of $f(x)$ is generally unknown, some assumptions regarding the boundedness of these derivatives are necessary in estimating the error term. The factor $\omega_n^{(i)}(x)$ is associated with the spacings of the pivotal points. Hence, it serves as an indicator of vanishing error with respect to diminishing pivotal spacings.

Expressions for Newton's interpolation polynomial and its associated error term (Equation 3.4) are derived in terms of divided differences for the interpolation interval $[a_0, a_n]$. They are equally applicable to both nonuniform and uniform spacings of the pivotal points. However, in practice, uniform spacing is widely used. Accordingly, we present some useful interpolation polynomials for equally spaced pivotal points, where the constant interval (spacing) is denoted by h.

Generally, there are three types of interpolation polynomials, depending on where the initial interpolating point a_0 is placed in the interval of interpolation $[a_0, a_n]$. For example, if we evaluate $P_n(x)$ near the right-hand region of the interpolation interval, it is natural to place a_0 close to the right-hand endpoint, so that greater emphasis may be placed on the pivotal function values near this end. Consequently, we introduce three sets of differently ordered pivotal points for *forward*, *backward*, and *central* interpolation polynomials, as shown in Figure 3.1a, b, and c, respectively. Figure 3.1 also indicates the definition of new

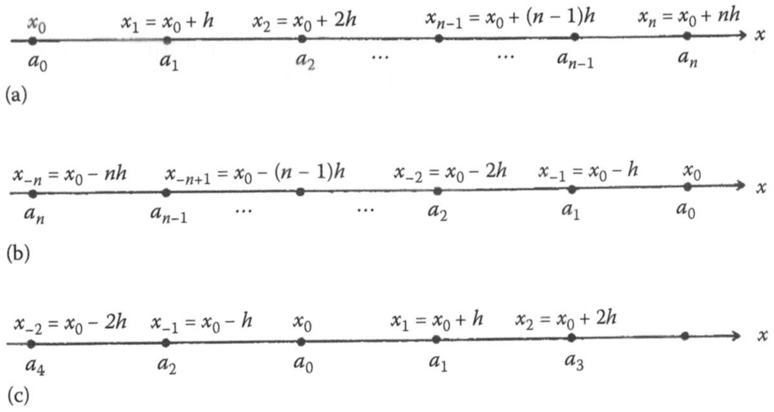

Figure 3.1 Equal pivotal spacings.

systems of pivotal coordinates $x_i = x_0 + ih$ $(i = 0, \pm 1, \pm 2, \ldots)$. This notation enables us to write any pivotal function value $f(x_i)$ in a convenient subscripted form [i.e., $f(x_i) = f(x_0 + ih) = f_{0+i}$].

Corresponding to the forward, backward, and central arrangements of the pivotal points shown in Figure 3.1, we introduce *forward*, *backward*, and *central differences* as follows:

$$\begin{aligned}
&\Delta f(x_i) = f(x_1 + h) - f(x_i) = f_{i+1} - f_i && \text{(forward)}\\
&\nabla f(x_i) = f(x_i) - f(x_i - h) = f_i - f_{i-1} && \text{(backward)}\\
&\delta f(x_i) = f\left(x_i + \frac{1}{2}h\right) - f\left(x_i + \frac{1}{2}h\right) = f_{i+1/2} - f_{i-1/2} && \text{(central)}
\end{aligned} \tag{3.21}$$

Note that the first central difference entails values of $f(x)$ other than those corresponding to values of x_i. In a manner similar to first-order differences, we write formulas for higher-order differences. For example, second-order differences are defined by

$$\begin{aligned}
&\Delta^2 f(x_i) = \Delta(\Delta f(x_i)) = \Delta(f_{i+1} - f_i) = f_{i+2} - 2f_{i+1} + f_i && \text{(forward)}\\
&\nabla^2 f(x_i) = \nabla(\nabla f(x_i)) = \nabla(f_i + f_{i-1}) = f_i - 2f_{i-1} + f_{i-2} && \text{(backward)}\\
&\delta^2 f(x_i) = \delta(\delta f(x_i)) = \delta(f_{i+1/2} - f_{i-1/2}) = f_{i+1} - 2f_i + f_{i-1} && \text{(central)}
\end{aligned} \tag{3.22}$$

We see that the second-order central difference depends on values of f_i. In general, central even differences depend on values of f_i and central odd differences depend on $f_{i+1/2}$. Generalization of Equations 3.21 and 3.22 yields for $r = 1, 2, 3, \ldots,$

$$\begin{aligned}
&\Delta^r f_i = \Delta(\Delta^{r-1} f_i) && \text{(forward)}\\
&\nabla^r f_i = \nabla(\nabla^{r-1} f_i) && \text{(backward)}\\
&\delta^r f_i = \delta(\delta^{r-1} f_i) && \text{(central)}
\end{aligned} \tag{3.23}$$

with $\Delta^0 f_i = \nabla^0 f_i = \delta^0 f_i = f_i$.

3.2.1 Newton's forward and backward interpolation polynomials

Consider the case of equally spaced x_i (Figure 3.1a). Newton's forward interpolation polynomial may be derived with respect to point x_0 using pivotal points to the right of point x_0. We first write the divided differences of Equation 3.1 directly in terms of the forward differences:

$$\begin{aligned} f[x_0,x_1] &= \frac{1}{h}(f_1 - f_0) = \frac{1}{h}\Delta f_0 \\ f[x_0,x_1,x_2] &= \frac{1}{2h}(f[x_1,x_2] - f[x_0,x_1]) = \frac{1}{2!h^2}\Delta^2 f_0 \\ &\vdots \\ f[x_0,x_1,x_2\ldots,x_n] &= \frac{1}{nh}(f[x_1,x_2,\ldots,x_n] - f[x_0,x_1,\ldots,x_{n-1}]) \\ &= \frac{1}{n!h^n}\Delta^n f_0 \end{aligned} \tag{3.24}$$

On denoting x by $x_0 + sh$, where s measures $x - x_0$ in units of h, and eliminating the divided differences by means of Equation 3.24 from the general term of Newton's interpolation polynomial defined in Equation 3.4, we obtain

$$(x - x_0)(x - x_1)\cdots(x - x_{n-1})f[x_0,x_1,\ldots,x_n] = s(s-1)(s-2)\cdots(s-n+1)\frac{\Delta^n f_0}{n!} \tag{3.25}$$

Hence, Newton's forward interpolation polynomial $P_n(x) = P_n(x_0 + sh)$ is expressible as

$$P_n(x_0 + sh) = f_0 + s\Delta f_0 + s(s-1)\frac{\Delta^2 f_0}{2!} + \cdots + s(s-1)\cdots(s-n+1)\frac{\Delta^n f_0}{n!} \tag{3.26}$$

The rth ($r \le n$) derivative of $P_n(x_0 + sh)$ with respect to s then yields the desired approximation to $f^{(r)}(x_0 + sh)$. The associated error can be easily estimated from Equation 3.20, as is demonstrated later in connection with finite difference equations.

If the order of x_i is reversed so that $n - 1$ interpolation points are to the left of x_0 as shown in Figure 3.1b, we may construct Newton's backward interpolation polynomial $P_n(x_0 - sh)$. In a manner analogous to the previous derivation, we obtain

$$P_n(x_0 - sh) = f_0 + (-s)\nabla f_0 + (-s)(-s+1)\frac{\nabla^2 f_0}{2!} + \cdots + (-s)(-s+1)\cdots(-s+n-1)\frac{\nabla^n f_0}{n!} \tag{3.27}$$

The forward and backward differences in Equations 3.26 and 3.27 can be conveniently expressed in terms of linear combinations of pivotal function values f_i. For example, note from Equation 3.1 that

$$f[x_0,x_1,x_2,\ldots,x_n]=\sum_{i=0}^{n}f_i\Bigg/\prod_{\substack{j=0\\j\neq 1}}^{n}(x_i-x_j) \tag{3.28}$$

However,

$$\prod_{\substack{j=0\\j\neq i}}^{n}(x_i-x_j)=\prod_{\substack{j=0\\j\neq i}}^{n}(i-j)h-h^n\prod_{j=0}^{i-1}(i-j)\prod_{r=i+1}^{n}(i-r)=h^n(i)!(n-i)!(-1)^{n-i}$$

Hence, Equation 3.28 may be written in the form

$$f[x_0,x_1,x_2,\ldots,x_n]=\frac{1}{n!h^n}\sum_{i=0}^{n}(-1)^{n-i}\frac{n!}{(n-i)!i!}f_i=\frac{1}{n!h^n}\sum_{i=0}^{n}(-1)^{n-i}\binom{n}{i}f_i \tag{3.29}$$

where

$$\binom{n}{i}=\frac{n!}{(n-i)!i!}$$

On comparing the last equation of Equation 3.24 with 3.29, we deduce that

$$\Delta^n f_0=\sum_{i=0}^{n}(-1)^{n-i}\binom{n}{i}f_i \tag{3.30}$$

In a similar manner, we find that

$$\nabla^n f_0=\sum_{i=0}^{n}(-1)^{i}\binom{n}{i}f_i \tag{3.31}$$

These formulas may be used to facilitate conversion of higher-order differences to their corresponding combinations of function values f_i.

3.2.2 Newton–Gauss interpolation polynomial

Many occasions arise in the application of finite difference techniques where it is required to approximate derivatives of $f(x)$ at interior pivotal points (Chapra and Canale, 2012; Stanton, 1961). To cope with these situations, we develop central interpolation polynomials. Let the arrangement of the pivotal points x_i be centrally distributed about the point x_0 as shown in Figure 3.1c. We derive first the *Newton–Gauss interpolation formula.*

From the definition of the divided differences of Equation 3.1 and the centrally arranged pivotal points x_i, we write

$$\begin{aligned}
f[x_0,x_1] &= \frac{1}{h}(f_1 - f_0) = \frac{1}{h}\delta f_{0+1/2} \\
f[x_{-1},x_0,x_1] &= \frac{f[x_0,x_1]-f[x_{-1},x_0]}{x_1 - x_{-1}} = \frac{1}{2h^2}(\delta f_{0+1/2} - \delta f_{0-1/2}) = \frac{\delta^2 f_0}{2!h^2} \\
&\vdots \\
f[x_{-m+1},\ldots,x_0,\ldots,x_m] &= \frac{\delta^{2m-1} f_{0+1/2}}{(2m-1)!h^{2m-1}} \\
f[x_{-m},\ldots,x_0,\ldots,x_m] &= \frac{\delta^{2m} f_0}{(2m)!h^{2m}}
\end{aligned} \tag{3.32}$$

Next, consider the factors associated with the last two divided differences of Equation 3.32 as they appear to Newton's interpolation polynomial (Equation 3.4). Noting that $x_i = x_0 + ih$ with $i = 0, \pm 1, \pm 2, \ldots, \pm m$, we write ($n = 2m - 1$)

$$\prod_{i=-m}^{m-1}(x - x_i) = (x - x_0)\prod_{i=-1}^{-m+1}(x - x_i)\prod_{i=1}^{m-1}(x - x_i) = sh^{2m-1}\prod_{i=1}^{m-1}(s^2 - i^2) \tag{3.33}$$

where $x = x_0 + sh$ is used. Similarly, we obtain

$$\prod_{i=-m+1}^{m}(x - x_i) = sh^{2m}(s - m)\prod_{i=1}^{m-1}(s^2 - i^2) \tag{3.34}$$

Using Equations 3.32, 3.33, and 3.34, we rewrite Equation 3.4 into the following form, known as the *Newton–Gauss central interpolation formula:*

$$\begin{aligned}
P_{2m}(x_0 + sh) &= f_0 + s\delta f_{0+1/2} + s(s-1)\frac{\delta^2 f_0}{2!} + s(s^2 - 1^2)\frac{\delta^3 f_{0+1/2}}{3!} + s(s^2 - 1^2)(s-2)\frac{\delta^4 f_0}{4!} + \cdots \\
&\quad + s\left[\prod_{i=1}^{m-1}(s^2 - i^2)\right]\frac{\delta^{2m-1} f_{0+1/2}}{(2m-1)!} + s\,(s-m)\left[\prod_{i=1}^{m-1}(s^2 - i^2)\right]\frac{\delta^{2m-1} f_0}{(2m)!}
\end{aligned} \tag{3.35}$$

If the last term is omitted, we obtain the expression for $P_{2m-1}(x_0 + sh)$.

3.2.3 Stirling's central interpretation polynomial

Because of the unsymmetrical nature of the odd-order central differences of Equation 3.35, its practical use is limited. However, we may rearrange the terms and effect a symmetrical representation, called *Stirling's central interpolation formula*, that is commonly used in finite difference methods. In practice, it is natural to select an equal number of pivotal points to the right and to the left of the point x_0. Then we limit ourselves to even-degree interpolation polynomials $P_{2m}(x_0 + sh)$. We then rewrite Equation 3.35 as

$$\begin{aligned} P_{2m}(x_0+sh) &= f_0 + \left[s\delta f_{0+1/2} - \frac{s}{2!}\delta^2 f_0\right] + \frac{s}{2\times 2!}[(s-1)+(s+1)\delta^2 f_0 \\ &+ \left[s(s^2-1^2)\frac{\delta^2 f_{0+1/2}}{3!} - \frac{2s}{4!}\delta^4 f_0\right] + \frac{s(s^2-1^2)}{2\times 4!}[(s-2)+(s+2)]\delta^4 f_0 + \cdots \\ &+ \left\{s\left[\prod_{i=1}^{m-1}(s^2-i^2)\right]\frac{\delta^{2m-1}f_{0+1/2}}{(2m-1)!} - \frac{sm}{(2m)!}\delta^{2m} f_0\right\} \\ &+ \left[\prod_{i=1}^{m-1}(s^2-i^2)\right]\frac{s[(s-m)+(s+m)]}{2\times(2m)!}\delta^{2m} f_0 \end{aligned}$$

By expressing all the even-order central differences in the brackets as combinations of their next-lower central odd differences and then combining the terms within each bracket, we obtain Stirling's formula:

$$\begin{aligned} P_{2m}(x_0+sh) &= f_0 + \frac{s}{2}(\delta f_{0+1/2} + \delta f_{0-1/2}) + \frac{s}{2\times 2!}[(s-1)+(s+1)\delta^2 f_0 \\ &+ \frac{s(s^2-1^2)}{2\times 3!}(\delta^3 f_{0+1/2} + \delta^3 f_{0-1/2}) + \frac{s(s^2-1^2)}{2\times 4!}[(s-2)+(s+2)]\delta^4 f_0 + \cdots \\ &+ \frac{s}{2(2m-1)!}\left[\prod_{i=1}^{m-1}(s^2-i^2)\right](\delta^{2m-1}f_{0+1/2} + \delta^{2m-1}f_{0-1/2}) \\ &+ \frac{s}{2\times(2m)!}\left[\prod_{i=1}^{m-1}(s^2-i^2)\right][(s-m)+(s+m)+(s+m)]\delta^{2m} f_0 \end{aligned} \tag{3.36}$$

This formula, being symmetric about x_0, is widely used in practice. The error term associated with Equation 3.36 is exactly that of Equation 3.35.

In the finite difference approximation, the central differences appearing in Equation 3.36 must be expanded into linear combinations of f_i. Theoretically, this can be accomplished by successively lowering the order of the particular central difference according to Equation 3.23, but it is a rather cumbersome procedure. Accordingly, we seek a more expedient representation of the central differences. The dual representations of the divided difference $f[x_{-m}, \ldots, x_0, \ldots, x_m]$ from Equations 3.1 and 3.32 allow us to write

$$\sum_{i=-m}^{m} f_i \Bigg/ \prod_{\substack{j=-m \\ j\neq 1}}^{m} (x_i - x_j) = \frac{\delta^{2m} f_0}{(2m)!\,h^{2m}} \tag{3.37}$$

To simplify this equation, note the following relation (see Appendix 3B):

$$\sum_{i=-m}^{m}\left[\prod_{\substack{j=-m \\ j\neq i}}^{m} (x_i - x_j)\right] = \sum_{i=1}^{m} (-1)^{m+i} (h)^{2m} (m-i)!(m+i)! + \sum_{i=0}^{m} (-1)^{m-i} (h)^{2m} (m-i)!(m+i)! \tag{3.38}$$

It follows from Equations 3.37 and 3.38 that

$$\delta^{2m} f_0 = \sum_{i=1}^{m} (-1)^{m+i} \frac{(2m)!}{(m-i)!(m+i)!} f_{-i} + \sum_{i=0}^{m} (-1)^{m-i} \frac{(2m)!}{(m-i)!(m+i)!} f_i \tag{3.39}$$

In similar fashion, we further obtain

$$\delta^{2m-1} f_{0+1/2} = \sum_{i=1}^{m-1} (-1)^{m+i} \frac{(2m-1)!}{(m-i-1)(m+i)!} f_{-i} + \sum_{i=0}^{m} (-1)^{m-i} \frac{(2m-1)!}{(m-i)(m+i-1)!} f_i \tag{3.40}$$

and

$$\delta^{2m-1} f_{0-1/2} = \sum_{i=1}^{m} (-1)^{m+i-1} \frac{(2m-1)!}{(m-i)!(m+i-1)!} f_{-i} + \sum_{i=0}^{m-1} (-1)^{m-i-1} \frac{(2m-1)!}{(m+i)!(m-i-1)!} f_i \tag{3.41}$$

These formulas permit us to write any-order central differences directly into corresponding linear combinations of pivotal function values f_i.

3.3 APPROXIMATE EXPRESSIONS FOR DERIVATIVES

To replace the governing differential equations of elastic systems approximately by the finite difference equations, we represent derivatives of the desired functions in terms of derivatives of corresponding interpolation polynomials. Depending on the desired location of the derivatives in the given interval of independent variable, we choose appropriate interpolation polynomials developed in Section 3.2. Suppose that it is required to approximate the first and second derivatives of $f(x)$ at one of the interior pivotal points, x_0, and express them in terms of pivotal function values f_i· near x_0. For this situation, naturally, we select Stirling's formula with pivotal points arranged as shown in Figure 3.1c. On limiting Stirling's polynomial to second degree in x and differentiating it twice with respect to x, we obtain at $x = x_0$

$$\begin{aligned}\frac{d}{dx}f_0 &= \frac{d}{dx}P_2(x_0)+\frac{d}{dx}R_2(x_0)\\ \frac{d^2}{dx^2}f_0 &= \frac{d^2}{dx^2}P_2(x_0)+\frac{d^2}{dx^2}R_2(x_0)\end{aligned} \tag{3.42}$$

Since

$$\frac{d}{dx}=\frac{1}{h}\frac{d}{ds}$$

where $x = x_0 + hs$ and $dx = h\ ds$, we find from Equation 3.36 that

$$\begin{aligned}\frac{d}{dx}P_2(x_0) &= \frac{1}{2h}(\delta f_{0+1/2}+\delta f_{0-1/2})=\frac{1}{2h}(-f_{-1}+f_1)\\ \frac{d^2}{dx^2}P_2(x_0) &= \frac{1}{h^2}\delta^2 f_0=\frac{1}{h^2}(f_{-1}-2f_0+f_1)\end{aligned} \tag{3.43}$$

Suppressing the derivatives of the error term and introducing the approximation $F_i \approx f_i$, we arrived at the desired approximations

$$\frac{d}{dx}f_0 \approx \frac{1}{2h}(-F_{-1}+F_1), \qquad \frac{d^2}{dx^2}f_0 \approx \frac{1}{h^2}(F_{-1}-2F_0+F_1) \tag{3.44}$$

To estimate the error terms, we recall Equation 3.20:

$$R_n^{(r)}(x)=\sum_{i=0}^{r}\frac{r!}{(n+r-i+1)!\,i!}\omega_n^{(i)}(x)f^{(n+r-i+1)}(\xi_i) \tag{3.45}$$

in which $\omega_n^{(i)}(x)$ for $n = 2$, $i = 0, 1, 2$ and $x = x_0$ are $\omega_2^{(0)}(x_0)=0$, $\omega_2^{(1)}(x_0)=-h^2$, and $\omega_2^{(2)}(x_0)=0$. Hence, Equation 3.45 yields the estimates

$$\begin{aligned}\frac{d}{dx}R_2(x_0) &= -\frac{h^2}{6}f^{(3)}(\xi_1)\\ \frac{d^2}{dx^2}R_2(x_0) &= \frac{h^2}{12}f^{(4)}(\xi_1)\end{aligned} \tag{3.46}$$

where ξ_1 is a point in the interval $x_{-1} < \xi_1 < x_1$. Since the approximations are carried out with the second-degree Stirling polynomial, we call these second-order approximations. Note that they entail three pivotal points, x_{-1}, x_0, and x_1. The associated errors are of $O(h^2)$.

Third- and fourth-order derivatives of $P_n(x)$ may be obtained by considering the fourth-degree Stirling polynomial $P_4(x)$. Derivatives of $P_4(x)$ at $x = x_0$ are

$$\begin{aligned}
\frac{d}{dx}P_4(x_0) &= \frac{1}{h}\left[\frac{1}{2}(\delta f_{0+1/2}+\delta f_{0-1/2}) - \frac{1}{12}(\delta^3 f_{0+1/2}+\delta^3 f_{0-1/2})\right] \\
&= \frac{1}{12h}(f_{-2}-8f_{-1}+8f_1-f_2) \\
\frac{d^2}{dx^2}P_4(x_0) &= \frac{1}{h^2}\left(\delta^2 f_0 - \frac{1}{12}\delta^4 f_0\right) \\
&= \frac{1}{12h^2}(-f_2+16f_{-1}-30f_0+16f_1-f_2) \\
\frac{d^3}{dx^3}P_4(x_0) &= \frac{1}{h^3}\left[\frac{1}{2}(\delta^3 f_{0+1/2}+\delta^3 f_{0-1/2})\right] \\
&= \frac{1}{2h^3}(-f_{-2}+2f_{-1}-2f_1+f_2) \\
\frac{d^4}{dx^4}P_4(x_0) &= \frac{1}{h^4}\delta^4 f_0 = \frac{1}{h^4}(f_{-2}-4f_{-1}+6f_0-4f_1+f_2)
\end{aligned} \tag{3.47}$$

The corresponding error terms become

$$\begin{aligned}
\frac{d}{dx}R_4(x_0) &= \frac{h^4}{30}f^{(5)}(\xi_1) \\
\frac{d^2}{dx^2}R_4(x_0) &= \frac{h^4}{90}f^{(6)}(\xi_1) \\
\frac{d^3}{dx^3}R_4(x_0) &= \frac{h^4}{210}f^{(7)}(\xi_1) - \frac{h^2}{4}f^{(5)}(\xi_2) \\
\frac{d^4}{dx^4}R_4(x_0) &= \frac{h^4}{840}f^{(8)}(\xi_1) - \frac{h^2}{6}f^{(6)}(\xi_2)
\end{aligned} \tag{3.48}$$

where ξ_1 and ξ_2 are points in the interval containing x_{-2}, x_{-1}, x_0, x_1, and x_2 pivotal points. It may be seen that the fourth-order $dP_4(x_0)/dx$ and $d^2P_4(x_0)/dx^2$ are more accurate than the second-order $dP_2(x_0)/dx$ and $d^2P_2(x_0)/dx^2$, since their errors are of order $O(h^4)$ instead of $O(h^2)$. However, the expressions for $d^3P_4(x_0)/dx^3$ and $d^4P_4(x_0)/dx^4$ are accurate within $O(h^2)$. Higher-order finite difference formulas may be derived in a similar manner.

If derivatives are required at the endpoints of a given interval of x_i, the forward and backward Newton polynomials (Equations 3.26 and 3.27) should be used. Tables 3.1, 3.2, and 3.3 show the second- and fourth-order approximations of $f^{(r)}(x_0)$ in forward, backward, and central finite differences, respectively, together with their error terms. On comparing Table 3.3 with Tables 3.1 and 3.2, we may conclude that in even-order derivatives, the central difference formulas possess errors that are smaller by the order $O(h)$ than those for forward and backward difference formulas. However, the order of errors are identical for all the odd-order derivatives.

Next, we derive finite difference formulas for unequal pivotal spacings, which may be used in the generation of finite difference equations near curved boundaries where unequal pivotal spacings are unavoidable. A detailed application is described in Section 3.4; at present, consider the second-order approximations for $f^{(r)}(x_0)$ in forward, backward, and central differences for the unequal pivotal spacings shown in Figure 3.2. We refer to Newton's basic interpolation polynomial in terms of divided differences (Equation 3.4). Differentiation of Equation 3.4 gives at $x = x_0$

$$
\begin{aligned}
\frac{d}{dx}f_0 &= f[x_0,x_1]+(x_0-x_1)f[x_0,x_1,x_2]+\frac{d}{dx}R_2(x_0) \qquad \text{(forward and backward)}\\
&= f[x_0,x_1]+(x_0-x_1)f[x_1,x_0,x_1]+\frac{d}{dx}R_2(x_0) \qquad \text{(central)}\\
\frac{d^2}{dx^2}f_0 &= 2f[x_0,x_1,x_2]+\frac{d^2}{dx^2}R_2(x_0) \qquad \text{(forward and backward)}\\
&= 2f[x_{-1},x_0,x_1]+\frac{d^2}{dx^2}R_2(x_0) \qquad \text{(central)}
\end{aligned} \tag{3.49}
$$

Table 3.1 Forward difference formulas[a]

Order of approximation	*Second-order forward approximation, n = 2*	*Fourth-order forward approximation, n = 4*
Intervals (h)	0 1 2 → x; x_0	0 1 2 3 4 → x; x_0
$\frac{d}{dx}P_n(x_0)$	$\frac{1}{2h}(-3f_0+4f_1-f_2)$	$\frac{1}{24h}(-50f_0+96f_1-72f_2+32f_3-6f_4)$
$\frac{d}{dx}R_n(x_0)$	$\frac{h^2}{3}f^{(3)}(\xi_1)$	$\frac{h^4}{5}f^{(5)}(\xi_1)$
$\frac{d^2}{dx^2}P_n(x_0)$	$\frac{1}{h^2}(f_0-2f_1+f_2)$	$\frac{1}{12h^2}(35f_0-104f_1+114f_2-56f_3+11f_4)$
$\frac{d^2}{dx^2}R_n(x_0)$	$\frac{h^2}{6}f^{(4)}(\xi_1)-hf^{(3)}(\xi_2)$	$\frac{h^4}{15}f^{(6)}(\xi_1)-\frac{5h^3}{6}f^{(5)}(\xi_2)$
$\frac{d^3}{dx^3}P_n(x_0)$		$\frac{1}{4h^3}(-10f_0+36f_1-48f_2+28_3f-6f_4)$
$\frac{d^3}{dx^3}R_n(x_0)$		$\frac{h^4}{35}f^{(7)}(\xi_1)-\frac{5h^3}{12}f^{(6)}(\xi_2)+\frac{7h^2}{4}f^{(5)}(\xi_3)$
$\frac{d^4}{dx^4}P_n(x_0)$		$\frac{1}{h^4}(f_0-4f_1+6f_2-4f_3+f_4)$
$\frac{d^4}{dx^4}R_n(x_0)$		$\frac{h^4}{70}f^{(8)}(\xi_1)-\frac{5h^3}{21}f^{(7)}(\xi_2)+\frac{7h^2}{6}f^{(6)}(\xi_3)$ $-2hf^{(5)}(\xi_4)$

[a] $\min(x_0, x_1, \ldots, x_n) < \xi_i < \max(x_0, x_1, \ldots, x_n)$.

Table 3.2 Backward difference formulas

Order of approximation	*Second-order backward approximation, n = 2*	*Fourth-order forward approximation, n = 4*
Intervals (h)	$-2 \;\; -1 \;\; 0 \rightarrow x$, x_0	$-4 \;\; -3 \;\; -2 \;\; -1 \;\; 0 \rightarrow x$, x_0
$\frac{d}{dx}P_n(x_0)$	$\frac{1}{2h}(f_{-2}-4f_{-1}+3f_0)$	$\frac{1}{24h}(6f_{-4}-32f_{-3}+72f_{-2}-96f_{-1}+50f_0)$
$\frac{d}{dx}R_n(x_0)$	$\frac{h^2}{3}f^{(3)}(\xi_1)$	$\frac{h^4}{5}f^{(5)}(\xi_1)$
$\frac{d^2}{dx^2}P_n(x_0)$	$\frac{1}{h^2}(f_{-2}-2f_{-1}+f_0)$	$\frac{1}{12h^2}(11f_{-4}-56f_{-3}+114f_{-2}-104f_{-1}+35f_0)$
$\frac{d^2}{dx^2}R_n(x_0)$	$\frac{h^2}{6}f^{(4)}(\xi_1)-hf^{(3)}(\xi_2)$	$\frac{h^4}{15}f^{(6)}(\xi_1)+\frac{5h^3}{6}f^{(5)}(\xi_2)$
$\frac{d^3}{dx^3}P_n(x_0)$		$\frac{1}{4h^3}(6f_{-4}-28f_{-3}+48f_{-2}-36f_{-1}+10f_0)$
$\frac{d^3}{dx^3}R_n(x_0)$		$\frac{h^4}{35}f^{(7)}(\xi_1)+\frac{5h^3}{12}f^{(6)}(\xi_2)+\frac{7h^2}{4}f^{(5)}(\xi_3)$
$\frac{d^4}{dx^4}P_n(x_0)$		$\frac{1}{h^4}(f_{-4}-4f_{-3}+6f_{-2}-4f_{-1}+f_0)$
$\frac{d^4}{dx^4}R_n(x_0)$		$\frac{h^4}{70}f^{(8)}(\xi_1)+\frac{5h^3}{21}f^{(7)}(\xi_1)+\frac{7h^2}{6}f^{(6)}(\xi_3)$ $-2hf^{(5)}(\xi_4)$

Evaluating the divided differences in Equation 3.49 by means of Equation 3.1 for the pivotal arrangements of Figure 3.2, we obtain

$$
\begin{aligned}
&\frac{d}{dx}P_2(x_0)=\frac{1}{h}\left[-\frac{2\alpha+1}{\alpha(\alpha+1)}f_0+\frac{\alpha+1}{\alpha}f_1-\frac{\alpha}{\alpha+1}f_2\right] \\
&\frac{d^2}{dx^2}P_2(x_0)=\frac{2}{h^2}\left[\frac{1}{\alpha(\alpha+1)}f_0-\frac{1}{\alpha}f_1+\frac{1}{(\alpha+1)}f_2\right] \qquad \text{(forward)} \\
&\frac{d}{dx}P_2(x_0)=\frac{1}{h}\left[\frac{\alpha}{(\alpha+1)}f_{-2}-\frac{(\alpha+1)}{\alpha}f_{-1}+\frac{2\alpha+1}{\alpha(\alpha+1)}f_0\right] \\
&\frac{d^2}{dx^2}P_2(x_0)=\frac{2}{h^2}\left[\frac{1}{(\alpha+1)}f_{-2}-\frac{1}{\alpha}f_{-1}+\frac{1}{\alpha(\alpha+1)}f_0\right] \qquad \text{(backward)} \\
&\frac{d}{dx}P_2(x_0)=\frac{1}{h}\left[-\frac{\alpha}{(\alpha+\beta)\beta}f_{-1}+\frac{\beta-\alpha}{\alpha\beta}f_0+\frac{\beta}{\alpha(\alpha+\beta)}f_1\right] \\
&\frac{d^2}{dx^2}P_2(x_0)=\frac{2}{h^2}\left[\frac{1}{(\alpha+\beta)\beta}f_{-1}-\frac{1}{\alpha\beta}f_0+\frac{1}{(\alpha+\beta)\alpha}f_1\right] \qquad \text{(central)}
\end{aligned}
\tag{3.50}
$$

Table 3.3 Central difference formulas

Order of approximation	*Second-order central approximation, n = 2*	*Fourth-order central approximation, n = 4*
Intervals (*h*)	−1 0 1 → x; x_0	−2 −1 0 1 2 → x; x_0
$\frac{d}{dx}P_n(x_0)$	$\frac{1}{2h}(-f_{-1}+f_1)$	$\frac{1}{12h}(f_{-2}-8f_{-1}+8f_1-f_2)$
$\frac{d}{dx}R_n(x_0)$	$-\frac{h^2}{6}f^{(3)}(\xi_1)$	$\frac{h^4}{30}f^{(5)}(\xi_1)$
$\frac{d^2}{dx^2}P_n(x_0)$	$\frac{1}{h^2}(f_{-1}-2f_0+f_1)$	$\frac{1}{12h^2}(-f_{-2}+16f_{-1}-30f_0+16f_1-f_2)$
$\frac{d^2}{dx^2}R_n(x_0)$	$-\frac{h^2}{12}f^{(4)}(\xi_1)$	$\frac{h^4}{90}f^{(6)}(\xi_1)$
$\frac{d^3}{dx^3}P_n(x_0)$		$\frac{1}{2h^3}(-f_{-2}+2f_{-1}-2f_1+f_2)$
$\frac{d^3}{dx^3}R_n(x_0)$		$\frac{h^4}{210}f^{(7)}(\xi_1)-\frac{h^2}{4}f^{(5)}(\xi_3)$
$\frac{d^4}{dx^4}P_n(x_0)$		$\frac{1}{h^4}(f_{-2}-4f_{-1}+6f_0-4f_1+f_2)$
$\frac{d^4}{dx^4}R_n(x_0)$		$\frac{h^4}{840}f^{(8)}(\xi_1)-\frac{h^2}{6}f^{(6)}(\xi_3)$

where $0 \le \alpha \le 1$ and $0 \le \beta \le 1$.

Application of Equation 3.45 then yields the errors associated with Equation 3.50. They are

$$
\begin{aligned}
&\frac{d}{dx}R_2(x_0) = \frac{\alpha(\alpha+1)h^2}{6}f^{(3)}(\xi_1) \\
&\frac{d^2}{dx^2}R_2(x_0) = \frac{\alpha(\alpha+1)h^2}{12}f^{(4)}(\xi_1) - \frac{(2\alpha+1)h}{3}f^{(3)}(\xi_2) \qquad \text{(forward)} \\
&\frac{d}{dx}R_2(x_0) = \frac{\alpha(\alpha+1)h^2}{6}f^{(3)}(\xi_1) \\
&\frac{d^2}{dx^2}R_2(x_0) = \frac{\alpha(\alpha+1)h^2}{12}f^{(4)}(\xi_1) + \frac{(2\alpha+1)h}{3}f^{(3)}(\xi_2) \qquad \text{(backward)} \\
&\frac{d^2}{dx^2}R_2(x_0) = -\frac{\alpha\beta h^2}{6}f^{(3)}(\xi_1) \qquad \text{(central)} \\
&\frac{d^2}{dx^2}R_2(x_0) = -\frac{\alpha\beta h^2}{12}f^{(4)}(\xi_1) + \frac{(\beta-\alpha)h}{3}f^{(3)}(\xi_2)
\end{aligned}
\tag{3.51}
$$

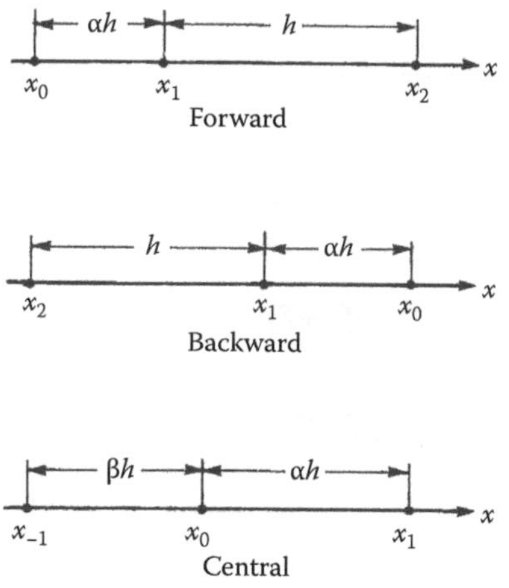

Figure 3.2 Unequal pivotal spacings.

When α and β are set equal to unity, Equations 3.50 and 3.51 reduce to formulas given in Tables 3.1, 3.2, and 3.3. From the error $d^2R_2(x_0)/dx^2$ for the central difference formula $d^2P_2(x_0)/dx^2$, note that the accuracy of $d^2P_2(x_0)/dx^2$ deteriorates with the use of unequal pivotal spacings.

3.4 TWO-DIMENSIONAL HARMONIC EQUATION, BIHARMONIC EQUATION, AND CURVED BOUNDARIES

In this section, we consider finite difference approximations of partial derivatives that appear in the partial differential equations of elasticity. The method is analogous to the techniques developed for one independent variable. Without loss of generality, we restrict the treatment to problems of two independent variables (x, y). The extension to three or more independent variables is apparent.

For two-dimensional problems, an equally space rectangular mesh of the type shown in Figure 3.3 is generally employed to define pivotal points as intersections of mesh lines. Thus, any pivotal point may be located by

$$x = x_0 + ih \text{ and } y = y_0 + jk, \qquad i, j = 0, \pm 1, \pm 2, \ldots \tag{3.52}$$

with respect to a conveniently chosen origin (x_0, y_0). The parameters h and k denote the spacings of mesh lines in the x and y directions, respectively. Whenever $h = k$, we say that the mesh is square.

Let $f(x, y)$ be a function of the independent variables (x, y). The function value of $f(x, y)$ at pivotal point $(x_0 + ih, y_0 + jk)$ is identified by appropriate subscripts. Thus, $f_{i,j}$ denotes $f(x_0 + ih, y_0 + jk)$. The first subscript is associated with the mesh line $x_0 + ih$, the second with the mesh line $y_0 + jk$.

Approximations to the derivatives $(\partial^r f/\partial x^r)_{i,j}$ and $(\partial^r f/\partial y^r)_{i,j}$ may be expressed by means of the forward, backward, or central difference formulas given in Tables 3.1, 3.2, and 3.3. On

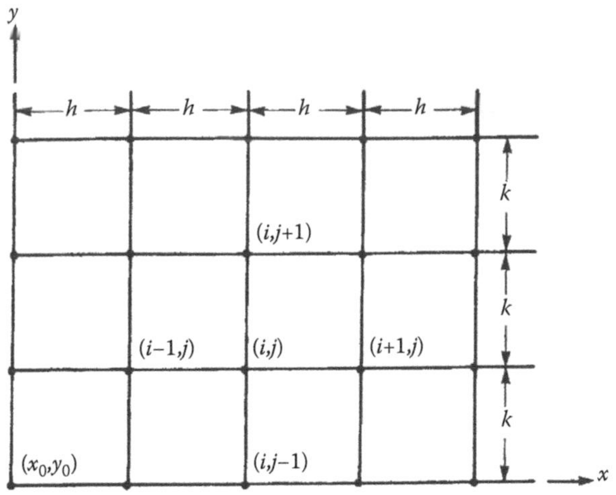

Figure 3.3 Rectangular mesh.

admitting errors of the order $O(h^2)$ or $O(k^2)$, for example, we write directly from Table 3.3 the second-order central finite difference formulas for partial derivatives:

$$\begin{aligned}
\left(\frac{\partial f}{\partial x}\right)_{i,j} &= \frac{1}{2h}(-f_{i-1,j} + f_{i+1,j}) + O(h^2) \\
\left(\frac{\partial^2 f}{\partial x^2}\right)_{i,j} &= \frac{1}{h^2}(f_{i-1,j} - 2f_{i,j} + f_{i+1,j}) + O(h^2) \\
\left(\frac{\partial f}{\partial y}\right)_{i,j} &= \frac{1}{2k}(-f_{i,j-1} + f_{i,j+1}) + O(k^2) \\
\left(\frac{\partial^2 f}{\partial y^2}\right)_{i,j} &= \frac{1}{k^2}(f_{i,j-1} - 2f_{i,j} + f_{i,j+1}) + O(k^2)
\end{aligned} \tag{3.53}$$

To compute the mixed partial derivative $(\partial^2 f/\partial x\, \partial y)_{i,j}$, we first hold the y variable fixed and write

$$\left(\frac{\partial f}{\partial x}\right)_i = \frac{1}{2h}(-f_{i-1,j} + f_{i+1,j}) - \frac{h^2}{6}\left(\frac{\partial^3 f}{\partial x^3}\right)_{x=\xi} \tag{3.54}$$

where the single subscript refers to the x variable. Noting that every term of Equation 3.54 is a function of y, we differentiate both sides with respect to y and evaluate the result at $y = y_0 + jk$. In particular, considering $(1/2h)(-f_{i-1} + f_{i+1})$ as a function of y, we apply $\partial/\partial y$ to obtain (see Table 3.3)

$$\left(\frac{\partial^2 f}{\partial x \partial y}\right)_{i,j} = \frac{1}{(2h)(2k)}(f_{i-1,j-1} - f_{i-1,j+1} - f_{i+1,j-1} + f_{i+1,j+1})$$
$$- \frac{k^2}{6}\left[\frac{\partial^3}{\partial y^3}\left(\frac{-f_{i-1} + f_{i+1}}{2h}\right)\right]_{y=\eta} - \frac{h^2}{6}\left(\frac{\partial^3 f}{\partial x^3}\right)_{\substack{x=\xi \\ y=y_0+jk}} \tag{3.55}$$

We transform the error term in parentheses by means of Equation 3.54:

$$-\frac{k^2}{6}\frac{\partial^3}{\partial y^3}\left(\frac{-f_{i-1} + f_{j+1}}{2h}\right)_{y=\eta} = -\frac{k^2}{6}\left\{\frac{\partial^3}{\partial y^3}\left[\left(\frac{\partial f}{\partial x}\right)_i + \frac{h^2}{6}\left(\frac{\partial^3 f}{\partial x^3}\right)_{x=\xi}\right]\right\}_{y=\eta}$$
$$= -\frac{k^2}{6}\left(\frac{\partial^4 f}{\partial x\, \partial y^3}\right)_{\substack{x=x_0+ih \\ y=\eta}} - \frac{h^2k^2}{6}\left(\frac{\partial^6 f}{\partial x^3 \partial y^3}\right)_{\substack{x=\xi \\ y=\eta}}$$

where ξ and η are bounded by $x_0 + (i - 1)h < \xi < x_0 + (i + 1)h$ and $y_0 + (j - 1)k < \eta < y_0 + (j + 1)$. Accordingly, we write

$$\left(\frac{\partial^2 f}{\partial x\, \partial y}\right)_{i,j} = \frac{1}{4hk}(f_{i-1,j-1} - f_{i-1,j+1} - f_{i+1,j-1} + f_{i+1,j+1}) + O(h^2 \text{ or } k^2) \tag{3.56}$$

Analogously, differentiating the second equation of Equation 3.53 with respect to y twice and estimating the error as earlier, we find that

$$\left(\frac{\partial^4 f}{\partial x^2\, \partial y^2}\right)_{i,j} = \frac{1}{h^2k^2}[f_{i-1,j-1} + f_{i-1,j+1} + f_{i+1,j-1} - f_{i+1,j+1}$$
$$- 2(f_{i-1,j} + f_{i,j+1} + f_{i+1,j} + f_{i,j-1}) + 4f_{i,j}] + O(h^2 \text{ or } k^2) \tag{3.57}$$

If the fourth derivatives $(\partial^4 f/\partial x^4)_{i,j}$ and $(\partial^4 f/\partial y^4)_{i,j}$ are required, we employ the *fourth-order central difference formulas* of Table 3.3 to obtain

$$\left(\frac{\partial^4 f}{\partial x^4}\right)_{i,j} = \frac{1}{h^4}(f_{i-2,j} - 4f_{i-1,j} + 6f_{i,j} - 4f_{i+1,j} + f_{i+2,j}) + O(h^2)$$
$$\left(\frac{\partial^4 f}{\partial y^4}\right)_{i,j} = \frac{1}{k^4}(f_{i,j-2} - 4f_{i,j-1} + 6f_{i,j} - 4f_{i,j+1} + f_{i,j+2}) + O(k^2) \tag{3.58}$$

Similarly, formulas for higher-order approximations for derivatives with respect to x, and for derivatives with respect to y, may be written.

3.4.1 Harmonic equation

The plane harmonic equation (Laplace equation) is $\nabla^2\phi = 0$, where $\phi = \phi(x, y)$ and ∇^2 denotes the Laplacian $\partial^2/\partial x^2 + \partial^2/\partial y^2$. By means of Equation 3.53, the harmonic equation may be replaced by a finite difference equation at the pivotal point $(x_0 + ih, y_0 + jk)$ or simply (i, j); that is,

$$(\nabla^2\phi)_{i,j} = \frac{1}{h^2}(\phi_{i-1,j} - 2\phi_{i,j} + \phi_{i+1,j}) + \frac{1}{k^2}(\phi_{i,j-1} - 2\phi_{i,j} + \phi_{i,j+1}) + O(h^2 \text{ or } k^2) = 0 \tag{3.59}$$

However, the representation of $\nabla^2\phi = 0$ in Equation 3.59 is not unique. For instance, by using fourth-order central formulas, we may write (see Table 3.3)

$$(\nabla^2\phi)_{i,j} = \frac{1}{12h^2}(-\phi_{i-2,j} + 16\,\phi_{i-1,j} - 30\phi_{i,j} + 16\phi_{i+1,j} - \phi_{i+2,j}) + \frac{1}{12k^2}(-\phi_{i,j-2} + 16\,\phi_{i,j-1} - 30\phi_{i,j} + 16\phi_{i,j+1} - \phi_{i,j+2}) + O(h^4 \text{or } k^4) \tag{3.60}$$

Extension of these formulas to three variables (x, y, z) follows directly by assigning a third index and the associated spacing.

3.4.2 Biharmonic equation

The plane biharmonic equation is $\nabla^4\phi = \partial^4\phi/\partial x^4 + 2\partial^4\phi/\partial x^2\,\partial y^2 + \partial^4\phi/\partial y^4 = 0$. On replacing the partial derivatives occurring in this equation by Equations 3.57 and 3.58, we arrive at

$$\begin{aligned}(\nabla^4\phi)_{i,j} &= \frac{1}{h^4}(\phi_{i-2,j} - 4\phi_{i-1,j} + 6\phi_{i,j} - 4\phi_{i+1,j} + \phi_{i+2,j}) + \frac{2}{h^2k^2}[\phi_{i-1,j-1} + \phi_{i-1,j+1}\\ &+ \phi_{i+1,j-1} + \phi_{i+1,j+1} - 2(\phi_{i-1,j} + \phi_{i,j+1} + \phi_{i+1,j} + \phi_{i,j-1}) + 4\phi_{i,j}]\\ &+ \frac{1}{k^4}(\phi_{i,j-2} - 4\phi_{i,j-1} + 6\phi_{i,j} - 4\phi_{i,j+1} + \phi_{i,j+2}) + O(h^2 \text{ or } k^2) = 0\end{aligned} \tag{3.61}$$

For a square mesh $k = h$, Equation 3.61 reduces to

$$\begin{aligned}(\nabla^4\phi)_{i,j} &= \frac{1}{h^4}[20\phi_{i,j} - 8(\phi_{i+1,j} + \phi_{i,j+1} + \phi_{i-1,j} + \phi_{i,j-1})\\ &+ 2(\phi_{i+1,j+1} + \phi_{i,-1,j+1} + \phi_{i-1,j-1} + \phi_{i+1,j-1})\\ &+ (\phi_{i+2,j} + \phi_{i,j+2} + \phi_{i-2,j} + \phi_{i,j-2})] + O(h^2) = 0\end{aligned} \tag{3.62}$$

The corresponding formula of Equation 3.62 with error of order $O(h^4)$ is

$$
\begin{aligned}
(\nabla^4\phi)_{i,j} = \frac{1}{6h^4}[184\phi_{i,j} &+ 20(\phi_{i+1,j+1} + \phi_{i-1,j+1} + \phi_{i-1,j-1} + \phi_{i+1,j-1}) - 77(\phi_{i,j+1} \\
&+ \phi_{i-1,j} + \phi_{i,j-1} + \phi_{i+1,j}) + 14(\phi_{i+2,j} + \phi_{i,j+2} + \phi_{i-2,j} + \phi_{i,j-2}) \\
&- (\phi_{i+3,j} + \phi_{i+2,j+1} + \phi_{i+1,j+2} + \phi_{i,j+3} + \phi_{i-1,j+2} + \phi_{i-2,j+1} + \phi_{i-3,j} \\
&+ \phi_{i-2,j-1} + \phi_{i-1,j-2} + \phi_{i,j-3} + \phi_{i+1,j-2} + \phi_{i+2,j-1})] + O(h^4) = 0
\end{aligned} \tag{3.63}
$$

3.4.3 Curved boundaries

When the pivotal point $(x_0 + ih, y_0 + jk)$ is near a curved or irregular boundary as shown in Figure 3.4, unequal mesh spacings may occur in both x and y directions. To demonstrate techniques to deal with this situation, consider the approximation of the plane harmonic equation $\nabla^2\phi = 0$ at such a pivotal point.

Basically, there are two types of boundary value problems associated with the equation, one with function values specified along the boundary (Dirichlet problem) and the other with normal derivative specified (Neumann problem). Accordingly, we present two versions of the finite difference approximations to $\nabla^2\phi = 0$.

Consider first the Dirichlet problem where the ϕ value along the boundary is prescribed. Let the unequal mesh near the pivotal point (i, j) be defined as in Figure 3.4, where $0 \le \alpha \le 1$ and $0 \le \gamma \le 1$. Applying the last formula of Equation 3.50 to the function

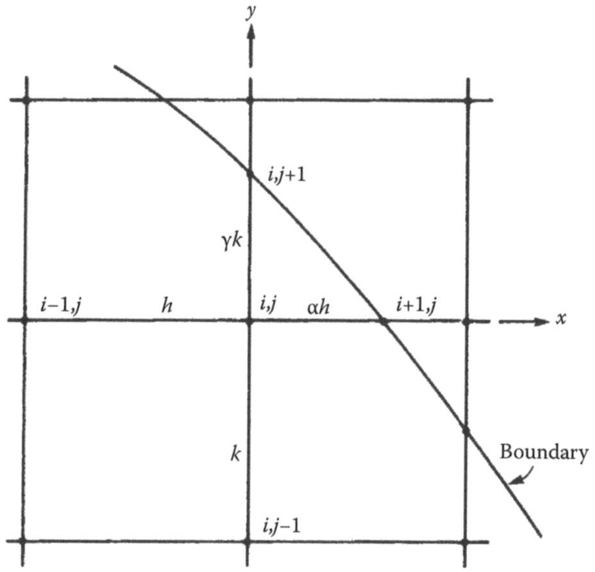

Figure 3.4 Unequal mesh near a curved boundary. Direchlet Problem.

ϕ once with respect to x with $\beta = 1$ and $\alpha = \alpha$, and once with respect to y with $\beta = 1$ and $\alpha = \gamma$, we obtain

$$(\nabla^2\phi)_{i,j} = \frac{2}{h^4}\left(\frac{1}{(\alpha+1)}\phi_{i-1,j} - \frac{1}{\alpha}\phi_{i,j} + \frac{1}{\alpha(\alpha+1)}\phi_{i+1,j}\right) + \frac{2}{k^2}\left(\frac{1}{(\gamma+1)}\phi_{i,j-1} - \frac{1}{\gamma}\phi_{i,j} + \frac{1}{\gamma(\gamma+1)}\phi_{i,j+1}\right) + O[(1-\alpha)h \text{ or } (1-\gamma)k] = 0 \tag{3.64}$$

where the magnitude of error is estimated by Equation 3.61.

For the Neumann problem, consider the situation pictured in Figure 3.5, where the regular pivotal points $(i, j + 1)$ and $(i + 1, j)$ fall outside the curved boundary. We construct two normals, n_1 and n_2, to the boundary through these two pivotal points. Let the intersections of these normals with the boundary be denoted by p and q and with the mesh lines, by r and s, as shown in the figure. The locations of r and s are defined by α and β parameters, respectively. For convenience, the distance between the pivotal point $(i, j + 1)$ and r is denoted by a, and the distance between the pivotal point $(i + 1, j)$ and s by b.

By means of the first-degree Newton interpolation polynomial of Equation 3.4, we write

$$\left(\frac{\partial\phi}{\partial n_1}\right)_p = \frac{1}{a}(\phi_{i,j+1} - \phi_r) \text{ and } \left(\frac{\partial\phi}{\partial n_2}\right)_q = \frac{1}{b}(\phi_{i+1,j} - \phi_s)$$

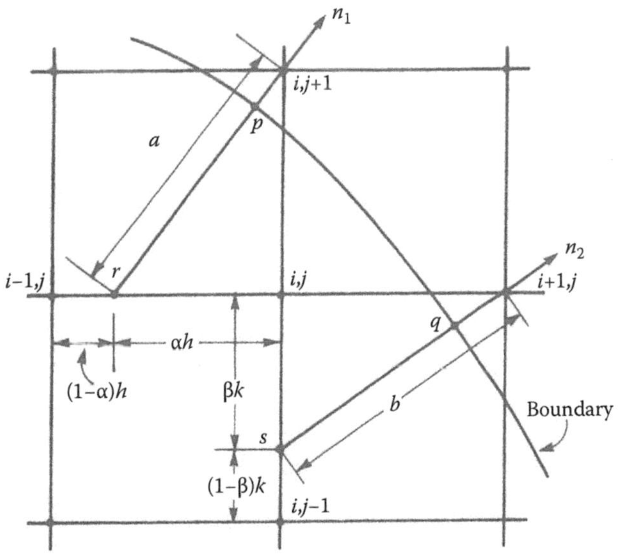

Figure 3.5 Unequal mesh near a curved boundary. Neumann Problem.

The solution for $\phi_{i,j+1}$ and $\phi_{i+1,j}$ from these equations yields

$$\phi_{i,j+1} = a\left(\frac{\partial\phi}{\partial n_1}\right)_p + \phi_r \qquad \phi_{i+1,j} = b\left(\frac{\partial\phi}{\partial n_2}\right)_q + \phi_s \tag{3.65}$$

Repeated application of the first-degree Newton interpolation polynomial along the *j*th and *i*th mesh lines, respectively, produces the following expressions for ϕ_r and ϕ_s:

$$\begin{aligned}\phi_r &= \phi_{i,j} + (-\alpha h)\frac{\phi_{i-1,j} - \phi_{i,j}}{(-h)} = \alpha\phi_{i-1,j} + (1-\alpha)\phi_{i,j} \\ \phi_s &= \beta\phi_{i,j-1} + (1-\beta)\phi_{i,j}\end{aligned} \tag{3.66}$$

Hence, from Equations 3.65 and 3.66, we have

$$\begin{aligned}\phi_{i,j+1} &= a\left(\frac{\partial\phi}{\partial n_1}\right)_p + \alpha\phi_{i-1,j} + (1-\alpha)\phi_{i,j} \\ \phi_{i+1,j} &= b\left(\frac{\partial\phi}{\partial n_2}\right)_q + \beta\phi_{i,j-1} + (1-\beta)\phi_{i,j}\end{aligned} \tag{3.67}$$

Next, elimination of $\phi_{i,j+1}$ and $\phi_{i+1,j}$ from Equation 3.59 by the use of Equation 3.67 leads to the desired approximate harmonic equation,

$$\begin{aligned}(\nabla^2\phi)_{i,j} = &\left[\frac{1}{h^2}\phi_{i-1,j} - (1+\beta)\phi_{i,j} + \beta\phi_{i,j-1} + b\left(\frac{\partial\phi}{\partial n_2}\right)_q\right] \\ &+ \frac{1}{k_2}\left[\phi_{i,j-1} - (1+\alpha)\phi_{i,j} + \alpha\phi_{i-1,j} + a\left(\frac{\partial\phi}{\partial n_1}\right)_p\right] \\ &+ O(h \text{ or } k)\end{aligned} \tag{3.68}$$

Note that the use of the first-degree Newton interpolation polynomial leads to error of $O(h \text{ or } k)$.

3.5 FINITE DIFFERENCE APPROXIMATION OF THE PLANE STRESS PROBLEM

The equations of equilibrium for the plane stress problem of elasticity may be formulated in two ways: one in terms of the Airy stress function, which leads to the biharmonic partial differential equation, and the other in terms of (x, y) displacement components (u, v), which leads to a set of two coupled second-order partial differential equations (Boresi et al., 2011, Chapter 5). Since the boundary conditions associated with the Airy stress function formulation contain higher-order derivatives than those connected with the displacement formulation, in view of the nature of finite difference methods, the displacement formulation is preferred.

Accordingly, we consider the governing differential equation for the plane stress problem in the following form (Boresi et al., 2011, Equation 5.8.3):

$$\begin{aligned}
&\frac{\partial^2 u}{\partial x^2}+\frac{1}{2}(1-v)\frac{\partial^2 v}{\partial y^2}+\frac{1}{2}(1+v)\frac{\partial^2 v}{\partial x\,\partial y}+\frac{1}{E}\left(\frac{\partial u}{\partial x}+v\frac{\partial v}{\partial y}\right)\frac{\partial E}{\partial x}\\
&\quad+\frac{(1-v)}{2E}\left(\frac{\partial u}{\partial y}+\frac{\partial v}{\partial x}\right)\frac{\partial E}{\partial y}=\frac{(1+v)}{E}\frac{\partial}{\partial x}(EKT)\\
&\frac{1}{2}(1+v)\frac{\partial^2 u}{\partial x\,\partial y}+\frac{1}{2}(1-v)\frac{\partial^2 v}{\partial x^2}+\frac{\partial^2 v}{\partial y^2}+\frac{(1-v)}{2E}\left(\frac{\partial u}{\partial y}+\frac{\partial v}{\partial x}\right)\frac{\partial E}{\partial x}\\
&\quad+\frac{1}{E}\left(\frac{\partial v}{\partial y}+v\frac{\partial u}{\partial x}\right)\frac{\partial E}{\partial y}=\frac{(1+v)}{E}\frac{\partial}{\partial y}(EKT)
\end{aligned} \tag{3.69}$$

where $T = T(x, y)$ is a known temperature field and where the modulus of elasticity E, and the coefficient of linear thermal expansion K depend on (x, y). Poisson's ratio v is taken to be constant.

The finite difference method may be applied to obtain an approximate solution to Equation 3.69, subject to appropriate boundary conditions. For simplicity of demonstration, let E = constant, although in general, this restriction is unnecessary. Then Equation 3.69 reduces to

$$\begin{aligned}
&\frac{\partial^2 u}{\partial x^2}+\frac{1}{2}(1-v)\frac{\partial^2 u}{\partial y^2}+\frac{1}{2}(1+v)\frac{\partial^2 v}{\partial x\,\partial y}=(1+v)\frac{\partial}{\partial x}(KT)\\
&\frac{1}{2}(1+v)\frac{\partial^2 u}{\partial x\,\partial y}+\frac{1}{2}(1-v)\frac{\partial^2 v}{\partial x^2}+\frac{\partial^2 v}{\partial y^2}=(1+v)\frac{\partial}{\partial y}(KT)
\end{aligned} \tag{3.70}$$

By way of illustration, consider the plane region R: $0 \le x \le a$, $0 \le y \le b$ (Figure 3.6). Let the mesh dimensions be h and k. Let the (i, j) coordinates run from (1 to m) and (1 to n), respectively. Observe that there are mn pivotal points in R; hence, we have $2mn$ unknown pivotal u and v displacements.

To supply the $2mn$ equations required for the solution, we first consider the approximate satisfaction of the governing differential equation within the region R. Thus, at the $2(m - 2)(n - 2)$ interior pivotal points, we require that the finite difference equations that replace Equation 3.70 be satisfied. Employing *second-order central difference formulas* (Table 3.3), we write the $2(m - 2)(n - 2)$ algebraic equations that approximate Equation 3.70:

$$\begin{aligned}
&\frac{1}{h^2}(U_{i+1,j}-2U_{i,j}+U_{i-1,j})+\frac{1-v}{2k^2}(U_{i,j+1}-2U_{i,j}+U_{i,j-1})+\frac{1+v}{8hk}(V_{i+1,j+1}-V_{i+1,j-1}-V_{i-1,j+1}+V_{i-1,j-1})\\
&-\frac{1+v}{2h}(KT_{i+1,j}-KT_{i-1,j})=0\,\frac{1+v}{8hk}(U_{i+1,j+1}-U_{i+1,j-1}-U_{i-1,j+1}-U_{i-1,j-1})\\
&+\frac{1-v}{2h^2}(V_{i+1,j}-2V_{i,j}+V_{i-1,j})+\frac{1}{k^2}(V_{i,j+1}-2V_{i,j}+V_{i,j-1})-\frac{1+v}{2k}(KT_{i,j+1}-KT_{i,j-1})=0
\end{aligned} \tag{3.71}$$

for $i = 2, 3, \ldots, m - 1$ and $j = 2, 3, \ldots, n - 1$. Here, (U, V) are approximations to (u, v).

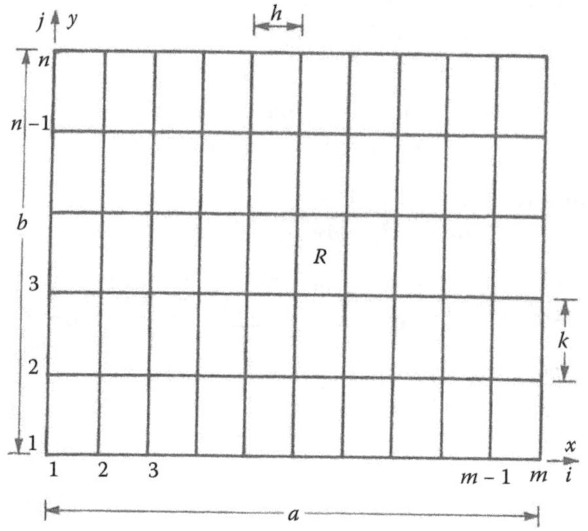

Figure 3.6 A rectangular region *R*.

Next, we derive the remaining required $4m + 4n - 8$ equations from consideration of boundary conditions. A variety of boundary conditions exist for the plane stress problem. However, the simplest one, in the sense of the present numerical scheme, is the case where the edges are prevented from displacing ($u = v = 0$ on the edges). In this case, we simply assign u and v values along the $2m + 2n - 4$ boundary pivotal points to obtain another $4m + 4n - 8$ equations.

Alternatively, we may consider the free edge boundary conditions where the boundary edges are free of any stress tractions. In this case, we must eliminate rigid-body motion (Boresi et al., 2011, Chapter 2). The translatory and rotatory rigid-body displacements of the region R can be prevented if the following conditions are satisfied:

$$U_{1,1} = V_{1,1} = V_{2,1} = 0 \tag{3.72}$$

In general, the stress-displacement relations for plane stress are (Boresi et al., 2011, Equation 5.8.2)

$$\begin{aligned}
\sigma_x &= \frac{E}{1-v^2}\left[\frac{\partial u}{\partial x} + v\frac{\partial u}{\partial y} - (1+v)KT\right] \\
\sigma_y &= \frac{E}{1-v^2}\left[\frac{\partial v}{\partial y} + v\frac{\partial u}{\partial x} - (1+v)KT\right] \\
\tau_{xy} &= G\left(\frac{\partial u}{\partial y} + \frac{\partial v}{\partial x}\right); \qquad G = \frac{E}{2(1+v)}
\end{aligned} \tag{3.73}$$

For the edge $y = 0$ free, $\sigma_y = \tau_{xy} = 0$. Hence, Equation 3.73 yields for $y = 0$

$$\begin{aligned} &\frac{\partial v}{\partial y} + v\frac{\partial u}{\partial x} - (1+v)KT = 0 \\ &\frac{\partial u}{\partial y} + \frac{\partial v}{\partial x} = 0 \end{aligned} \tag{3.74}$$

Since the second-order central differences with error of $O(h^2 \text{ or } k^2)$ have been employed in deriving Equation 3.71, the same order of difference formulas should be used in replacing Equation 3.74. Otherwise, the approximation may be inconsistent. Approximating Equation 3.73, we use the following finite difference approximations of the derivatives of u and v along the edge $y = 0$:

$$\left(\frac{\partial v}{\partial y}\right)_{i,1} = \frac{1}{2k}(-3V_{i,1} + 4V_{i,2} - V_{i,3}) \qquad i = 3,\ldots,m$$

$$\left(\frac{\partial u}{\partial x}\right)_{i,1} = \frac{1}{2h}(-U_{i-1,1} + U_{i+1,1}) \qquad i = 2,\ldots,m-1$$

$$\left(\frac{\partial u}{\partial x}\right)_{i,1} = \frac{1}{2h}(U_{i-2,1} - 4U_{i-1,1} + 3U_{i,1}) \qquad i = m$$

$$\left(\frac{\partial u}{\partial y}\right)_{i,1} = \frac{1}{2k}(-3U_{i,1} + 4U_{i,2} - U_{i,3}) \qquad i = 2,\ldots,m$$

$$\left(\frac{\partial v}{\partial x}\right)_{i,1} = \frac{1}{2h}(-V_{i-1,1} + V_{i+1,1}) \qquad i = 2,\ldots,m-1$$

$$\left(\frac{\partial v}{\partial x}\right)_{i,1} = \frac{1}{2h}(V_{i-2,1} - 4V_{i-1,1} + 3V_{i,1}) \qquad i = m$$

where the forward, backward, and central difference formulas of Tables 3.1, 3.2, and 3.3 are used. These expressions permit us to approximate Equation 3.74 as

$$\begin{aligned} &\frac{1}{2k}(-3V_{i,1} + 4V_{i,2} - V_{i,3}) + \frac{v}{2h}(-U_{i-1,1} + U_{i+1,1}) - (1+v)KT_{i,1} = 0 \qquad i = 3,\ldots,m-1 \\ &\frac{1}{2k}(-3U_{i,1} + 4U_{i,2} - U_{i,3}) + \frac{1}{2h}(-V_{i-1,1} + V_{i+1,1}) = 0 \qquad i = 2,\ldots,m-1 \\ &\frac{1}{2k}(-3V_{i,1} + 4V_{i,2} - V_{i,3}) + \frac{v}{2h}(U_{i-2,1} - 4U_{i-1,1} + 3U_{i,1}) - (1+v)KT_{i,1} = 0 \qquad i = m \\ &\frac{1}{2k}(-3U_{i,1} + 4U_{i,2} - U_{i,3}) + \frac{1}{2h}(V_{i-2,1} - 4V_{i-1,1} + 3V_{i,1}) = 0 \qquad i = m \end{aligned} \tag{3.75}$$

Equation 3.75 represents $(m - 2) + (m - 1)$ algebraic equations.

Along a free edge $y = b$, again $\sigma_y = \tau_{xy} = 0$. Hence, in an analogous manner, we write another $2(m - 1)$ equations:

$$
\begin{aligned}
&\frac{1}{2k}(V_{i,n-2} - 4V_{i,n-1} + 3V_{i,n}) + \frac{v}{2h}(-U_{i-1,n} + U_{i+1,n}) - (1+v)KT_{i,n} = 0 \qquad i = 2,\ldots,m-1\\
&\frac{1}{2k}(U_{i,n-2} - 4U_{i,n-1} + 3U_{i,n}) + \frac{1}{2h}(-V_{i-1,n} + V_{i+1,n}) = 0 \qquad i = 2,\ldots,m-1\\
&\frac{1}{2k}(V_{i,n-2} - 4V_{i,n-1} + 3V_{i,n}) + \frac{v}{2h}(U_{i-2,n} - 4U_{i-1,n} + 3U_{i,n}) - (1+v)KT_{i,n} = 0 \qquad i = m\\
&\frac{1}{2k}(U_{i,n-2} - 4U_{i,n-1} + 3U_{i,n}) + \frac{1}{2h}(V_{i-2,n} - 4V_{i-1,n} + 3V_{i,n}) = 0 \qquad i = m
\end{aligned}
\tag{3.76}
$$

Along a free edge $x = 0$, $\sigma_x = \tau_{xy} = 0$. Hence, by Equation 3.73,

$$
\begin{aligned}
&\frac{\partial u}{\partial x} + v\frac{\partial v}{\partial y} - (1+v)KT = 0\\
&\frac{\partial u}{\partial y} + \frac{\partial v}{\partial x} = 0
\end{aligned}
\tag{3.77}
$$

Equation 3.77 can be cast into corresponding finite difference equations in a manner similar to previous approximations. Upon writing the resulting difference equations at $(n - 1)$ pivotal points ($j = 2, 3, \ldots, n$) along the free edge $x = 0$, and $(n - 2)$ pivotal points ($j = 2, 3, \ldots, n - 1$) along the free edge $x = a$, we further obtain $2(n - 1) + 2(n - 2)$ equations for the case of a rectangular region with free edges.

Accordingly, the boundary conditions provide a total of $4m + 4n - 8$ equations. Equation 3.71 combined with these boundary equations form $2mn$ linear algebraic equations for the $2mn$ unknowns $(U_{i,j}, V_{i,j})$. The solution of these equations subject to Equation 3.72 represents an approximate solution of the plane stress problem of elasticity for the rectangular region with stress-free edges. Various methods of solving such equations in conjunction with digital computers have been developed (Chapra and Canale, 2012; Collatz, 2012; Shoup, 1979).

3.6 TORSION PROBLEM

In Chapter 7 of Boresi et al. (2011), we formulated the torsion of elastic prismatic bars in terms of Saint-Venant's warping function and also in terms of Prandtl's torsion function. Prandtl's formulation leads to a simpler Dirichlet-type boundary value problem (Boresi et al., 2011, Equation 7.3.16):

$$\begin{aligned} \nabla^2\phi &= -2G\beta \quad \text{over region } R \\ \phi &= 0 \quad \text{on contour } C \end{aligned} \tag{3.78}$$

where C is the bounding curve of the bar cross-section R and β is the angle of twist per unit length of the bar. Accordingly, we employ Equation 3.78 in the finite difference analysis of the torsion problem.

3.6.1 Square cross-section

Let the bar cross-section R be square (Figure 3.7). Then, by symmetry, we need consider only a quarter of the region R. Figure 3.7 shows a 4 × 4 square mesh subdivision with mesh width $h = a/4$. Numbering the pivotal points as shown in Figure 3.7, we need consider only the three values $F_{1,1}$, $F_{2,1}$, and $F_{2,2}$, where $F_{i,j}$ are approximations to $\phi_{i,j}$. Note that the $\phi_{i,j}$ on the contour C is identically zero.

Employing the second-order central difference formula for $\nabla^2\phi$ (Equation 3.59), we write the governing differential equation in finite difference form for the three interior pivotal points to obtain (since $\phi_{i,j} = 0$ on the boundary)

$$\begin{aligned} -4F_{1,1} + 4F_{2,1} &= -2G\beta h^2 \\ F_{1,1} - 4F_{2,1} + 2F_{2,2} &= -2G\beta h^2 \\ 2F_{2,1} + 4F_{2,2} &= -2G\beta h^2 \end{aligned} \tag{3.79}$$

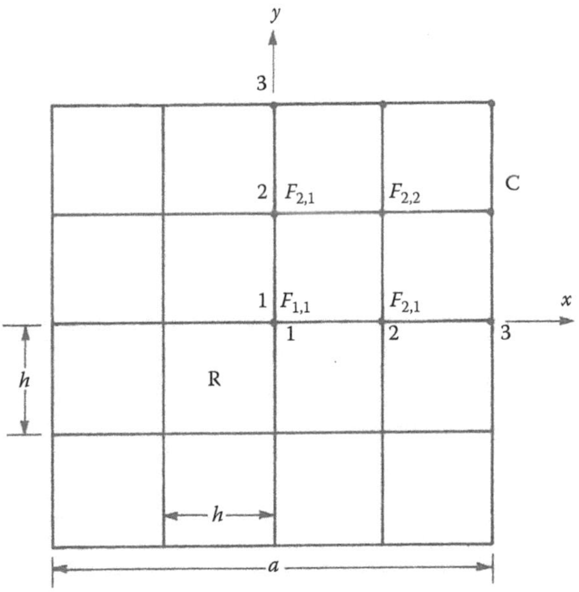

Figure 3.7 Square cross-section.

The solution of Equation 3.79 yields

$$\begin{aligned} F_{1,1} &= 2.250G\beta h^2 \\ F_{2,1} &= 1.750G\beta h^2 \\ F_{2,2} &= 1.375G\beta h^2 \end{aligned} \tag{3.80}$$

Next, we determine the maximum shearing stress $\tau_{zy} = \partial\phi/\partial x$ that occurs at the point $(x, y) = (a/2, 0)$. On taking account of the symmetry and using fourth-order backward difference formulas (Tables 3.2), we approximate $\partial\phi/\partial x$ at $(x, y) = (a/2, 0)$ as

$$\left(\frac{\partial\phi}{\partial x}\right)_{\substack{x=a/2\\y=0}} = \frac{1}{24h}(0 - 32F_{2,1} + 72F_{1,1} - 96F_{2,1} + 0) \tag{3.81}$$

Substitution of Equation 3.80 into 3.81 gives

$$(\tau_{zy})_{\max} = -2.583G\beta h = -0.646G\beta a$$

The exact solution (Boresi et al., 2011, Section 7.10) yields $(\tau_{zy})_{\max} = -0.675G\beta a$. If we take a finer mesh, we obtain a better approximation. For example, with $h = a/8$, we find that $(\tau_{zy})_{\max} = -0.666G\beta a$.

3.6.2 Bar with elliptical cross-section

An explicit closed-form analytical solution of the torsion problem with curved or irregular boundary contours generally is not available, except for a few special cases. However, approximate solutions by the finite difference method are still feasible, although generation of the approximating finite difference equations becomes numerically more involved.* To illustrate the numerical procedure, we consider torsion of a bar with elliptical cross-section for which an exact solution is available (Boresi et al., 2011, Section 7.4). By comparison with the exact solution, we may verify the accuracy of the finite difference approximate solution.

By way of illustration, let us subdivide the elliptical cross-section ($a = 10$ and $b = 7$), as shown in Figure 3.8. Only 11 unknown interior pivotal function values F_i are required because of the symmetric nature of the problem. For simplicity, we have denoted each pivotal point by a single number. Hence, F_i represents the approximate value of the Prandtl torsion function ϕ at the pivotal point i (Figure 3.8). We see from Figure 3.8 that the finite difference approximation (Equation 3.59) that replaces $\nabla^2\phi = -2G\beta$ may be written at interior pivotal points 1, 2, …, 7 with constant mesh interval of width $h = 2.5$. However, unequal pivotal spacings occur adjacent to pivotal points 8, 9, 10, and 11 near the boundary. Therefore, for these pivotal points, Equation 3.64 is used to approximate $\nabla^2\phi = -2G\beta$. In conjunction with the application of Equation 3.64, the unequal pivotal spacings adjacent to

* It is shown in Chapter 4 that the inherent difficulties of the finite difference method in dealing with curved boundaries or in satisfying complicated boundary conditions can be circumvented by use of the finite element method.

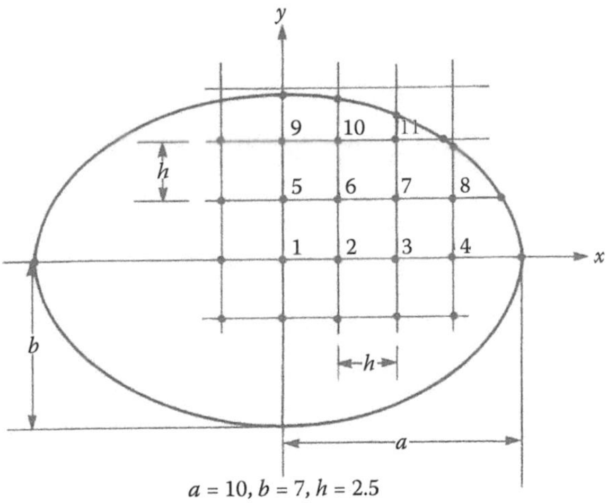

Figure 3.8 Subdivided elliptical cross-section (11 unknowns).

points 8, 9, 10, and 11 are shown in Figure 3.9. Accordingly, by Equations 3.59 and 3.64, we obtain 11 algebraic equations for the 11 unknowns, F_i. They are

$$\begin{aligned}
-4F_1 + 2F_2 + 2F_5 &= -2G\beta h^2 \\
F_1 - 4F_2 + F_3 + 2F_6 &= -2G\beta h^2 \\
F_2 - 4F_3 + F_4 + 2F_7 &= -2G\beta h^2 \\
F_3 - 4F_4 + F_8 &= -2G\beta h^2 \\
F_1 - 4F_5 + 2F_6 + F_9 &= -2G\beta h^2 \\
F_2 + F_5 - 4F_6 + F_7 + F_{10} &= -2G\beta h^2 \\
F_3 + F_6 - 4F_7 + F_8 + F_{11} &= -2G\beta h^2 \\
1.1111F_5 - 4.5000F_9 + 2F_{10} &= -2G\beta h^2 \\
1.1688F_6 + F_9 - 4.8126F_{10} + F_{11} &= -2G\beta h^2 \\
1.4036F_7 + 1.1114F_{10} - 7.2090F_{11} &= -2G\beta h^2 \\
1.0799F_4 + 1.1519F_7 - 5.0639F_8 &= -2G\beta h^2
\end{aligned} \tag{3.82}$$

where the condition $\phi = 0$ on the boundary has been used.

The solution of Equation 3.82 yields the 11 unknowns, F_i. The F_i values along the y axis, designated as approximate ϕ values, are listed in Table 3.4 for $h = 2.5$ and $h = 1.5$, along with corresponding exact values. From the F_i values for $h = 2.5$, we calculate $(\tau_{zx})_{max} = \partial\phi/\partial y$ at $(x, y) = (0, 7)$ by means of the second-degree Newton interpolation formula of Equation 3.4. The approximate value $(\tau_{zx})_{max} = -9.0475G\beta$ is 3.7% less than the corresponding exact value $(\tau_{zx})_{max} = -9.3960G\beta$.

For $h = 1.5$, we obtain a finer mesh subdivision with 30 unknown F_i values (Figure 3.10). The approximate maximum shearing stress $(\tau_{zx})_{max} = -9.3702G\beta$ differs by only 0.28% from the exact value (Table 3.4).

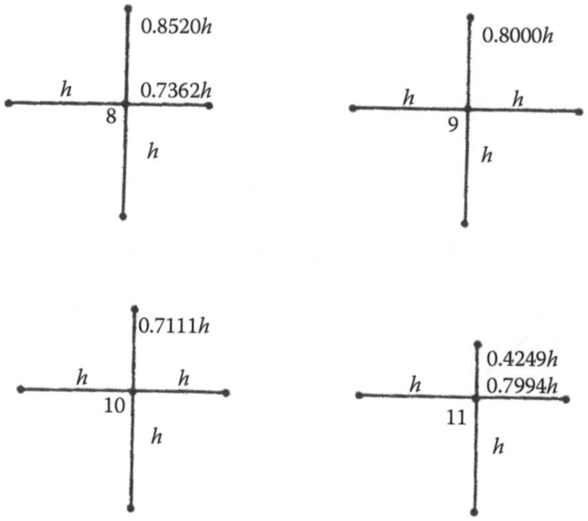

Figure 3.9 Unequal pivotal spacings (h = 2.5).

Table 3.4 Comparison of exact and approximate values of torsion function and maximum shearing stress

	11 unknowns with h = 2.5			*30 unknowns with h = 1.5*	
y	*Exact* $\phi/G\beta h^2$	*Approximate* $\phi/G\beta h^2$	*y*	*Exact* $\phi/G\beta h^2$	*Approximate* $\phi/G\beta h^2$
0.0	5.2617	5.2616	0.0	14.6159	14.5861
2.5	4.5906	4.5905	1.5	13.9448	13.9155
5.0	2.5772	2.5771	3.0	11.9314	11.9044
			4.5	8.5756	8.5542
			6.0	3.8777	3.8673
	Exact	Approximate		Exact	Approximate
$(\tau_{zx})_{max}$	$-9.3960G\beta$	$-9.0475G\beta$		$-9.3960G\beta$	$-9.3702G\beta$

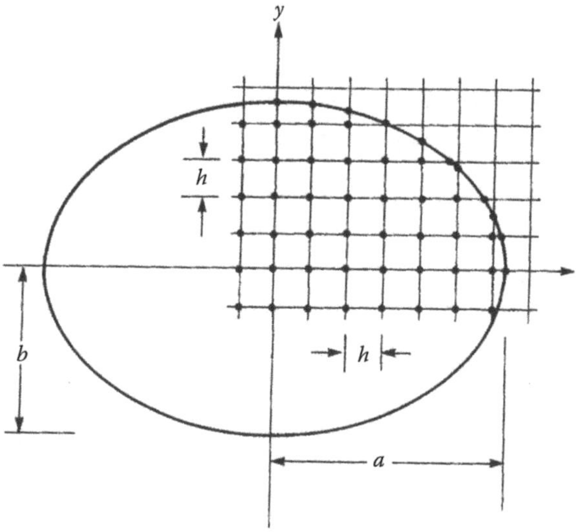

Figure 3.10 Subdivided elliptical cross-section (30 unknowns).

We note that the torsional function ϕ can be approximated fairly accurately even with a rough mesh subdivision. However, since the determination of shearing stress requires numerical differentiation of the approximate torsion function, for a given mesh size, the approximation of the shear stress is less accurate than that of the stress function (Table 3.4). Accordingly, in general, accurate approximation of the shear stress requires a finer mesh subdivision than does accurate approximation of the torsion function.

APPENDIX 3A: DERIVATION OF EQUATION 3.16

By Equation 3.14, with $p = 1$ and $n = 1$, we have with Equation 3.1,

$$g[b_0,b_1]=\frac{g(b_1)-g(b_0)}{b_1-b_0} \tag{3A.1}$$

Let $b_0 = x$ and $b_1 = x + \varepsilon$. Then, Equation 3A.1 becomes

$$g[x,x+\varepsilon]=\frac{g(x+\varepsilon)-g(x)}{\varepsilon} \tag{3A.2}$$

Consider the limit of Equation 3A.2 as $\varepsilon \to 0$. Then,

$$g[x,x+\varepsilon]\to g[x,x]=\lim_{\varepsilon\to 0}\frac{g(x+\varepsilon)-g(x)}{\varepsilon}=g'(x) \tag{3A.3}$$

where the prime denotes derivative with respect to x. Thus, although $g\,[x, x]$ cannot be evaluated directly from the values of g and the definition of Equation 3.1, it can be expressed in terms of the derivative of $g(x)$.

Similarly, consider

$$g[b_0,b_1,b_2]=\frac{g[b_1,b_2]-g[b_0,b_1]}{b_2-b_0} \tag{3A.4}$$

Let $b_0 = x$, $b_1 = b_0 + \varepsilon$, $b_2 = b_1 + \varepsilon$, or $b_0 = x$, $b_1 = x + \varepsilon$, $b_2 = x + 2\varepsilon$. Then, Equation 3A.4 may be written as

$$\begin{aligned}
g[b_0,b_1,b_2] &= g[x,x+\varepsilon,x+2\varepsilon]\\
&= \frac{g[x+\varepsilon,x+2\varepsilon]-g[x,x+\varepsilon]}{2\varepsilon}\\
&= \frac{g[x+2\varepsilon]-g[x+\varepsilon]}{2\varepsilon\cdot\varepsilon}-\frac{g[x+\varepsilon]-g[x]}{2\varepsilon\cdot\varepsilon}\\
&= \frac{g[x+2\varepsilon]-2g[x+\varepsilon]+g[x]}{2\varepsilon^2}
\end{aligned}$$

Taking the limit as $\varepsilon \to 0$, we have

$$g[b_0, b_1, b_2] \xrightarrow[\varepsilon \to 0]{} g[x, x, x] = \lim_{\varepsilon \to 0} \frac{g[x+2\varepsilon] - 2g[x+\varepsilon] + g[x]}{2\varepsilon^2} \to \frac{1}{2!} g''(x)$$

Continuing in this manner, we obtain

$$g[x, x, \ldots, x] = \frac{1}{n!} g^n(x) \tag{3A.5}$$

where x is repeated $n + 1$ times. (See Equation 3.16.)

APPENDIX 3B: DERIVATION OF EQUATION 3.38

Since $x_i = x_0 + ih$ and $x_j = x_0 + jh$, we have $x_i - x_j = h(i - j)$. Now considering the left-hand side of Equation 3.38, we expand in the following manner:

$$\sum_{i=-m}^{m} \left[\prod_{\substack{j=-m \\ j \neq 1}}^{m} (x_i - x_j) \right] = \sum_{i=0}^{m} \left[\prod_{j=-1}^{-m} (x_i - x_j) \prod_{\substack{j=0 \\ j \neq i}}^{m} (x_i - x_j) \right] + \sum_{i=-1}^{-m} \left[\prod_{j=-1}^{-m} (x_i - x_j) \prod_{j=0}^{m} (x_i - x_j) \right] \tag{3B.1}$$

With the condition $x_i - x_j = h(i - j)$, the right-hand side of Equation 3B.1 can be written in the form

$$\sum_{i=0}^{m} [h^m (i+1)(i+2) \cdots (i+m) h^m (i) \times (i-1) \cdots (i-j-1)(i-j+1) \cdots (i-m)]$$
$$+ \sum_{i=-1}^{-m} [h^m (i+1)(i+2) \cdots (-1)(1) \cdots (i+m) h^m (i)(i-1) \cdots (i-m)] \tag{3B.2}$$

Collecting terms, we find that Equation 3B.2 can be written in the form

$$\sum_{i=0}^{m} h^{2m} (-1)^{m-i} (m+i)!(m-i)! + \sum_{i=1}^{m} h^{2m} (-1)^{m+i} (m+i)!(m-i)! \tag{3B.3}$$

where in the last summation, we have taken $i = -i$ and $m = -m$ in order to sum with positive values of i. Substitution of Equation 3B.3 for the right-hand side of Equation 3B.1 yields Equation 3.38.

REFERENCES

Boresi, A. P., Chong, K. P., and Lee, J. D. 2011. *Elasticity in Engineering Mechanics*, Wiley, New York.

Bushnell, D. 1973. Finite difference energy models versus finite element models: Two variational approaches in one computer program, *ONR Symposium*, Academic Press, San Diego, CA, pp. 291–336.

Chapra, S. C., and Canale, R. P. 2012. *Numerical Methods for Engineers* (Vol. 2), McGraw-Hill, New York.

Collatz, L. 2012. *The Numerical Treatment of Differential Equations* (Vol. 60), Springer-Verlag, Berlin, Germany.

Hartree, D. R. 1961. *Numerical Analysis*, 2nd ed., Oxford University Press, Oxford.

Isaacson, E., and Keller, H. B. 1994. *Analysis of Numerical Methods*, Courier Corp., North Chelmsford, MA.

Shoup, T. E. 1979. *A Practical Guide to Computer Methods for Engineers*, Prentice-Hall, Upper Saddle River, NJ.

Stanton, R. G. 1961. *Numerical Methods for Science and Engineering*, Prentice-Hall, Upper Saddle River, NJ.

Thomas, G. B., and Finney, R. L. 1984. *Calculus and Analytical Geometry*, Addison-Wesley, Reading, MA.

Zienkiewicz, O. C., and Morgan, K. 2006. *Finite Elements and Approximation*, Courier Corp., North Chelmsford, MA.

BIBLIOGRAPHY

Allaire, P. E. 1999. *Basics of the Finite Element Method*, PWS Publishers, Dubuque, IA.

Bertsekas, D. P., and Tsitsiklis, J. 1997. *Parallel and Distributed Computation: Numerical Methods*, Athena Scientific, Belmont, MA.

Boresi, A. P., and Chong, K. P. (eds.) 1984. *Engineering Mechanics in Civil Engineering* (Vols. 1 and 2), American Society of Civil Engineers (Engineering Mechanics Division), Reston, VA.

Chau, K. T. 2018. *Theory of Differential Equations in Engineering and Mechanics*, CRC Press, Boca Raton, FL.

Chong, K. P., and Miller, M. M., Jr. 1979. Behavior of thin hyperbolic shells on rectangular enclosures, in *Proc. IASS World Congress on Shell and Spatial Structures*, IASS, Madrid, Spain, pp. 421–435.

Fillerup, J. M., and Boresi, A. P. 1982. Excitation of finite viscoelastic solid on springs, *Nucl. Eng. Des.*, **71**, 179–193.

Graves, R. W., 1996. Simulating seismic wave propagation in 3D elastic media using staggered-grid finite differences, *Bull. Seismological Soc. Am.*, *86*(4), pp. 1091–1106.

Kikuchi, N. 1986. *Finite Element Methods in Mechanics*, Cambridge University Press, Cambridge.

La Fara, R. L. 1973. *Computer Methods for Science and Engineering*, Hayden, Rochelle Park, NJ.

Langhaar, H. L. 1969. Two methods that converge to the method of least squares, *J. Franklin Inst.*, **288**, 165–173.

Langhaar, H. L., and Chu, S. C. 1970. Piecewise polynomials and the partition method for ordinary differential equations, in *Developments in Theoretical and Applied Mechanics* (Vol. 8), Frederick, D. (ed.), Pergamon Press, New York, pp. 553–564.

Langhaar, H. L., Boresi, A. P., and Miller, R. E. 1974. Periodic excitation of a finite linear viscoelastic solid, *Nucl. Eng. Des.*, **30**, 349–368.

Lapidus, L., and Pinder, G. F., 2011. *Numerical Solution of Partial Differential Equations in Science and Engineering*, Wiley, New York.

Larson, R., Hostetler, R. P., Edwards, B. H., and Heyd, D. E., 2002. *Calculus with Analytic Geometry*, Houghton Mifflin Co., Boston, MA.

Mitchell, A. R., and Griffiths, D. F. 1980. *The Finite Difference Method in Partial Differential Equations*, Wiley, New York.

Ralston, A., and Rabinowitz, P. 2012. *A First Course in Numerical Analysis*, Courier Corp., North Chelmsford, MA.

Sharma, S. K., and Boresi, A. P. 1978. Finite element weighted residual methods: Axisymmetric shells, *J. Eng. Mech. Div. ASCE*, **104**(EM4), 895–909.

Smith, G. D., 1985. *Numerical Solution of Partial Differential Equations: Finite Difference Methods*, Oxford University Press, New York.

Taflove, A., and Hagness, S. C. 1995. *Computational Electrodynamics: The Finite–Difference Time–Domain Method*, 2nd ed., Artech House, Dedham, MA.

Vandergraft, J. S., 2014. *Introduction to Numerical Computations*, Academic Press, San Diego, CA.

Virieux, J., 1986. P-SV wave propagation in heterogeneous media: Velocity-stress finite-difference method, *Geophysics,* **51**(4), 889–901.

Williams, P. W. 1972. *Numerical Computation*, Harper & Row, New York.

Zienkiewicz, O., C. Chan, A. H. C., Pastor, M., and Schrefler, B. A. 1999. *Computational Geomechanics: With Special Reference to Earthquake Engineering*, Wiley, New York.

Chapter 4

The finite element method

4.1 INTRODUCTION

The finite element method,* the most powerful numerical technique available today for the analysis of complex structural and mechanical systems, can be used to obtain numerical solutions to a wide range of problems. The method can be used to analyze both linear and nonlinear systems. Nonlinear analysis can include material yielding, creep, or cracking; aeroelastic response; buckling and postbuckling response; and contact and friction. It can be used for both static and dynamic analysis. In its most general form, the finite element method is not restricted to structural (or mechanical) systems. It has been applied to problems in fluid flow, heat transfer, electric potential, and multiscale problems. Its versatility is a major reason for the popularity of the method.

A complete study of finite element methods is beyond the scope of this book. So the objective of this chapter is to outline the basic formulation for problems in linear elasticity. The formulation for plane elasticity is presented first. Then use of the method to analyze framed structures is examined. Finally, accuracy, convergence, and proper modeling techniques are discussed. Advanced topics such as analysis of plate bending and shell problems, three-dimensional problems, and dynamic and nonlinear analysis are left to more specialized texts.

4.1.1 Analytical perspective

The classical method of analysis in elasticity involves the study of an infinitesimal element of an elastic body (continuum or domain). Relationships among stress, strain, and displacement for the infinitesimal element are developed (Boresi et al., 2011) that are usually in the form of differential (or integral) equations that apply to each point in the body. These equations must be solved subject to appropriate boundary conditions. In other words, the approach is to define and solve a classical boundary value problem in mathematics (Boresi et al., 2011). Problems in engineering usually involve very complex shapes and boundary conditions. Consequently, for such cases, the equations cannot be solved exactly but must finally be solved by approximate methods: for example, by truncated series, finite differences, numerical integration, and so on. All of these approximate methods require some form of discretization of the solution.

* The discovery of the method is often attributed to Courant (1943). The use of the method in structural (aircraft) analysis was first reported by Turner et al. (1956). The method received its name from Clough (1960).

By contrast, the formulation of finite element solutions recognizes at the outset that discretization is likely to be required. The first step in application of the method is to discretize the domain into an assemblage of a finite number of finite-size *elements* (or subregions) that are connected at specified node points. The quantities of interest (usually just displacements) are assumed to vary in a particular fashion over the element. This *assumed* element behavior leads to relatively simple integral equations for the individual elements. The integral equations for an element are evaluated to produce algebraic equations (in the case of static loading) in terms of the displacements of the node points. The algebraic equations for all elements are assembled to achieve a system of equations for the structure as a whole. Appropriate numerical methods are then used to solve this system of equations.

In summary, using the classical approach, we often are confronted with partial differential (or integral) equations that cannot be solved in closed form. This is due to the complexity of the geometry of the domain or of the boundary conditions. Consequently, we are forced to use numerical methods to obtain an approximate solution. These numerical methods always involve some type of discretization. In the finite element method, the discretization is performed at the outset. Then further approximation in either the formulation or the solution may not be necessary.

4.1.2 Sources of error

There are three sources of error in the finite element method: errors due to approximation of the domain (discretization error), errors due to approximation of the element behavior (formulation error), and errors due to use of finite-precision arithmetic (numerical error).

Discretization error is due to the approximation of the domain with a finite number of elements of fixed geometry. For instance, consider the analysis of a rectangular plate with a centrally located hole, Figure 4.1a. Due to symmetry, it is sufficient to model only one-fourth of the plate. If the region is subdivided into triangular elements (a triangular mesh or grid), the circular hole is approximated by a series of straight lines. If a few large triangles are used in a coarse mesh (Figure 4.1b), a greater discretization error results than if a large number of small elements is used in a fine mesh (Figure 4.1c). Other geometric shapes may be chosen. For example, with quadrilateral elements that can represent curved sides, the circular hole is more accurately approximated (Figure 4.1d). Hence, discretization error may be reduced by grid refinement. The grid can be refined by using more elements of the same type but of smaller size (*h-refinement*; Cook et al., 2007) or by using elements of different type (*p-refinement*).

Formulation error results from use of finite elements that do not precisely describe the behavior of the continuum. For instance, a particular element might be formulated on the assumption that displacements vary linearly over the domain. Such an element would contain no formulation error when used to model a prismatic bar under constant tensile load; in this case, the assumed displacement matches the actual displacement. If the same bar were subjected to uniformly distributed body force, the actual displacements vary quadratically and formulation error would exist. Formulation error can be minimized by proper selection of element type and by appropriate grid refinement.

Numerical error is a consequence of round-off during floating-point computations and the error associated with numerical integration procedures. This source of error is dependent on the order in which computations are performed in the program and on the use of double or extended-precision variables and functions. The use of bandwidth minimization* can help

* See Section 4.6 for a discussion of bandwidth minimization.

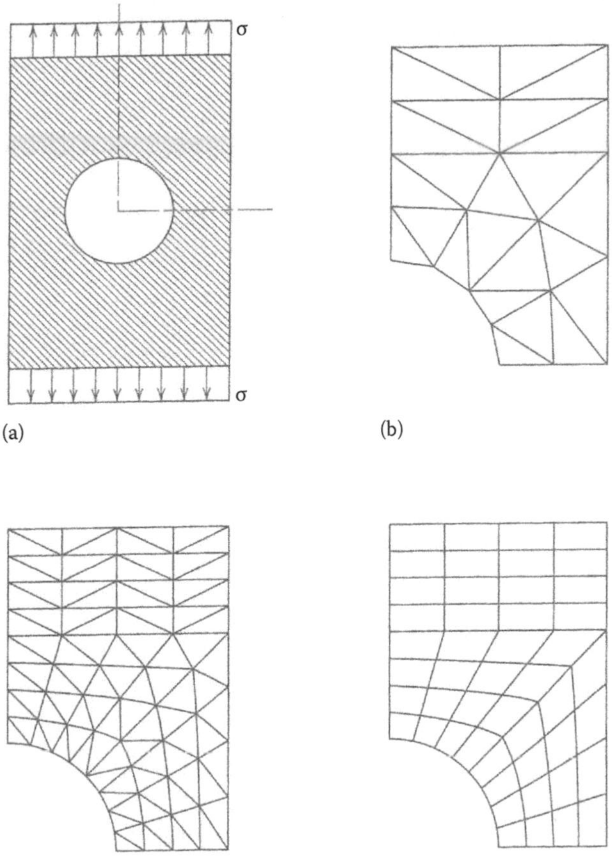

Figure 4.1 Finite element models of plate with centrally located hole. (a) Plate geometry and loading. (b) Coarse mesh of triangles. (c) Fine mesh of triangles. (d) Mesh of quadrilaterals with curved edges.

control numerical error. Generally, in a well-designed finite element program, numerical error is small relative to formulation error.

4.2 FORMULATION FOR PLANE ELASTICITY

4.2.1 Elasticity concepts

One approach for developing the algebraic equations of the finite element method is to use energy principles. Fundamental energy expressions for an elastic solid can be found in standard texts (e.g., Boresi et al., 2011). For plane elasticity, these expressions are simplified appropriately. The first law of thermodynamics states that for a two-dimensional body in equilibrium and subjected to arbitrary virtual displacements (δu, δv), the variation in work of the external forces W is equal to the variation of internal energy δU. Since *virtual displacements* are imposed, we define δW as the *virtual work of the external loads* and δU as the *virtual work of the internal forces*. The virtual work of the external forces δW can be divided into the work δW_S of the surface tractions, the work δW_B of the body forces, and the work δW_C* of the concentrated forces. For a two-dimensional body, these quantities are (Boresi and Schmidt, 2003, Chapters 3 and 5)

$$\delta W - \delta U = \delta W_S + \delta W_B + \delta W_C - \delta U = 0 \tag{4.1}$$

$$\delta W_S = \int_S (\sigma_{Px}\, \delta u + \sigma_{Py}\, \delta v)\, dS \tag{4.2}$$

$$\delta W_B = \int_V (B_x\, \delta u + B_y\, \delta v)\, dV \tag{4.3}$$

$$\delta W_C = \sum F_{ix} \delta u_i + \sum F_{iy} \delta v_i \tag{4.4}$$

and

$$\delta U = \int_V (\sigma_{xx}\, \delta\varepsilon_{xx} + \sigma_{yy}\, \delta\varepsilon_{yy} + \sigma_{xy}\, \delta\gamma_{xy})\, dV \tag{4.5}$$

where σ_{Px} and σ_{Py} are projections of the stress vector along axes x and y, respectively; B_x and B_y are body force components per unit volume; F_{ix} and F_{iy} are the x and y components of the concentrated force F_i at point i; δu_i and δv_i are the x and y components of the virtual displacement at point i; and $\gamma_{xy} = 2\varepsilon_{xy}$. In matrix notation,

$$\delta W_S = \int_S \{\delta u\}^T \{F_S\}\, dS \tag{4.6}$$

$$\delta W_B = \int_V \{\delta u\}^T \{F_B\}\, dV \tag{4.7}$$

$$\delta W_C = \sum \{\delta u_i\}^T \{F_i\} \tag{4.8}$$

and

$$\delta U = \int_V \{\delta \varepsilon\}^T \{\sigma\}\, dV \tag{4.9}$$

* Concentrated forces were not discussed in Chapter 3 or 5 of Boresi and Schmidt, 2003, but are included here for completeness.

where

$$\{\delta u\} = \{\delta u\ \delta v\}^{\mathrm{T}}$$

$$\{\delta u_i\} = \{\delta u_i\ \delta v_i\}^{\mathrm{T}}$$

$$\{F_S\} = [\sigma_{Px}\ \sigma_{Py}]^{\mathrm{T}}$$

$$\{F_B\} = [B_x\ B_y]^{\mathrm{T}}$$

$$\{F_i\} = [F_{ix}\ F_{iy}]^{\mathrm{T}}$$

$$\{\delta\varepsilon\} = [\delta\varepsilon_{xx}\ \delta\varepsilon_{yy}\ \delta\varepsilon_{xy}]^{\mathrm{T}}$$

$$\{\sigma\} = [\sigma_{xx}\ \sigma_{yy}\ \sigma_{xy}]^{\mathrm{T}}$$

In matrix form, the two-dimensional stress–strain relations are, by appropriate simplification (Boresi and Schmidt, 2003, Chapter 3),

$$\{\sigma\} = [D]\{\varepsilon\} \tag{4.10}$$

where $\{\varepsilon\} = [\varepsilon_{xx}\ \varepsilon_{yy}\ \gamma_{xy}]^{\mathrm{T}}$ and $[D]$ is the matrix of elastic coefficients. For plane stress,

$$[D] = \frac{E}{1-v^2}\begin{bmatrix} 1 & v & 0 \\ v & 1 & 0 \\ 0 & 0 & 1-\dfrac{v}{2} \end{bmatrix} \tag{4.11}$$

and for plane strain,

$$[D] = \frac{E}{(1+v)(1-2v)}\begin{bmatrix} 1-v & v & 0 \\ v & 1-v & 0 \\ 0 & 0 & \dfrac{1-2v}{2} \end{bmatrix} \tag{4.12}$$

Similarly, the two-dimensional, small displacement, strain–displacement relations are (see Chapter 2; Boresi and Schmidt, 2003)

$$\{\varepsilon\} = [L]\{u\} \tag{4.13}$$

where $\{u\} = [u(x, y)\ v(x, y)]^{\mathrm{T}}$ and $[L]$ is a matrix of linear differential operators:

$$[L] = \begin{bmatrix} \dfrac{\partial}{\partial x} & 0 \\ 0 & \dfrac{\partial}{\partial y} \\ \dfrac{\partial}{\partial y} & \dfrac{\partial}{\partial x} \end{bmatrix} \tag{4.14}$$

4.2.2 Displacement interpolation: Constant-strain triangle

Consider a plane elasticity problem such as that shown in Figure 4.1a. As discussed previously, the first step in applying the finite element method is the discretization of the domain into a finite number of elements. Consider triangular elements as shown in Figure 4.1b and c. If the entire domain is in equilibrium, so too is each of the elements. Hence, the virtual work concepts earlier can be applied to an individual triangular element.

A typical triangular element is shown in Figure 4.2 with corner nodes 1, 2, and 3 numbered in a counterclockwise order. The (x, y) displacement components at the nodes are (u_1, v_1), (u_2, v_2), and (u_3, v_3), as shown. The *nodal* displacements are the primary variables (unknowns) that are to be determined by the finite element method. In general, for plane elasticity elements, node i has two degrees of freedom (DOFs), u_i and v_i, where the subscript identifies the node at which the DOF exists. Quantities that are continuous over the element (those not associated with a particular node) are denoted without a subscript. A single triangular element with three nodes has six nodal DOFs These DOFs are ordered according to the node numbering as

$$\{u_i\} = [u_1 \quad v_1 \quad u_2 \quad v_2 \quad u_3 \quad v_3]^T \tag{4.15}$$

Displacements (u, v) at any point P within the element are continuous functions of the spatial coordinates (x, y).

A fundamental approximation in the finite element method (that leads to formulation error) is that the displacement (u, v) at any point P in the element can be written in terms of the nodal displacements. Specifically, the displacement (u, v) at point P within the element is *interpolated* from the displacements of the nodes using interpolation polynomials. The order of the interpolation depends on the number of DOFs in the element. For the three-node triangular element, the displacement is assumed to vary linearly over the element.

$$\begin{aligned} u(x, y) &= a_1 + a_2 x + a_3 y \\ v(x, y) &= a_4 + a_5 x + a_6 y \end{aligned} \tag{4.16}$$

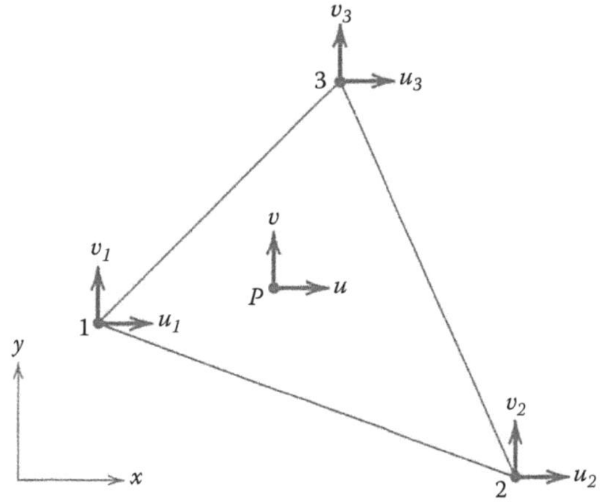

Figure 4.2 CST element.

The coefficients a_i are constants (sometimes called *generalized displacement coordinates*) that are evaluated in terms of the nodal displacements. Before making this evaluation, we consider some properties of the linear displacement approximation.

1. Substitution of Equation 4.16 into Equation 4.13 yields $\varepsilon_{xx} = a_2$, $\varepsilon_{yy} = a_6$, and $\gamma_{xy} = a_3 + a_5$. Thus, the strain components in the element are constant: hence, the name *constant-strain triangle* (CST) element. Since the stress–strain relations are linear (Equation 4.10), stress components are also constant in the element.
2. If $a_2 = a_3 = a_5 = a_6 = 0$, then $u(x, y) = a_1$ and $v(x, y) = a_4$. Constant values of u and v displacement indicate that the element can represent rigid-body translation.
3. If $a_1 = a_2 = a_4 = a_6 = 0$ and $a_3 = -a_5$, then $u(x, y) = a_3 y$ and $v(x, y) = -a_3 x$. Thus, the element can represent rigid-body rotation.

These three element characteristics ensure that the solution will converge monotonically as the mesh is refined (see Section 4.6 for a discussion of convergence).

To express the continuous displacement field in terms of the nodal displacements, Equation 4.16 is evaluated at each node. The resulting equations are then solved for the coefficients a_i. Consider first the u displacement.

$$\begin{aligned} u(x_1, y_1) &= a_1 + a_2 x_1 + a_3 y_1 = u_1 \\ u(x_2, y_2) &= a_1 + a_2 x_2 + a_3 y_2 = u_2 \\ u(x_3, y_3) &= a_1 + a_2 x_3 + a_3 y_3 = u_3 \end{aligned}$$

In matrix form, these equations are written as

$$[A]\{a\} = \{u_i\} \tag{4.17}$$

where

$$[A] = \begin{bmatrix} 1 & x_1 & y_1 \\ 1 & x_2 & y_2 \\ 1 & x_3 & y_3 \end{bmatrix} \quad \{a\} = \begin{Bmatrix} a_1 \\ a_2 \\ a_3 \end{Bmatrix} \quad \{u_i\} = \begin{Bmatrix} u_1 \\ u_2 \\ u_3 \end{Bmatrix}$$

The solution of Equation 4.17 for $\{a\}$ and substitution into Equation 4.16 yield

$$u(x, y) = \frac{1}{2A}(\alpha_1 + \beta_1 x + \gamma_1 y)u_1 + \frac{1}{2A}(\alpha_2 + \beta_2 x + \gamma_2 y)u_2 + \frac{1}{2A}(\alpha_3 + \beta_3 x + \gamma_3 y)u_3 \tag{4.18}$$

where A is the area of the triangle:

$$A = \frac{1}{2}\left[x_1(y_2 - y_3) + x_2(y_3 - y_1) + x_3(y_1 - y_2)\right] \tag{4.19}$$

and

$$\begin{array}{lll} \alpha_1 = x_2y_3 - x_3y_2 & \beta_1 = y_2 - y_3 & \gamma_1 = x_3 - x_2 \\ \alpha_2 = x_3y_1 - x_1y_3 & \beta_2 = y_3 - y_1 & \gamma_2 = x_1 - x_3 \\ \alpha_3 = x_1y_2 - x_2y_1 & \beta_3 = y_1 - y_2 & \gamma_3 = x_2 - x_1 \end{array} \tag{4.20}$$

Similarly, for the v displacement,

$$v(x,y) = \frac{1}{2A}(\alpha_1 + \beta_1 x + \gamma_1 y)v_1 + \frac{1}{2A}(\alpha_2 + \beta_2 x + \gamma_2 y)v_2 + \frac{1}{2A}(\alpha_3 + \beta_3 x + \gamma_3 y)v_3 \tag{4.21}$$

The functions that multiply the nodal displacements in Equations 4.18 and 4.21 are known as *shape functions* (other common names are *interpolation functions* and *basis functions*). The shape functions for the CST element are

$$\begin{aligned} N_1(x,y) &= \frac{1}{2A}(\alpha_1 + \beta_1 x + y_1 y) \\ N_2(x,y) &= \frac{1}{2A}(\alpha_2 + \beta_2 x + y_2 y) \\ N_3(x,y) &= \frac{1}{2A}(\alpha_3 + \beta_3 x + y_3 y) \end{aligned} \tag{4.22}$$

Then Equations 4.18 and 4.21 take the form

$$u(x,y) = \sum_{i=1}^{3} N_i u_i \qquad v(x,y) = \sum_{i=1}^{3} N_i v_i$$

In matrix notation,

$$\{u\} = [N]\{u_i\} \tag{4.23}$$

where

$$[N] = \begin{bmatrix} N_1 & 0 & N_2 & 0 & N_3 & 0 \\ 0 & N_1 & 0 & N_2 & 0 & N_3 \end{bmatrix} \tag{4.24}$$

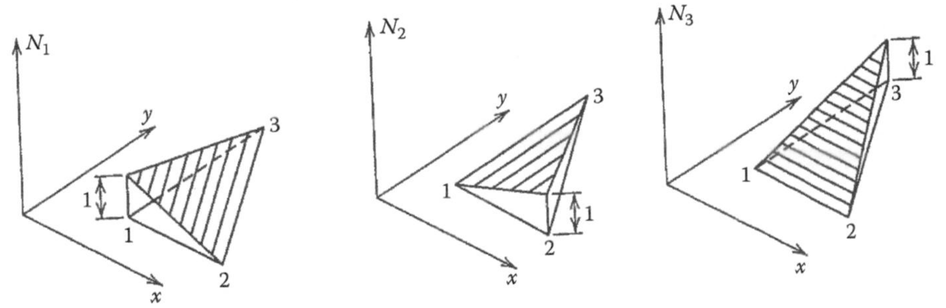

Figure 4.3 Graphical representation of shape functions for the CST element.

The shape functions are illustrated in Figure 4.3, where $N_1 = 1$ at node 1 and $N_1 = 0$ at nodes 2 and 3. Shape functions N_2 and N_3 behave similarly. Another important characteristic of the shape functions is that

$$\sum_{i=1}^{3} N_i(x, y) = 1.0$$

which is a consequence of the fact that the element can represent rigid-body motion.

4.2.3 Element stiffness matrix: CST

With the displacement field for the element expressed in terms of the nodal displacements, the remainder of the formulation involves relatively straightforward manipulation of the virtual work expressions, Equations 4.1 and 4.6 through 4.9. Consider first the strain–displacement relations. Substitution of Equation 4.23 into Equation 4.13 gives the relationship between *continuous element strains* and *nodal displacements*,

$$\{\varepsilon\} = [L][N]\{u_i\} = [B]\{u_i\} \tag{4.25}$$

where, by Equations 4.14 and 4.24,

$$[B] = \left[\begin{array}{cc|cc|cc} \dfrac{\partial N_1}{\partial x} & 0 & \dfrac{\partial N_2}{\partial x} & 0 & \dfrac{\partial N_3}{\partial x} & 0 \\ 0 & \dfrac{\partial N_1}{\partial y} & 0 & \dfrac{\partial N_2}{\partial y} & 0 & \dfrac{\partial N_3}{\partial y} \\ \dfrac{\partial N_1}{\partial y} & \dfrac{\partial N_1}{\partial x} & \dfrac{\partial N_2}{\partial y} & \dfrac{\partial N_2}{\partial x} & \dfrac{\partial N_3}{\partial y} & \dfrac{\partial N_3}{\partial x} \end{array}\right] \tag{4.26}$$

where $[B]$ is partitioned into nodal submatrices. The matrix $[B]$ is sometimes called the *semidiscretized gradient operator*. Since the shape functions are linear in x and y, $[B]$ contains only constants that depend on the nodal coordinates.

For simplicity, let us temporarily assume that no body forces or surface tractions are applied to the element. However, concentrated loads at node points are permitted. The virtual work of these loads is (see Equation 4.8)

$$\delta W = \delta W_C = \{\delta u_i\}^T \{F_i\} \tag{4.27}$$

Substitution of Equations 4.27 and 4.9 into 4.1 leads to

$$\{\delta u_i\}^T \{F_i\} - \int_V \{\delta \varepsilon\}^T \{\sigma\}\, dV = 0 \tag{4.28}$$

Note that $\{\sigma\} = [D]\{\varepsilon\}$, $\{\varepsilon\} = [B]\{u_i\}$, and $\{\delta\varepsilon\}^T = \{\delta u_i\}^T[B]^T$. Substitution of these expressions into Equation 4.28 gives

$$\{\delta u_i\}^T \{F_i\} - \int_V \{\delta u_i\}^T \{B\}^T [D][B]\{u_i\}\, dV = 0$$

Since $\{u_i\}$ and $\{\delta u_i\}$ are *nodal* quantities, they can be removed from the integral. Thus,

$$\{\delta u_i\}^T \left(\{F_i\} - \left[\int_V \{B\}^T [D][B] dV \right] \{u_i\} \right) = 0 \tag{4.29}$$

Since $\{\delta u_i\}$ is arbitrary, Equation 4.29 becomes

$$\{F_i\} = \left[\int_V [B]^T [D][B] dV \right] \{u_i\}$$

or

$$\{F_i\} = [K]\{u_i\} \tag{4.30}$$

where

$$[K] = \int_V [B]^T [D][B] dV \tag{4.31}$$

The element stiffness matrix $[K]$ relates nodal loads to nodal displacements in a system of linear algebraic equations, Equation 4.30 For the CST element, all terms in the integral are constants. Hence, for an element of constant thickness t, the element stiffness matrix is

$$[K] = At[B]^T [D][B] \tag{4.32}$$

The individual terms in $[K]$ are denoted k_{ij}, where $i, j = 1, 2, \ldots, 6$ are the row and column positions, respectively. Since the element has six nodal DOF, $[K]$ has order (6×6). The explicit form of the CST element stiffness matrix for a plane stress condition is given in Table 4.1.

Table 4.1 CST element stiffness matrix, plane stress case (partitioned into 2 × 2 nodal submatrices)

	Column index						Row index
$j \rightarrow$	1	2	3	4	5	6	$i \downarrow$
C	$y_{23}^2 + \frac{1-v}{2} x_{32}^2$	$\frac{1+v}{2} x_{32}y_{23}$	$y_{13}y_{23} + \frac{1-v}{2} x_{13}y_{32}$	$vx_{13}y_{23} + \frac{1-v}{2} x_{32}y_{31}$	$y_{12}y_{23} + \frac{1-v}{2} x_{21}x_{32}$	$vx_{21}y_{23} + \frac{1-v}{2} x_{32}x_{12}$	1
	$\frac{1+v}{2} x_{32}y_{23}$	$x_{32}^2 + \frac{1-v}{2} y_{23}^2$	$vx_{32}y_{31} + \frac{1-v}{2} x_{13}y_{23}$	$x_{13}x_{32} + \frac{1-v}{2} y_{23}y_{31}$	$vx_{32}y_{12} + \frac{1-v}{2} x_{21}y_{23}$	$x_{21}y_{32} + \frac{1-v}{2} y_{12}y_{23}$	2
	$y_{31}y_{23} + \frac{1+v}{2} x_{13}x_{32}$	$vx_{32}x_{31} + \frac{1-v}{2} x_{13}y_{23}$	$y_{31}^2 + \frac{1-v}{2} x_{13}^2$	$\frac{1-v}{2} x_{13}y_{31}$	$y_{12}y_{31} + \frac{1-v}{2} x_{13}y_{21}$	$vx_{21}y_{31} + \frac{1-v}{2} x_{13}y_{12}$	3
	$vx_{13}y_{23} + \frac{1-v}{2} x_{32}y_{31}$	$x_{13}x_{32} + \frac{1-v}{2} y_{23}y_{31}$	$\frac{1-v}{2} x_{13}y_{31}$	$x_{13}^2 + \frac{1-v}{2} y_{31}^2$	$vx_{13}y_{12} + \frac{1-v}{2} x_{21}y_{31}$	$x_{13}y_{21} + \frac{1-v}{2} y_{12}y_{31}$	4
	$y_{12}y_{23} + \frac{1-v}{2} x_{21}x_{32}$	$vx_{32}y_{12} + \frac{1-v}{2} x_{21}y_{23}$	$y_{12}y_{31} + \frac{1-v}{2} x_{13}x_{21}$	$vx_{13}y_{12} + \frac{1-v}{2} x_{21}y_{31}$	$y_{12}^2 + \frac{1-v}{2} x_{21}^2$	$\frac{1+v}{2} x_{21}y_{12}$	5
	$vx_{21}y_{23} + \frac{1-v}{2} x_{32}y_{12}$	$x_{21}x_{32} + \frac{1-v}{2} y_{12}y_{23}$	$vx_{21}y_{31} + \frac{1-v}{2} x_{13}y_{12}$	$x_{13}y_{21} + \frac{1-v}{2} x_{12}y_{31}$	$\frac{1-v}{2} x_{21}y_{12}$	$x_{21}^2 + \frac{1-v}{2} y_{12}^2$	6

$$C = \frac{Et}{4A(1-v^2)};\ x_{ij} = x_i - x_j;\ y_{ij} = y_i - y_j$$

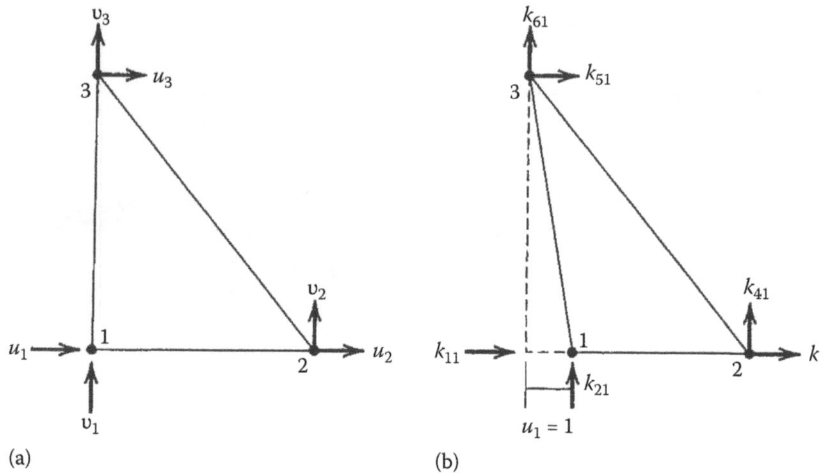

Figure 4.4 Physical interpretation of k_{ij}. (a) Undeformed element, (b) deformed element, forces k_{il} required to maintain u_l, = 1.

Examination of Equation 4.30 helps to establish a physical interpretation of the stiffness coefficients (the individual terms in $[K]$). Let a unit displacement be assigned to u_1 and take all other DOFs to be zero. The resulting displacement vector is

$$\{u_i\} = [1 \quad 0 \quad 0 \quad 0 \quad 0 \quad 0]^{\mathrm{T}}$$

Substitution of this displacement vector into Equation 4.30 gives the force vector required to maintain the deformed shape:

$$\{F_i\} = [k_{11} \quad k_{21} \quad k_{31} \quad k_{41} \quad k_{51} \quad k_{61}]^{\mathrm{T}}$$

Hence, an individual stiffness coefficient k_{ij} can be interpreted as the nodal force in the direction of DOF i that results from a unit displacement in the direction of DOF j while all other DOFs are set equal to zero. The physical system is illustrated in Figure 4.4.

4.2.4 Equivalent nodal load vector: CST

Assume that body forces are applied to the CST element (surface tractions will be considered subsequently). The virtual work δW_B of the body forces on the element during an arbitrary virtual displacement $\{\delta u\}$ is given by Equation 4.7. Substitution of Equation 4.23 into Equation 4.7 gives

$$\delta W_B = \{\delta u_i\}^{\mathrm{T}} \int_V [N]^{\mathrm{T}} \{F_B\}\, dV \tag{4.33}$$

The total external virtual work δW is the sum of the virtual work of the body forces and the virtual work of the concentrated forces, so that Equation 4.29 becomes

$$\{\delta u_i\}^{\mathrm{T}} \left(\{F_i\} + \int_V [N]^{\mathrm{T}} \{F_B\}\, dV - \left[\int_V [B]^{\mathrm{T}} [D][B]\, dV \right] \{u_i\} \right) = 0 \tag{4.34}$$

Comparison of Equation 4.34 with Equation 4.29 shows that with the addition of body forces, the load vector for the element becomes

$$\{P_i\} = \{F_i\} + \int_V [N]^T \{F_B\}\, dV \tag{4.35}$$

The vector $\{P_i\}$ is the equivalent nodal load vector for the element. That is, the work of the loads $\{P_i\}$ under the virtual displacement $\{\delta u_i\}$ of the nodes is equivalent to the work of the actual concentrated loads and body forces under the virtual displacement $\{\delta u\}$.

In Equation 4.35, the body force $\{F_B\}$ is expressed as a continuous function of the spatial coordinates. However, when constructing a finite element model, it is customary for the analyst to define element loads in terms of the intensity of the load at the nodes rather than in functional form. So, for convenience, assume that the body force distribution may be expressed in terms of the force intensities at the nodes according to the relation

$$\{F_B\} = [N]\{f_{Bi}\}$$

where $\{f_{Bi}\}$ is the vector of nodal force intensities. Substitution of this relation into Equation 4.35 gives

$$\{P_i\} = \{F_i\} + \int_V [N]^T [N]\{f_{Bi}\}\, dV$$

Since $\{f_{Bi}\}$ does not vary over the element,

$$\{P_i\} = \{F_i\} + [Q]\{f_{Bi}\} \tag{4.36}$$

where

$$[Q] = \int_V [N]^T [N]\, dV$$

Thus, for a CST element,

$$[Q] = \frac{At}{9}\begin{bmatrix} 1 & 0 & 1 & 0 & 1 & 0 \\ 0 & 1 & 0 & 1 & 0 & 1 \\ 1 & 0 & 1 & 0 & 1 & 0 \\ 0 & 1 & 0 & 1 & 0 & 1 \\ 1 & 0 & 1 & 0 & 1 & 0 \\ 0 & 1 & 0 & 1 & 0 & 1 \end{bmatrix}$$

Now suppose that in addition to concentrated nodal loads and body forces, the element is subjected to surface tractions along a single edge and that the continuous load function $\{F_S\}$ is expressed in terms of the nodal force intensities $\{f_{Si}\}$ by use of the shape functions.

Since only one edge is loaded, only two of the nodes have nodal intensities and only these two nodes have equivalent nodal load components. Hence, for these two nodes, the interpolation equation is

$$\{\bar{F}_S\} = [\bar{N}]\{\bar{f}_{Si}\}$$

where the overbar indicates that only these two element nodes are included in the equation.

By the same approach as for body forces, the equivalent nodal loads due to surface traction on one edge are

$$\{\bar{P}_i\} = [\bar{Q}]\{\bar{f}_{Si}\} \tag{4.37}$$

where

$$[\bar{Q}] = \int_S [\bar{N}]^{\mathrm{T}}[\bar{N}]\, dS \tag{4.38}$$

and the integral is evaluated over the loaded edge only, where $dS = t\, ds$, t = thickness, and s is a coordinate along the loaded edge. The equivalent nodal load vector $\{\bar{P}_i\}$ in Equation 4.37 is then added to $\{P_i\}$ from Equation 4.36, but first it must be expanded from four to six terms to account for the fact that one node does not participate in the loading.

Example 4.1: Equivalent Nodal Loads for Linear Surface Traction

A horizontally directed, linearly varying surface traction is applied to edge 1–3 of the CST element with nodal intensities as shown in Figure E4.1. Determine the vector of equivalent nodal loads for the element.

Solution: The surface traction function is interpolated from the nodal intensities at nodes 1 and 3 and from the corresponding shape functions:

$$f_x(y) = N_1 f_{1x} + N_2 f_{3x} \tag{4E1.1}$$

With the coordinates of the nodes, the shape functions are simplified to

$$N_1 = 1 - \frac{y}{b} \qquad N_3 = \frac{y}{b} \tag{4E1.2}$$

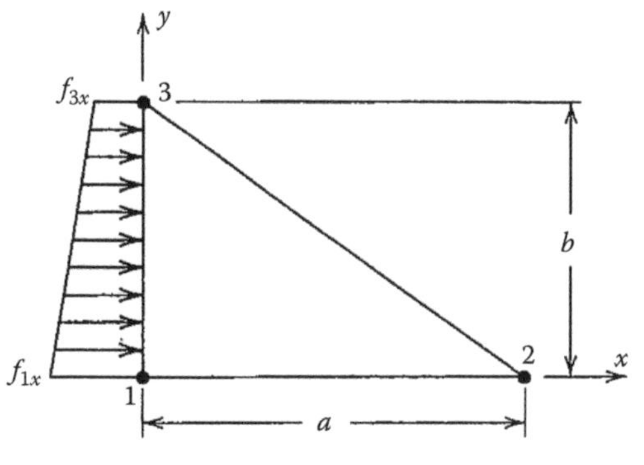

Figure E4.1 Surface traction applied to an edge of a CST element.

By Equation 4.38, with $ds = dy$,

$$[\bar{Q}] = t\int_0^b \begin{bmatrix} N_1^2 & 0 & N_1N_3 & 0 \\ 0 & N_1^2 & 0 & N_1N_3 \\ N_1N_3 & 0 & N_3^2 & 0 \\ 0 & N_1N_3 & 0 & N_3^2 \end{bmatrix} dy \tag{4E1.3}$$

By Equations 4E1.2 and 4E1.3

$$[\bar{Q}] = t\begin{bmatrix} b/3 & 0 & b/6 & 0 \\ 0 & b/3 & 0 & b/6 \\ b/6 & 0 & b/3 & 0 \\ 0 & b/6 & 0 & b/3 \end{bmatrix} \tag{4E1.4}$$

And by Equation 4E1.1, the vector of nodal intensities $\{\bar{f}_{Si}\}$ is

$$\{\bar{f}_{Si}\} = [f_{1x} \;\; 0 \;\; f_{3x} \;\; 0]^T \tag{4E1.5}$$

With Equations 4E1.4 and 4E1.5, the equivalent nodal load vector $\{\bar{P}_i\}$ is obtained from Equation 4.37 as

$$\{\bar{P}_i\} = \left\{ \begin{array}{c} tb\left(\dfrac{f_{1x}}{3} + \dfrac{f_{3x}}{6}\right) \\ 0 \\ \hline tb\left(\dfrac{f_{1x}}{6} + \dfrac{f_{3x}}{3}\right) \\ 0 \end{array} \right\} \tag{4E1.6}$$

Equation 4E1.5 is partitioned to identify the equivalent nodal loads associated with nodes 1 and 3.

If the vector $\{\bar{P}_i\}$ is expanded to include positions for node 2, it becomes

$$\{\bar{P}_i\} = \left\{ \begin{array}{c} tb\left(\dfrac{f_{1x}}{3} + \dfrac{f_{3x}}{6}\right) \\ 0 \\ \hline 0 \\ 0 \\ tb\left(\dfrac{f_{1x}}{6} + \dfrac{f_{3x}}{3}\right) \\ 0 \end{array} \right\} \tag{4E1.7}$$

4.2.5 Assembly of the structure stiffness matrix and load vector

To solve a plane elasticity problem by the finite element method, it is necessary to combine the individual element stiffness matrices $[K]_j$ and load vectors $\{P\}_j$ to form the *structure stiffness matrix* $[\mathbf{K}]$ and the *structure load vector* $\{\mathbf{P}\}$, respectively. To demonstrate the logic associated

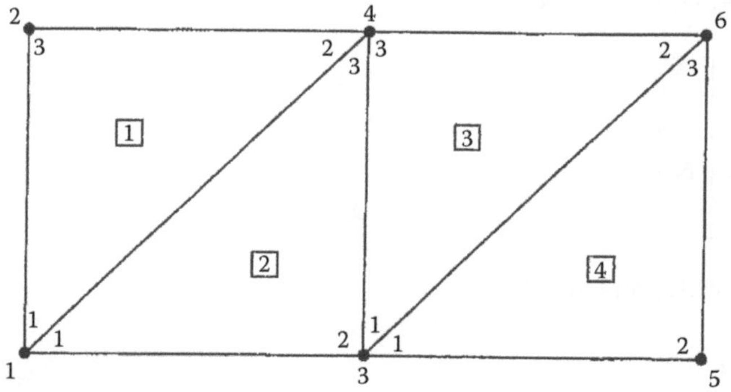

$NN = 6, \quad NE = 4$

Figure 4.5 Assembly of CST elements.

with the assembly process, two node numbering systems for the nodes are used. Let numerals in boldface refer to the nodes of the structural system and numerals in lightface refer to the nodes for a particular element. Similarly, lightface $[K]$, $\{u\}$, and $\{P\}$ refer to element quantities and boldface $[\mathbf{K}]$, $\{\mathbf{u}\}$, and $\{\mathbf{P}\}$ refer to structure quantities. A specific two-dimensional discretization is shown in Figure 4.5 to illustrate the node numbering. For this model, there are six structure nodes but a total of 12 separate element nodes. The assembly process involves assigning unique identifiers to each of the nodes in the model, using the structure node numbering and then combining element stiffness matrices and load vectors according to the numbering.

For purposes of demonstration, we consider first a mathematically precise but computationally inefficient approach for this assembly. Then we discuss an approach that is more appropriate for computer implementation.

For element j, define a matrix $[M]_j$ with order $(6 \times 2NN)$,* where NN is the number of nodes in the structure, to define the mapping from the element DOF vector $\{u\}_j$, with order (6×1), to the structure DOF vector $\{\mathbf{u}\}$, with order $(2NN \times 1)$:

$$\{u_i\} = [M]_i \{\boldsymbol{u}_i\} \tag{4.39}$$

By Figure 4.5, the mapping for element 1 takes the form

$$\{u\}_1 = [u_1 \quad v_1 \quad u_2 \quad v_2 \quad u_3 \quad v_3]_1^T$$

$$\{\mathbf{u}\} = [u_1 \quad v_1 \quad u_2 \quad v_2 \quad u_3 \quad \ldots v_6]^T$$

and

$$[M]_1 = \begin{bmatrix} 1 & 0 & 0 & 0 & 0 & 0 & 0 & 0 & 0 & 0 & 0 & 0 \\ 0 & 1 & 0 & 0 & 0 & 0 & 0 & 0 & 0 & 0 & 0 & 0 \\ 0 & 0 & 0 & 0 & 0 & 0 & 1 & 0 & 0 & 0 & 0 & 0 \\ 0 & 0 & 0 & 0 & 0 & 0 & 0 & 1 & 0 & 0 & 0 & 0 \\ 0 & 0 & 1 & 0 & 0 & 0 & 0 & 0 & 0 & 0 & 0 & 0 \\ 0 & 0 & 0 & 1 & 0 & 0 & 0 & 0 & 0 & 0 & 0 & 0 \end{bmatrix}$$

* The matrix $[M]_j$ is known as a *Boolean connectivity matrix* since it contains only ones and zeros.

By inspection or by Equation 4.39, the DOF mapping for element 1 is

$$[u_1 \quad v_1 \quad u_2 \quad v_2 \quad u_3 \quad v_3]_1^{\mathrm{T}} \leftrightarrow [u_1 \quad v_1 \quad u_4 \quad v_4 \quad u_2 \quad v_2]^{\mathrm{T}}$$

The double-headed arrow indicates the reversibility of the mapping of the quantities on the left to the quantities on the right. Nodal forces and stiffness coefficients for element 1 follow the same mapping.

Next, the virtual work expressions for the entire structure are written as the sum of the virtual work for all elements.

$$\sum_{j=1}^{NE} \{\delta u_i\}_j^{\mathrm{T}}\{F_i\} + \sum_{j=1}^{NE} \{\delta u_i\}_j^{\mathrm{T}} \int_S [\bar{N}]_j^{\mathrm{T}}\{F_S\}_j \, dS + \sum_{j=1}^{NE} \{\delta u_i\}_j^{\mathrm{T}} + \int_V [N]_j^{\mathrm{T}}\{F_B\}_j \, dV - \sum_{j=1}^{NE} \{\delta u_i\}_j^{\mathrm{T}}[K]_j\{u_i\}_j = 0 \tag{4.40}$$

where NE is the number of elements in the model. Substitution of Equation 4.39 into Equation 4.40 for each element gives

$$\{\delta \mathbf{u}_i\}^{\mathrm{T}} \sum_{j=1}^{NE} [M]_j^{\mathrm{T}}\{F_i\}_j + \{\delta \mathbf{u}_i\}^{\mathrm{T}} \sum_{j=1}^{NE} [\bar{M}]_j^{\mathrm{T}} \int_S [\bar{N}]_j^{\mathrm{T}}\{F_S\}_j \, dS + \{\delta \mathbf{u}_i\}^{\mathrm{T}} \sum_{j=1}^{NE} [M]_j^{\mathrm{T}} \int_V [\bar{N}]_j^{\mathrm{T}}\{F_B\}_j \, dV - \{\delta \mathbf{u}_i\}^{\mathrm{T}} \left[\sum_{j=1}^{NE} [M]_j^{\mathrm{T}}[K]_j[M]_j \right] \{\mathbf{u}_i\} = 0 \tag{4.41}$$

Since $\{\delta \mathbf{u}_i\}$ is arbitrary, it is eliminated from Equation 4.41 to obtain

$$[\mathbf{K}]\{\mathbf{u}_i\} = \{\mathbf{P}_i\} \tag{4.42a}$$

where

$$[\mathbf{K}] = \left[\sum_{j=1}^{NE} [M]_j^{\mathrm{T}}[K]_j[M]_j \right] \tag{4.42b}$$

and

$$\{\mathbf{P}_i\} = \sum_{j=1}^{NE} [M]_j^{\mathrm{T}} \{F_i\}_j + \sum_{j=1}^{NE} [\bar{M}]_j^T \int_S [\bar{N}]_j^T \{F_S\}_j \, dS + \sum_{j=1}^{NE} [M]_j^T \int_V [N]_j^T \{F_B\}_j \, dV \tag{4.42c}$$

In Equations 4.41 and 4.42c, matrix $[\bar{M}]_i$, of order ($4 \times 2NN$), accounts for the mapping to the structure nodes of the two nodes in element j that participate in the surface tractions. If more than one edge on an element is loaded, Equations 4.41 and 4.42b are extended accordingly.

The forms of $[\mathbf{K}]$ and $\{\mathbf{P}_i\}$ in Equation 4.42 are precise, but they are not used in practice. The matrix products involving $[M]_j$, which involve multiplying by 0 or 1, do nothing more than move individual quantities from one position in the element stiffness matrix or load vector to another in the structure stiffness matrix or load vector. Although the development noted previously is not practical, it does demonstrate that the structure stiffness matrix is assembled by successively adding the stiffness terms from each element into appropriate locations of the structure matrix, and similarly for the structure load vector. A more direct approach to assembly is demonstrated in Example 4.2.

Example 4.2: Assembly of the Structure Stiffness Matrix

For the model shown in Figure 4.5, illustrate the assembly of the stiffness matrix for element 1 into the structure stiffness matrix.

Solution: Since the structure has six nodes, each of which has two DOFs, the structure stiffness matrix is of order 12×12. The individual stiffness coefficients are designated k_{ij}^e, where the superscript identifies the element number. With this notation, the stiffness matrix for element 1 is shown in Figure E4.2. The mapping of element node numbers to structure node numbers is determined by inspection of the model in Figure 4.5 and is summarized in Table E4.2. The list of structure node numbers that define the nodes for each element, commonly known as the incidence list, is one of the input requirements for finite element programs. Using the incidence list, one can obtain the mapping of the element 1 nodal submatrices into the structure stiffness matrix (Figure E4.2). Markers have been added to the nodal submatrices as an aid to visualization of the placement of element stiffness coefficients into the structure stiffness matrix.

As described in Example 4.2, the incidence list is used to drive the assembly process. Suppose that the node numbers that comprise the incidence list, for instance, Table E4.2, are placed into a matrix [INCID] that contains one column for each element. The i, j term in the matrix is defined as the structure node number that corresponds to element node number i of element j. Then, using the incidence matrix [INCID], each term from the element stiffness matrix $[K]_j$ is moved into the structure stiffness matrix $[\mathbf{K}]$ in a prescribed manner. The method is illustrated in Table 4.2 by a FORTRAN subroutine. The subroutine moves one nodal submatrix at a time. Note that this code is for illustrative purposes only. Because of the symmetry and sparsity of the structure stiffness, it is usually stored in some form other than a square matrix.

4.2.6 Application of constraints

The model shown in Figure 4.5 is not fastened to supports. Hence, it represents an unstable structure, a structure that is not capable of resisting external loads. The assembled stiffness matrix for an unstable structure is singular; it has a rank deficiency of 3, due to the three

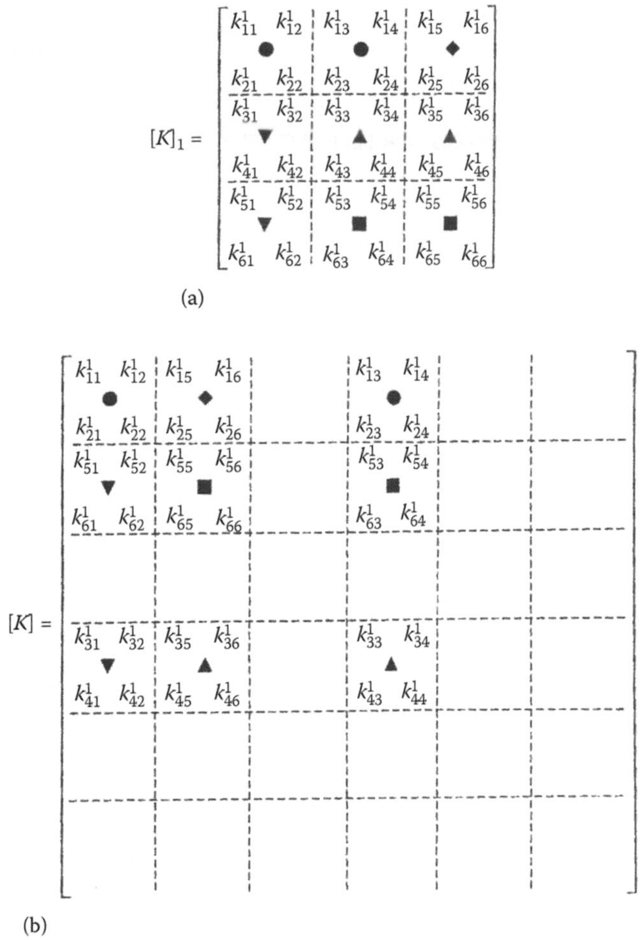

Figure E4.2 Assembly of element 1 stiffness matrix (with 2 × 2 nodal submatrix partitions). (a) Stiffness matrix for element 1. (b) Structure stiffness matrix with element 1 assembled.

Table E4.2 Element node-to-structure node mapping

Element node number	*Structure node numbers*			
	Element 1	*Element 2*	*Element 3*	*Element 4*
1	1	1	3	3
2	4	3	6	5
3	2	4	4	6

Table 4.2 FORTRAN subroutine for stiffness assembly

```
      SUBROUTINE ASSMBL (KS, KE, NNE, NDOF, INCID, IE, NRS, NRE, NE) (DIMENSION KS (NRS,
       NRS), KE (NRE, NR), INCID (NNE, NE) REAL KS, KE
C
C          ASSEMBLE THE STIFFNESS FOR ELEMENT 'IE' INTO THE
C          STRUCTURE STIFFNESS.
C
C     CONTROL VARIABLES:
C
C          KS, KE       = STRUCTURE & ELEMENT STIFFNESS MATRICES.
C          NNE          = NUMBER OF NODES IN AN ELEMENT.
C          NDOF         = NUMBER OF DOF AT EACH NODE.
C          INCID        = INCIDENCE MATRIX.
C          IE           = CURRENT ELEMENT NUMBER.
C          NRS, NRE     = NUMBER OF ROWS IN STRUCTURE & ELEMENT STIFFNESS.
C          NE           = NUMBER OF ELEMENTS IN THE MODEL.
C
C     LOCAL VARIABLES:
C
C          INE          = CURRENT ELEMENT SUBMATRIX ROW NUMBER.
C          JNE          = CURRENT ELEMENT SUBMATRIX COLUMN NUMBER.
C          INS          = CURRENT STRUCTURE SUBMATRIX ROW NUMBER.
C          JNS          = CURRENT STRUCTURE SUBMATRIX COLUMN NUMBER.
C          IDOF         = CURRENT DOF NUMBER IN SUBMATRIX ROW.
C          JDOF         = CURRENT DOF NUMBER IN SUBMATRIX COLUMN.
C          IKE          = ROW ENTRY IN THE ELEMENT STIFFNESS.
C          JKE          = COLUMN ENTRY IN THE ELEMENT STIFFNESS.
C          IKS          = ROW ENTRY IN THE STRUCTURE STIFFNESS.
C          JKS          = COLUMN ENTRY IN THE STRUCTURE STIFFNESS.
C
      DO 10 INE = 1, NNE
           INS = INCID (INE, IE)
      DO 10 JNE = 1, NNE
           JNS = INCID (JNE, IE)
C
C          ASSEMBLE ELEMENT SUBMATRIX (INE, JNE) INTO STRUCTURE SUBMATRIX (INS, JNS)
C
             DO 10 IDOF = 1, NDOF
               IKE = (INE - 1) * NDOF + IDOF
               IKS = (INS - 1) * NDOF + IDOF
             DO 10 JDOF = 1, NDOF
                       JKE = (JNE - 1) * NDOF + JDOF
                       JKS = (JNS - 1) * NDOF + JDOF
C
             KS(IKS, JKS) = KS(IKS, JKS) + KE(IKE, JKE)
C
      10     CONTINUE
             RETURN
             END
```

rigid-body modes that the model possesses. Physically, the structure must be supported to prevent rigid-body motion. In a similar fashion, if the structure stiffness matrix is modified to reflect the support conditions (commonly known as *constraints*), it becomes nonsingular. Several methods may be used to apply constraints to the structure stiffness matrix. Only one, the *equation modification method*, is discussed here.

To demonstrate the equation modification method, consider a model that contains only a single element (Figure 4.6a). The first step is to switch appropriate rows and columns of the stiffness such that those DOFs that are constrained are grouped together. The rearranged stiffness matrix, displacement vector, and load vector for the one-element model are shown in Figure 4.6b. For simplicity, the rearranged equations are represented in the symbolic form

$$\left[\begin{array}{c|c} \mathbf{K}_{cc} & \mathbf{K}_{cu} \\ \hline \mathbf{K}_{uc} & \mathbf{K}_{uu} \end{array}\right]\left\{\begin{array}{c} \mathbf{u}_c \\ \hline \mathbf{u}_u \end{array}\right\} = \left\{\begin{array}{c} \mathbf{P}_c \\ \hline \mathbf{P}_u \end{array}\right\} \tag{4.43}$$

where the subscript C represents the constrained DOF and the subscript u represents the unconstrained DOF.* The relationship between the submatrices and subvectors in Equation 4.43 and those in Figure 4.6b is determined by their respective positions in the equations.

The unknown quantities are the displacements $\{\mathbf{u}_u\}$ of the unconstrained DOF and the forces $\{\mathbf{P}_c\}$ at the constrained DOF. Rewrite Equation 4.43 as two separate submatrix/subvector equations.

$$[\mathbf{K}_{cc}]\{\mathbf{u}_c\} + [\mathbf{K}_{cu}]\{\mathbf{u}_u\} = \{\mathbf{P}_c\} \tag{4.44a}$$

$$[\mathbf{K}_{uc}]\{\mathbf{u}_c\} + [\mathbf{K}_{uu}]\{\mathbf{u}_u\} = \{\mathbf{P}_u\} \tag{4.44b}$$

Since $\{\mathbf{u}_c\}$ is known, it is moved to the load side of Equation 19.44b, to obtain

$$[\mathbf{K}_{uu}]\{\mathbf{u}_u\} = \{\mathbf{P}_u\} - [\mathbf{K}_{uc}]\{\mathbf{u}_c\} \tag{4.45}$$

Equation 4.45 is the constrained system of equations. If the imposed constraints $\{\mathbf{u}_c\}$ are nonzero, they serve to modify the load vector. If the constraints are all zero, such as in Figure 4.6a, then the second term on the right side of Equation 4.45 vanishes. In either case, the system of equations is reduced in order by the number of constrained DOFs. If appropriate constraints are applied to render the structure stable, $[\mathbf{K}_{uu}]$ will be nonsingular.

4.2.7 Solution of the system of equations

After assembly of the stiffness matrix and load vector and application of constraints, the system of linear algebraic equations may be solved. It is common to represent the solution of Equation 4.45 in the symbolic form

* The rearrangement of the equations and subsequent partitioning is done for convenience in representing the method. Computer implementation of this approach does not require that the equations be rearranged. Nor would such rearrangement be computationally efficient.

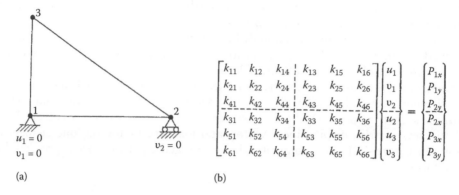

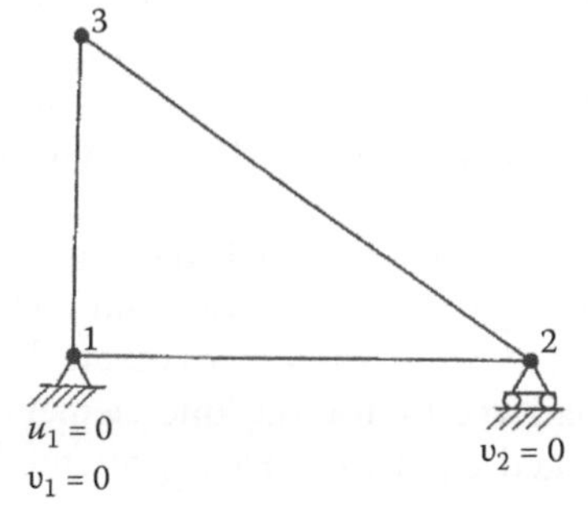

(a)

$$\left[\begin{array}{ccc|ccc} k_{11} & k_{12} & k_{14} & k_{13} & k_{15} & k_{16} \\ k_{21} & k_{22} & k_{24} & k_{23} & k_{25} & k_{26} \\ k_{41} & k_{42} & k_{44} & k_{43} & k_{45} & k_{46} \\ \hline k_{31} & k_{32} & k_{34} & k_{33} & k_{35} & k_{36} \\ k_{51} & k_{52} & k_{54} & k_{53} & k_{55} & k_{56} \\ k_{61} & k_{62} & k_{64} & k_{63} & k_{65} & k_{66} \end{array}\right] \left\{\begin{array}{c} u_1 \\ v_1 \\ v_2 \\ \hline u_2 \\ u_3 \\ v_3 \end{array}\right\} = \left\{\begin{array}{c} P_{1x} \\ P_{1y} \\ P_{2y} \\ \hline P_{2x} \\ P_{3x} \\ P_{3y} \end{array}\right\}$$

(b)

Figure 4.6 Application of constraints by the equation modification method. (a) One-element model with constraints. (b) Rearranged structure equations.

$$\{\mathbf{u}_u\} = [\mathbf{K}_{uu}]^{-1} = \left(\{\mathbf{P}_u\} - [\mathbf{K}_{uc}]\{\mathbf{u}_c\}\right)$$

However, inversion of the stiffness matrix $[\mathbf{K}_{uu}]$ is computationally expensive and can lead to significant numerical error. A more efficient approach, known as *Choleski decomposition*, involves triangular factorization of the stiffness matrix:

$$[\mathbf{K}_{uu}] = [\mathbf{U}]^{\mathrm{T}}[\mathbf{U}]$$

where $[\mathbf{U}]$ is an upper triangular matrix; that is, each term in the lower triangle of $[\mathbf{U}]$ is zero ($u_{ij} = 0,\ i > j$). Factorization of $[\mathbf{K}_{uu}]$ into this form permits direct solution for displacements via two *load-pass* operations. The first of these, known as the *forward load-pass*, yields an intermediate solution vector $\{\mathbf{y}\}$:

$$[\mathbf{U}]^{\mathrm{T}}\{\mathbf{y}\} = \{\mathbf{P}_u\} - [\mathbf{K}_{uc}]\{\mathbf{u}_c\}$$

The second operation, known as the *backward load pass*, yields the final displacement vector $\{\mathbf{u}_u\}$:

$$[\mathbf{U}]\{\mathbf{u}_u\} = \{\mathbf{y}\}$$

Upon solution for $\{\mathbf{u}_u\}$, the reactions that result from deformation of the structure can be found from Equation 4.44a. The total reactions are obtained by subtracting any nodal loads that are applied to the constrained DOF. Such loads frequently exist when element loads, in the form of body forces or surface tractions, are resolved into equivalent nodal loads.

Details of the equation-solving methods and discussion of their advantages and disadvantages can be found in books that specialize in the finite element method (see the Reference section at the end of this chapter).

4.3 BILINEAR RECTANGLE

The CST is the simplest element that can be used for plane elasticity problems. As such, it is an attractive choice for demonstration of the basic formulation of the finite element method.

However, because of its simplicity, the CST element exhibits relatively poor performance in a coarse mesh (a few large elements). To obtain satisfactory results with the CST element, a very highly refined mesh (many small elements) is generally needed for all but the most trivial problems. Alternatively, one may use a different element that is based on different displacement interpolation functions and that yields better results. The number of alternatives to the CST element is quite large, and no attempt is made to discuss all of them here. Instead, we examine two alternatives: the bilinear rectangle and the linear isoparametric quadrilateral. The development of bilinear rectangle follows. The linear isoparametric quadrilateral is presented in Section 4.4.

Consider a rectangular element of width $2a$, height $2b$, and with corner nodes numbered in counterclockwise order. The (x, y) coordinate axes for the element are parallel to the 1–2 and 1–4 edges of the element, respectively, and the origin of the coordinate system is at the centroid of the element (see Figure 4.7). As with the CST element, the displacement components (u, v) at any point P are expressed in terms of the nodal displacements. Since there are four nodes in the element, each with two nodal DOFs, the displacement functions for $u(x, y)$ and $v(x, y)$ each have four coefficients. Hence, we choose the bilinear functions*

$$u(x,y) = a_1 + a_2 x + a_3 y + a_4 xy$$
$$v(x,y) = a_5 + a_6 x + a_7 y + a_8 xy$$

As does the CST element, the *bilinear rectangle* can properly represent rigid-body translation, rigid-body rotation, and constant strain. The bilinear displacement components ($a_4 xy$ and $a_8 xy$) result in strain components such that ε_{xx} is linear in y, ε_{yy} is linear in x, and γ_{xy} is linear in both x and y. This higher-order response, compared to the CST element, results in more efficient and accurate numerical solutions.

The development of the stiffness matrix and load vector proceeds in a manner similar to that for the CST element. The shape functions are expressed as products of two one-dimensional Lagrange interpolation functions

$$\begin{aligned} N_1(x,y) &= \frac{(a-x)(b-y)}{4ab} \\ N_2(x,y) &= \frac{(a+x)(b-y)}{4ab} \\ N_3(x,y) &= \frac{(a+x)(b+y)}{4ab} \\ N_4(x,y) &= \frac{(a-x)(b+y)}{4ab} \end{aligned} \tag{4.46}$$

Since the shape function for node i is zero along any element edge that does not include node i, the shape function can be derived directly as the product of the equations of the lines that define these edges (see Figure 4.7). The shape functions for the bilinear rectangle are

* These functions are said to be bilinear functions of (x, y) because the dependency on x and y comes from the product of two linear expressions, one in x and one in y. The corresponding rectangular element is said to be bilinear. With the given functions (u, v), the straight edges of the bilinear rectangle remain straight under deformation (like the CST element). However, the strain components in the bilinear rectangle element are not constant.

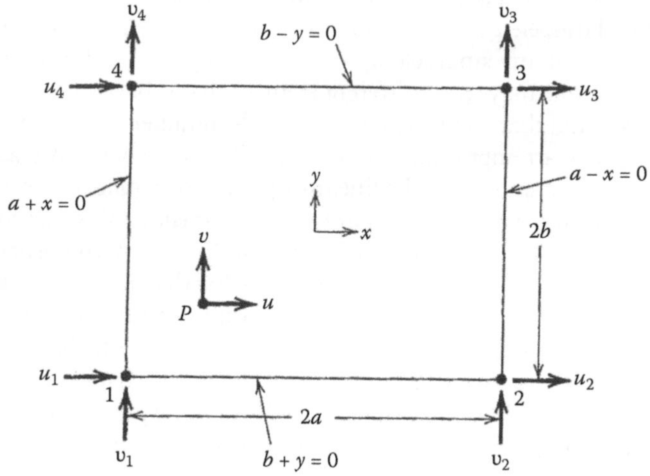

Figure 4.7 **Bilinear rectangle element.**

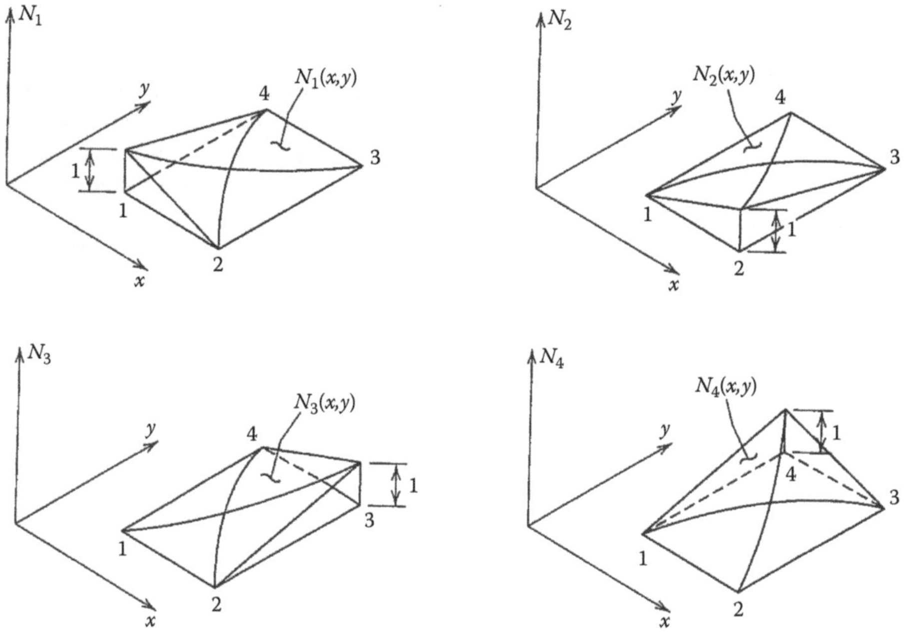

Figure 4.8 **Graphical representation of shape functions for the bilinear rectangle element.**

illustrated in Figure 4.8, where they form straight lines along the element edges. However, over the interior of the element, the functions form curved surfaces, with linearly varying slopes in the x and y directions.

The strain–displacement relations are written in the form of Equation 4.25, with the nodal displacement vector

$$\{u_i\} = [\, u_1 \;\; v_1 \;\; u_2 \;\; v_2 \;\; u_3 \;\; v_3 \;\; u_4 \;\; v_4 \,]^T$$

and [**B**] matrix

$$[B]=[\,B_1 \quad B_2 \quad B_3 \quad B_4\,] \tag{4.47}$$

where for node i

$$[B_i]=\begin{bmatrix} \frac{\partial N_i}{\partial x} & 0 \\ 0 & \frac{\partial N_i}{\partial y} \\ \frac{\partial N_i}{\partial y} & \frac{\partial N_i}{\partial x} \end{bmatrix} \tag{4.48}$$

The element stiffness matrix is found from Equation 4.31. That equation is repeated here with the order of each matrix shown as a subscript:

$$[K]_{8\times 8} = \int_V [B]^{\mathrm{T}}_{8\times 3}[D]_{3\times 3}[B]_{3\times 8}\, dV$$

The stiffness matrix can be written in terms of (2 × 2) nodal submatrices as

$$[K_{ij}]_{2\times 2} = \int_V [B_i]^{\mathrm{T}}_{2\times 3}[D]_{3\times 3}[B_j]_{3\times 2}\, dV$$

where i and j are element node numbers. The explicit form of the bilinear rectangle element stiffness matrix for a plane stress condition is given in Table 4.3.

By itself, the bilinear rectangle element is limited to rectangular domains. This is potentially a rather severe restriction. However, nonrectangular domains can be modeled with a combination of bilinear rectangle elements and CST elements. Since both elements represent linear displacement variation along their edges, they are compatible; that is, displacements will be continuous across element boundaries.

Example 4.3: Performance of the Bilinear Rectangle and CST Elements

Compare the ability of the bilinear rectangle and the CST elements to model in-plane bending of a thin, square plate.

Solution: A square plate of width a and thickness t is considered. For simplicity, Poisson's ratio is taken as zero, $v = 0$. To impose a state of pure bending, displacements $u = \pm\,\delta$ are imposed on the corners of the plate as shown in Figure E4.3a. From the theory of elasticity, the displacements are

$$u=-\frac{4xy\delta}{a^2} \tag{4E3.1}$$

$$v=\left(\frac{4x^2}{a}-1\right)\frac{\delta}{2} \tag{4E3.2}$$

Table 4.3 Bilinear rectangle stiffness matrix, plane stress case (partitioned into 2 × 2 nodal submatrices)

$j \rightarrow$	Column index 1	2	3	4	5	6	7	8	Row index $i \downarrow$
C	$4\beta+\frac{2(1-v)}{\beta}$	$\frac{3}{2}(1+v)$	$-4\beta+\frac{1-v}{\beta}$	$-\frac{3}{2}(1-3v)$	$-2\beta-\frac{1-v}{\beta}$	$-\frac{3}{2}(1+v)$	$2\beta-\frac{2(1-v)}{v}$	$\frac{3}{2}(1-3v)$	1
	$\frac{3}{2}(1+v)$	$\frac{4}{\beta}+2(1-v)\beta$	$\frac{3}{2}(1-3v)$	$\frac{2}{\beta}-2(1-v)\beta$	$-\frac{3}{2}(1+v)$	$-\frac{2}{\beta}-(1-v)\beta$	$-\frac{3}{2}(1-3v)$	$-\frac{4}{\beta}+(1-v)\beta$	2
	$-4\beta+\frac{1-v}{\beta}$	$\frac{3}{2}(1-3v)$	$4\beta+\frac{2(1-v)}{\beta}$	$-\frac{3}{2}(1+v)$	$2\beta-\frac{2(1-v)}{\beta}$	$-\frac{3}{2}(1-3v)$	$-2\beta+\frac{1-v}{\beta}$	$\frac{3}{2}(1-3v)$	3
	$-\frac{3}{2}(1-3v)$	$\frac{2}{\beta}-2(1-v)\beta$	$-\frac{3}{2}(1+v)$	$\frac{4}{\beta}+2(1-v)\beta$	$\frac{3}{2}(1-3v)$	$-\frac{4}{\beta}+(1-v)\beta$	$\frac{3}{2}(1+v)$	$-\frac{2}{\beta}-(1-v)\beta$	4
	$-2\beta-\frac{1-v}{\beta}$	$-\frac{3}{2}(1+v)$	$2\beta-\frac{2(1-v)}{\beta}$	$\frac{3}{2}(1-3v)$	$4\beta+\frac{2(1-v)}{\beta}$	$\frac{3}{2}(1+v)$	$-4\beta+\frac{1-v}{\beta}$	$-\frac{3}{2}(1-3v)$	5
	$-\frac{3}{2}(1+v)$	$-\frac{2}{\beta}-(1-v)\beta$	$-\frac{3}{2}(1-3v)$	$-\frac{4}{\beta}+(1-v)\beta$	$\frac{3}{2}(1+v)$	$\frac{4}{\beta}+2(1-v)\beta$	$\frac{3}{2}(1-3v)$	$\frac{2}{\beta}-2(1-v)\beta$	6
	$2\beta-\frac{2(1-v)}{\beta}$	$-\frac{3}{2}(1-3v)$	$-2\beta-\frac{1-v}{\beta}$	$\frac{3}{2}(1+v)$	$-4\beta+\frac{1-v}{\beta}$	$\frac{3}{2}(1-3v)$	$4\beta+\frac{2(1-v)}{\beta}$	$-\frac{3}{2}(1+v)$	7
	$\frac{3}{2}(1+3v)$	$-\frac{4}{\beta}+(1-v)\beta$	$\frac{3}{2}(1+v)$	$-\frac{2}{\beta}-(1-v)\beta$	$-\frac{3}{2}(1-3v)$	$\frac{2}{\beta}-2(1-v)\beta$	$-\frac{3}{2}(1+v)$	$\frac{4}{\beta}+2(1-v)\beta$	8

$$C=\frac{Et}{12(1-v^2)};\quad \beta=\frac{b}{a}$$

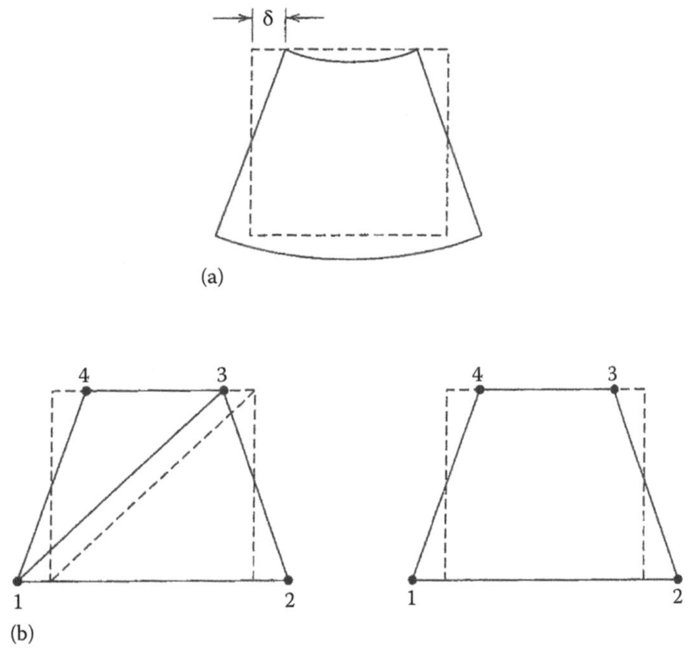

Figure E4.3 (a) Deformed shape, elasticity solution. (b) Deformed shape, finite element models.

Differentiation of Equations 4E3.1 and 4E3.2 gives the strains

$$\varepsilon_{xx} = -\frac{4y\delta}{a^2} \qquad \varepsilon_{yy} = 0 \qquad \gamma_{xy} = 0 \tag{4E3.3}$$

The strain energy in the plate is

$$U = \int_V U_0\, dV = \int_V \frac{E\varepsilon_{xx}^2}{2}\, dV = \frac{2}{3} Et\delta^2 \tag{4E3.4}$$

Two finite element models of the square plate are considered. The first uses two CST elements and the second uses a single bilinear rectangle. As for the elasticity solution, nodal displacements of $u_i = \pm\delta$ are imposed. The models and their deformed shapes are shown in Figure E4.3b. The model of two CST elements is considered first. Strains in the CST elements can be determined first, since the displacement vector is known.

$$\{u_i\} = [-\delta \;\; 0 \,|\, \delta \;\; 0 \,|\, -\delta \;\; 0 \,|\, \delta \;\; 0]^{\mathrm{T}} \tag{4E3.5}$$

The $[B]$ matrices for the two CST elements are defined by Equation 4.26. For the element geometries in Figure E4.3b, these matrices are

$$[B]_{\mathrm{CST-1}} = \frac{1}{a}\begin{bmatrix} 0 & 0 & 1 & 0 & -1 & 0 \\ 0 & -1 & 0 & 0 & 0 & 1 \\ 1 & 0 & 0 & 1 & 1 & 1 \end{bmatrix} \tag{4E3.6}$$

and

$$[B]_{CST-2} = \frac{1}{a}\left[\begin{array}{cc|cc|cc} -1 & 0 & 1 & 0 & 0 & 0 \\ 0 & 0 & 0 & -1 & 0 & 1 \\ 0 & 1 & -1 & 1 & 1 & 0 \end{array}\right] \tag{4E3.7}$$

Thus, the strains in the two elements are obtained by Equation 4.25 as

$$\{\varepsilon\}_{CST-1} = \left[-\frac{2\delta}{a} \quad 0 \quad \frac{2\delta}{a}\right]^T \tag{4E3.8}$$

and

$$\{\varepsilon\}_{CST-2} = \left[\frac{2\delta}{a} \quad 0 \quad -\frac{2\delta}{a}\right]^T \tag{4E3.9}$$

The structure stiffness for an assembly of two CST elements is

$$[K]_{CST} = Et\left[\begin{array}{cc|cc|cc|cc} 0.75 & 0.0 & -0.5 & 0.0 & 0.0 & -0.25 & -0.25 & 0.25 \\ 0.0 & 0.75 & 0.25 & -0.25 & -0.25 & 0.0 & 0.0 & -0.5 \\ \hline -0.5 & 0.25 & 0.75 & -0.25 & -0.25 & 0.0 & 0.0 & 0.0 \\ 0.0 & -0.25 & -0.25 & 0.75 & 0.25 & -0.5 & 0.0 & 0.0 \\ \hline 0.0 & -0.25 & -0.25 & 0.25 & 0.75 & 0.0 & -0.5 & 0.0 \\ -0.25 & 0.0 & 0.0 & -0.5 & 0.0 & 0.75 & 0.25 & -0.25 \\ \hline -0.25 & 0.0 & 0.0 & 0.0 & -0.5 & 0.25 & 0.75 & -0.25 \\ 0.25 & -0.5 & 0.0 & 0.0 & 0.0 & -0.25 & -0.25 & 0.75 \end{array}\right] \tag{4E3.10}$$

from which the product $[K]_{CST}\{\boldsymbol{u}_i\}$ gives the nodal forces $\{P_i\}_{CST}$ as

$$\{P_i\}_{CST} = Et\delta[-1.5 \;\; 0.5 \,|\, 1.5 \;\; -0.5 \,|\, -1.5 \;\; 0.5 \,|\, 1.5 \;\; -0.5]^T \tag{4E3.11}$$

The strain energy in the structure is

$$U_{CST} = \frac{1}{2}\{u_i\}^T\{P_i\}_{CST} = 3Et\delta^2 \tag{4E3.12}$$

Next, consider the model with only a single bilinear rectangle shown in Figure E4.3b. The $[B]$ matrix for the element is given by Equations 4.47 and 4.48 as

$$[B]_{BR} = \frac{1}{a^2}\left[\begin{array}{cc|cc|cc|cc} -\frac{a}{2}+y & 0 & \frac{a}{2}+y & 0 & \frac{a}{2}+y & 0 & -\frac{a}{2}-y & 0 \\ 0 & -\frac{a}{2}+x & 0 & -\frac{a}{2}-x & 0 & \frac{a}{2}+x & 0 & \frac{a}{2}-x \\ -\frac{a}{2}+x & -\frac{a}{2}+y & -\frac{a}{2}-x & \frac{a}{2}-y & \frac{a}{2}+x & \frac{a}{2}+y & \frac{a}{2}-x & -\frac{a}{2}-y \end{array}\right] \tag{4E3.13}$$

The strains are obtained by Equation 4.25 as

$$\{\varepsilon\}_{BR} = \left[-\frac{4y\delta}{a^2} \quad 0 \quad \frac{4x\delta}{a^2} \right]^T \tag{4E3.14}$$

The stiffness matrix for the bilinear rectangle is obtained from Table 4.3, which, for this problem, becomes

$$[K]_{BR} = Et\left[\begin{array}{cc|cc|cc|cc} 0.5 & 0.125 & -0.25 & -0.125 & -0.25 & -0.125 & 0.0 & 0.125 \\ 0.125 & 0.5 & 0.125 & 0.0 & -0.125 & -0.25 & -0.125 & -0.25 \\ \hline -0.25 & 0.125 & 0.5 & -0.125 & 0.0 & -0.125 & -0.25 & 0.125 \\ -0.125 & 0.0 & -0.125 & 0.5 & 0.125 & -0.25 & 0.125 & -0.25 \\ \hline -0.25 & -0.125 & 0.0 & 0.125 & 0.5 & 0.125 & -0.25 & -0.125 \\ -0.125 & -0.25 & -0.125 & -0.25 & 0.125 & 0.5 & 0.125 & 0.0 \\ \hline 0.0 & -0.125 & -0.25 & 0.125 & -0.25 & 0.125 & 0.5 & -0.125 \\ 0.125 & -0.25 & 0.125 & -0.25 & -0.125 & 0.0 & -0.125 & 0.5 \end{array}\right] \tag{4E3.15}$$

from which the product $[K]_{BR}\{u_i\}$ gives the nodal forces $\{P_i\}_{BR}$ as

$$\{P_i\}_{BR} = Et\delta[-0.5 \quad 0 \,|\, 0.5 \quad 0 \,|\, -0.5 \quad 0 \,|\, 0.5 \quad 0]^T \tag{4E3.16}$$

The strain energy in the element is

$$U_{BR} = \frac{1}{2}\{u_i\}^T\{P_i\}_{BR} = Et\delta^2 \tag{4E3.17}$$

This example clearly demonstrates that the bilinear rectangle is superior to the CST element. The bilinear rectangle correctly predicts the normal strains ε_{xx} and ε_{yy}. In addition, the bilinear rectangle model stores less strain energy than the CST model. Using the elasticity solution as the *exact* solution, $U_{BR} = 1.5U_{exact}$, while $U_{CST} = 4.5U_{exact}$. Notice, though, that both the CST and bilinear rectangle possess nonzero shear stress where none should exist. This defect, known as *parasitic shear*, contributes to excess strain energy in the elements. Although little can be done to improve the performance of the CST element, a more general formulation of the bilinear rectangle, known as the *linear isoparametric quadrilateral* (Section 4.4), can be used to control parasitic shear.

4.4 LINEAR ISOPARAMETRIC QUADRILATERAL

Suppose that an analyst wishes to model an irregular domain but wants to avoid the use of CST elements because of their relatively poor performance. Since the domain is irregular, the bilinear rectangle element would be inappropriate. Instead, arbitrarily shaped quadrilateral (four-sided) elements are selected to better fit boundaries. A quadrilateral element may be formulated directly, as was done previously for the CST and bilinear rectangle elements. However, the necessary integrations are quite complex. This is due, in part, to the difficulty in defining the limits of integration. Use of isoparametric elements eliminates this difficulty. Isoparametric elements are formulated in *natural* coordinates as square elements and then are *mapped* to physical coordinates via coordinate interpolation functions, similar to displacement interpolation functions. Depending on the type of isoparametric element that is used, the configuration of the element in physical coordinates can be nonrectangular and can have curved sides. If the shape functions used for coordinate interpolation are identical to those used for displacement interpolation, the element is said to be *isoparametric*. If coordinate interpolation is of higher order than displacement interpolation (i.e., more nodes are used to represent the variation in geometry than to represent the variation in displacements), the element is called *super-parametric*. If coordinate interpolation is of lower order than displacement interpolation (fewer nodes are used to represent the variation in geometry than to represent the variation in displacements), the element is called *subparametric* (Zienkiewicz and Taylor, 2005). Because of their versatility and accuracy, isoparametric elements have become the mainstay of modern finite element programs.

4.4.1 Isoparametric mapping

Consider the mapping of the four-node quadrilateral element from a *natural* (ξ, η) coordinate system (Figure 4.9*a*) to a physical (x, y) coordinate system (Figure 4.9*b*). In natural coordinates, the element is a 2 × 2 square and the origin of the coordinate system is at its center. In physical coordinates, the element is distorted from a rectangular shape. With shape functions in terms of the (ξ, η) coordinate system, the coordinates of any point P can be expressed in terms of the (x, y) coordinates of the nodes.

$$x(\xi,\eta)=\sum_{i=1}^{4}N_i(\xi,\eta)x_i \qquad y(\xi,\eta)=\sum_{i=1}^{4}N_i(\xi,\eta)y_i \tag{4.49a}$$

In matrix form, Equation 4.49a is

$$\begin{Bmatrix} x(\xi,\eta) \\ y(\xi,\eta) \end{Bmatrix}=[N][x_i] \tag{4.49b}$$

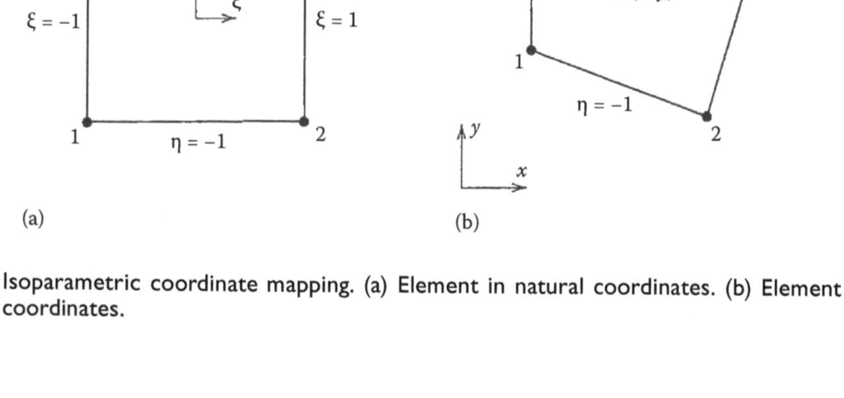

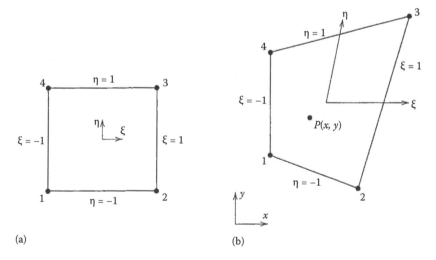

Figure 4.9 Isoparametric coordinate mapping. (a) Element in natural coordinates. (b) Element in physical coordinates.

where $\{x_i\}$ is the vector of nodal coordinates

$$\{x_i\} = [x_1 \;\; y_1 \;\; x_2 \;\; y_2 \;\; x_3 \;\; y_3 \;\; x_4 \;\; y_4]^T$$

and $[N]$ is the shape function matrix

$$[N] = \begin{bmatrix} N_1 & 0 & N_2 & 0 & N_3 & 0 & N_4 & 0 \\ 0 & N_1 & 0 & N_2 & 0 & N_3 & 0 & N_4 \end{bmatrix} \tag{4.50}$$

The shape functions are the Lagrange interpolation functions (refer to Equation 4.46) in dimensionless (ξ, η) coordinates

$$\begin{aligned} N_1(\xi,\eta) &= \frac{(1-\xi)(1-\eta)}{4} \\ N_2(\xi,\eta) &= \frac{(1+\xi)(1+\eta)}{4} \\ N_3(\xi,\eta) &= \frac{(1+\xi)(1+\eta)}{4} \\ N_4(\xi,\eta) &= \frac{(1-\xi)(1+\eta)}{4} \end{aligned} \tag{4.51}$$

After the element is mapped from natural to physical coordinates, the ξ- and η-axes need not remain orthogonal.

The principal reason for using isoparametric elements is to avoid integrating in physical coordinates. However, the general expression for the stiffness matrix, Equation 4.31, is expressed in terms of physical coordinates. Therefore, the differential lengths dx and dy must be expressed in terms of the natural coordinate differentials $d\xi$ and $d\eta$. In addition, strain is defined in terms of the derivatives of the shape functions with respect to physical coordinates. These derivatives are the elements in the $[B]$ matrix, and they must be converted to derivatives with respect to natural coordinates.

The differentials (dx, dy) are related to the differentials $(d\xi, d\eta)$ by means of Equation 4.49a. Thus,

$$\begin{aligned} dx &= \frac{\partial x}{\partial \xi} d\xi + \frac{\partial x}{\partial \eta} d\eta \\ dy &= \frac{\partial y}{\partial \xi} d\xi + \frac{\partial y}{\partial \eta} d\eta \end{aligned} \tag{4.52}$$

where

$$\frac{\partial x}{\partial \xi} = \sum \frac{\partial N_i}{\partial \xi} x_i \qquad \frac{\partial y}{\partial \xi} = \sum \frac{\partial N_i}{\partial \xi} y_i$$

$$\frac{\partial x}{\partial \eta} = \sum \frac{\partial N_i}{\partial \eta} x_i \qquad \frac{\partial y}{\partial \eta} = \sum \frac{\partial N_i}{\partial \eta} y_i$$

The coordinate derivatives are combined in matrix form as

$$[J] = \begin{bmatrix} \dfrac{\partial x}{\partial \xi} & \dfrac{\partial y}{\partial \xi} \\ \dfrac{\partial x}{\partial \eta} & \dfrac{\partial y}{\partial \eta} \end{bmatrix} \tag{4.53}$$

where $[J]$ is the *Jacobian* of the transformation (Courant, 2011).

Equations 4.52 and 4.53 relate the differentials of the two coordinate systems as

$$\begin{Bmatrix} dx \\ dy \end{Bmatrix} = [J]^T \begin{Bmatrix} d\xi \\ d\eta \end{Bmatrix} \tag{4.54}$$

In a similar manner, derivatives of the shape function for node i are related by

$$\begin{Bmatrix} \dfrac{\partial N_i}{\partial x} \\ \dfrac{\partial N_i}{\partial y} \end{Bmatrix} = [J]^{-1} \begin{Bmatrix} \dfrac{\partial N_i}{\partial \xi} \\ \dfrac{\partial N_i}{\partial \eta} \end{Bmatrix} \tag{4.55}$$

If $[J]^{-1}$ exists, the area mapping from (ξ, η) coordinates to (x, y) coordinates is unique and reversible. A physical interpretation of $[J]$ can be obtained by comparing the area of the element in (x, y) coordinates to that in (ξ, η) coordinates. If $|J| > 0$, the area of the element is preserved and the mapping is physically meaningful. In precise terms, $|J|$ is the instantaneous area ratio $A_{xy}/A_{\xi\eta}$ at any point in the element.

This physical interpretation of $[J]$ leads to a change in the differential volume for a constant-thickness plane elasticity element from $t\, dx\, dy$ to $t|J|\, d\xi\, d\eta$. The limits of integration are -1 to 1 in ξ and -1 to 1 in η. So the integral of any function $F(x, y)$ can be transformed to natural coordinates in the manner

$$\int_A F(x,y)\, dx\, dy = \int_{-1}^{1} \int_{-1}^{1} F(x(\xi,\eta), y(\xi,\eta)) |J|\, d\xi\, d\eta$$

4.4.2 Element stiffness matrix

Equation 4.31 defines the element stiffness matrix for any elasticity element (using displacement DOFs), including the isoparametric linear quadrilateral. A change in coordinate system from (x, y) to (ξ, η), with the modified limits of integration, leads to the stiffness matrix

$$[K] = t \int_{-1}^{1} \int_{-1}^{1} [B]^T [D][B] |J|\, d\xi\, d\eta \tag{4.56}$$

where $[B]$ is given by Equation 4.47 and $[B_i]$ by Equation 4.48. From Equations 4.48 and 4.55, the individual terms in $[B_i]$, in terms of (ξ, η), are

$$[B_i(\xi,\eta)] = \begin{bmatrix} J_{11}^* \dfrac{\partial N_i}{\partial \xi} + J_{12}^* \dfrac{\partial N_i}{\partial \eta} & 0 \\ 0 & J_{21}^* \dfrac{\partial N_i}{\partial \xi} + J_{22}^* \dfrac{\partial N_i}{\partial \eta} \\ J_{21}^* \dfrac{\partial N_i}{\partial \xi} + J_{22}^* \dfrac{\partial N_i}{\partial \eta} & J_{11}^* \dfrac{\partial N_i}{\partial \xi} + J_{12}^* \dfrac{\partial N_i}{\partial \eta} \end{bmatrix} \tag{4.57}$$

where J_{ij}^* is the i, j term from $[J]^{-1}$.

It is usually more convenient to work with just a single (2 × 2) nodal submatrix of $[K]$ at one time. So we write

$$[K_{ij}] = t\int_{-1}^{1}\int_{-1}^{1}[B_i]^{\mathrm{T}}[D][B_j]\left|J\right| d\xi\, d\eta \tag{4.58}$$

where i and j are node numbers for the element.

4.4.3 Numerical integration

While analytical expressions for the individual terms in Equation 4.58 can be developed, they are quite complex and thus prone to errors in algebra or in computer programming. As an alternative to direct integration, the integrals required are usually evaluated numerically within the finite element program. The most commonly used numerical integration method is *Gauss quadrature*. The Gauss quadrature method is more efficient than many other methods, such as the Newton–Cotes methods, since fewer sampling points are required to obtain a given accuracy. In fact, in one dimension, the use of n sampling points in Gauss quadrature results in exact integration of a polynomial of order $(2n - 1)$. However, the integration of functions that are not polynomials is approximate.

Consider a function $F(\xi, \eta)$ that is to be integrated over the limits of −1 to 1 in ξ and −1 to 1 in η. The integral is evaluated numerically by the form

$$I = \int_{-1}^{1}\int_{-1}^{1} F(\xi,\eta)\, d\xi\, d\eta = \sum_{k=1}^{m}\sum_{l=1}^{n} w_k w_l F(\xi_k,\eta_l)$$

where m and n are the number of sampling points in the ξ and η directions, respectively. Also, ξ_k and η_l are the locations of the kth and lth sampling points, and w_k and w_l are weights applied to $F(\xi, \eta)$ when it is evaluated at the sampling points. Usually, m and n are taken equal, in which case the numerical scheme is symmetric.

If Gauss quadrature is used to evaluate the nodal submatrix $[K_{ij}]$ in Equation 4.58, the integral becomes

$$[K_{ij}] = t\sum_{k=1}^{m}\sum_{l=1}^{n} w_k w_l [B_i(\xi_k,\eta_l)]^{\mathrm{T}}[D][B_j(\xi_k,\eta_l)]\left|J(\xi_k,\eta_l)\right| \tag{4.59}$$

The accuracy achieved with Gauss quadrature is dependent on proper selection of sampling point locations and weights. For elements in natural coordinates, the optimal sampling point locations and weights are given in Figure 4.10. Only symmetric integration and the one-, two-, and three-point rules are considered. Nonsymmetric integration and higher-order integration rules are discussed elsewhere.

The number of integration points that are used to evaluate Equation 4.59 influences the ultimate performance of the element. *Full integration* is the integration order needed to integrate the stiffness exactly for an undistorted element. For the linear quadrilateral, a two-point rule provides full integration. An integration rule below that required for full

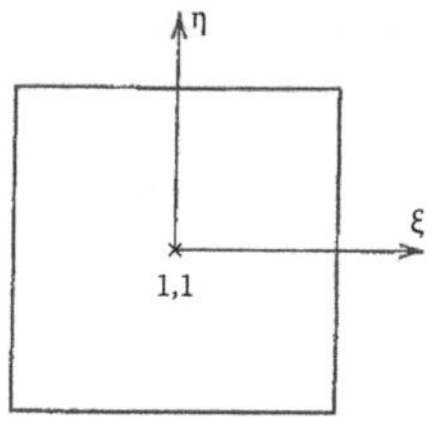

Point no.	ξ_i	η_j	w_i	w_j
1,1	0.0	0.0	1.0	1.0

(a)

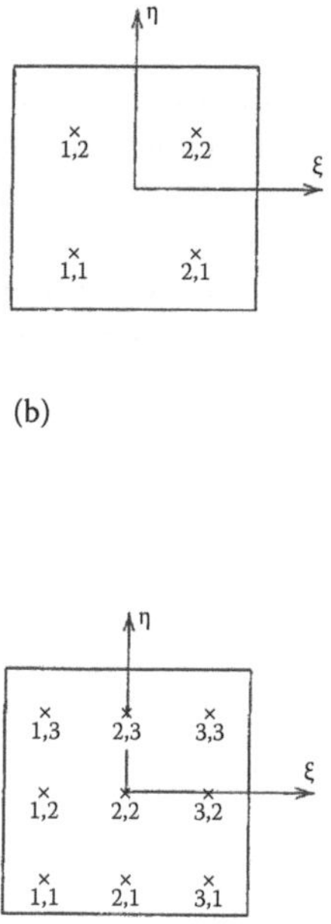

Point no.	ξ_i	η_j	w_i	w_j
1,1	$-1/\sqrt{3}$	$-1/\sqrt{3}$	1.0	1.0
2,1	$1/\sqrt{3}$	$-1/\sqrt{3}$	1.0	1.0
1,2	$-1/\sqrt{3}$	$1/\sqrt{3}$	1.0	1.0
2,2	$1/\sqrt{3}$	$1/\sqrt{3}$	1.0	1.0

(b)

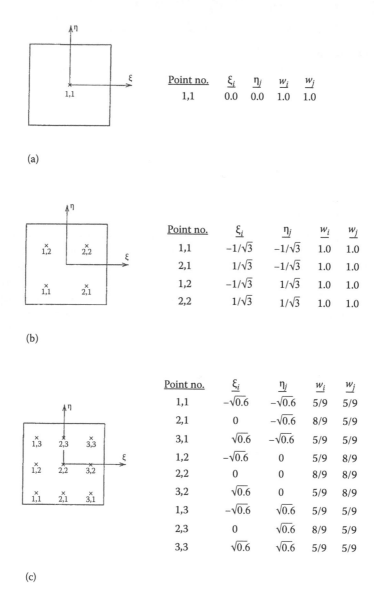

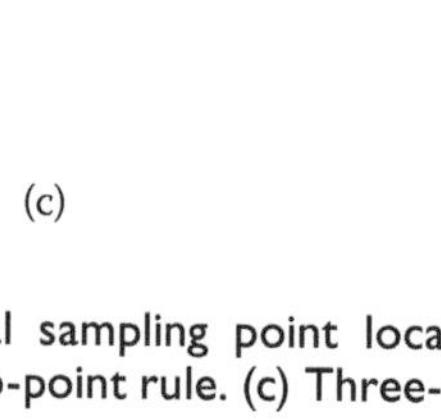

Point no.	ξ_i	η_j	w_i	w_j
1,1	$-\sqrt{0.6}$	$-\sqrt{0.6}$	5/9	5/9
2,1	0	$-\sqrt{0.6}$	8/9	5/9
3,1	$\sqrt{0.6}$	$-\sqrt{0.6}$	5/9	5/9
1,2	$-\sqrt{0.6}$	0	5/9	8/9
2,2	0	0	8/9	8/9
3,2	$\sqrt{0.6}$	0	5/9	8/9
1,3	$-\sqrt{0.6}$	$\sqrt{0.6}$	5/9	5/9
2,3	0	$\sqrt{0.6}$	8/9	5/9
3,3	$\sqrt{0.6}$	$\sqrt{0.6}$	5/9	5/9

(c)

Figure 4.10 Optimal sampling point locations and weights for Gauss quadrature. (a) One-point rule. (b) Two-point rule. (c) Three-point rule.

integration is termed *reduced integration*. Reduced integration, although not evaluating Equation 4.59 exactly, can often lead to improved performance of an element, relative to full integration. For instance, reduced integration of the linear quadrilateral can eliminate the parasitic shear that is a common defect in the element (see Example 4.3). A more complete discussion of reduced integration, including justification for its use, can be found in most finite element textbooks.

4.4.4 High-order isoparametric elements

The concept of isoparametric mapping has been applied to a broad list of element geometries. Within the scope of plane elasticity problems, elements with more than four nodes permit greater flexibility in element shape (including curved edges) and are capable of representing greater variation in displacements. Perhaps the most popular of all isoparametric elements is the eight-node quadrilateral. This element has four corner nodes, like the linear quadrilateral, but it also has four midside nodes, one midway along the length of each edge (Figure 4.11a). With three nodes along each edge, the element can have curved (parabolic) sides. Another popular high-order isoparametric element is the nine-node quadrilateral (Figure 4.11b). This element has four corner nodes, four midside nodes, and one interior node. Both the eight- and nine-node elements represent complete quadratic displacement fields. The generalization of these elements to cubic interpolation is straightforward (see Figure 4.11c and d).

The eight-node quadrilateral and other high-order elements that contain only boundary nodes are known as *serendipity elements*. The term *serendipity* is used because shape functions for this family of elements were initially developed by inspection. The nine-node quadrilateral and other high-order elements that contain a regular pattern of nodes are known as *Lagrangian elements* since their shape functions are based on Lagrange interpolation functions.

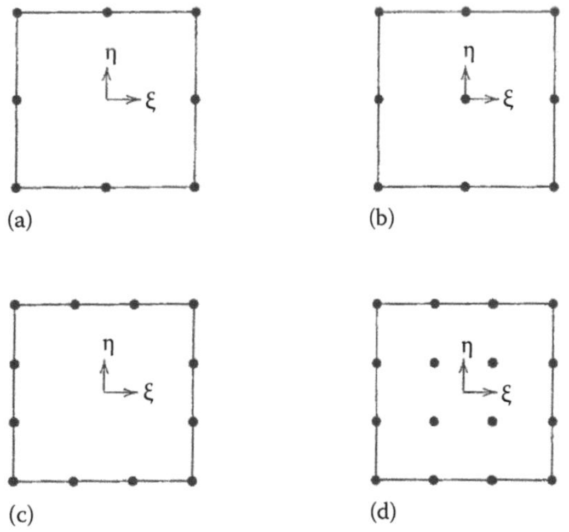

Figure 4.11 Higher-order isoparametric elements. (a) Quadratic serendipity element. (b) Quadratic Lagrange element. (c) Cubic serendipity element. (d) Cubic Lagrange element.

4.5 PLANE FRAME ELEMENT

Analysis of framed structures by the *stiffness method* (also known as *matrix analysis*) was fairly well established at the time of development of the finite element method. The stiffness method for frame analysis can be developed entirely from basic mechanics of materials principles, without the need to consider virtual work formulations and interpolation polynomials. As a result, many engineers view the two methods as distinct. However, it is clear that the stiffness method for frames is simply a special case of the finite element method. Hence, in this section, we develop a finite element that represents a plane frame member, using the same approach that was used for plane elasticity problems.

4.5.1 Element stiffness matrix

The classical plane frame element has two nodes, it is straight and prismatic, and it has three DOFs and three corresponding end actions at each node (see Figure 4.12a). The element has cross-sectional area A, moment of inertia I, and modulus of elasticity E. We assume that the axial response of the member is independent of the bending response. Consequently, the frame element stiffness is formulated as a superposition of the stiffness for an axial rod and that for a beam (Figure 4.12b). In the following, a local $(\bar{x}, \bar{y})$ coordinate system is established for the element. The local $\bar{x}$-axis is aligned with the longitudinal axis of the member and the $\bar{y}$-axis lies in the plane of the element cross-section. The stiffness matrix for the frame element is derived in terms of this local coordinate system. When the element

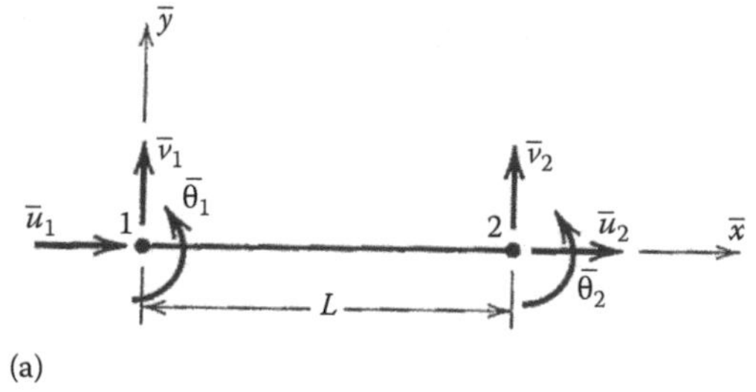

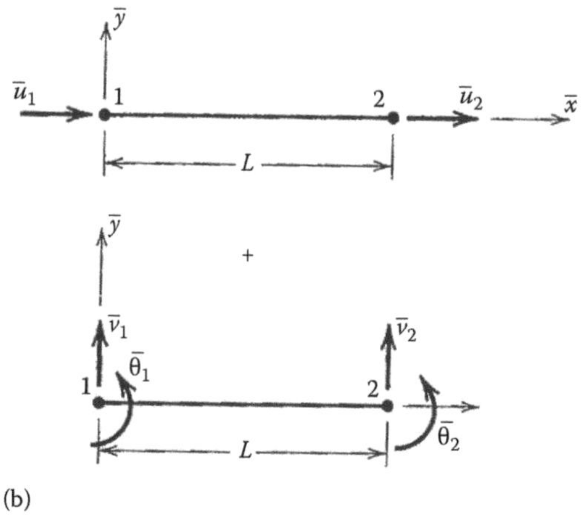

Figure 4.12 Plane frame element. (a) Element with combined axial and bending DOFs. (b) Axial and bending DOFs treated separately.

is oriented at some angle ϕ with respect to the *global* (x, y) coordinates for the structure, the nodal DOFs of the element must be related to the global coordinate system. Thus, a coordinate rotation from local to global coordinates is required for the displacements, loads, and stiffness. This rotation is discussed following Equation 4.75.

Consider first the case of the axial rod. There are two nodal DOFs associated with axial response, so the displacement variation is taken as a linear function:

$$\bar{u}(x) = a_0 + a_1\bar{x}$$

The coefficients a_0 and a_1 are evaluated based on the boundary conditions $\bar{u}(0) = \bar{u}_1$ and $\bar{u}(L) = \bar{u}_2$, where L is the element length. The displacement function, in terms of the nodal displacements, becomes

$$\bar{u}(x) = [N]\{\bar{u}_i\}$$

where $[N] = [1 - \bar{x}L \;\; \bar{x}/L]$ and $\{\bar{u}_i\} = [\bar{u}_1 \;\; \bar{u}_2]^{\mathrm{T}}$

The only nonzero strain component is ε_{xx}, which is written in terms of the nodal displacements as

$$\varepsilon_{xx} = [B_A]\{\bar{u}_i\}$$

in which the subscript A indicates *axial* response and

$$[B_A] = \begin{bmatrix} \dfrac{\partial N_1}{\partial \bar{x}} & \dfrac{\partial N_2}{\partial \bar{x}} \end{bmatrix} = \begin{bmatrix} -\dfrac{1}{L} & \dfrac{1}{L} \end{bmatrix}$$

The axial stress is written as $\sigma_{xx} = E\varepsilon_{xx}$ and the variation of internal energy is

$$\delta U = \int_V \delta\varepsilon_{xx}\sigma_{xx}\, dV \tag{4.60}$$

Assume that only concentrated nodal loads are applied. Then, substitution for σ_{xx} and $\delta\varepsilon_{xx}$ in Equation 4.60, and then substitution of Equation 4.60 into 4.1, yields

$$\{\delta\bar{u}_i\}^{\mathrm{T}}\{\bar{F}_i\} - \{\delta\bar{u}_i\}^{\mathrm{T}}\left[A\int_0^L [B_A]^{\mathrm{T}} E[B_A]\, d\bar{x}\right]\{\bar{u}_i\} = 0$$

Since $\{\delta\bar{u}_i\}$ is arbitrary,

$$\{\bar{F}_i\} = \left[A\int_0^L [B_A]^{\mathrm{T}} E[B_A]\, d\bar{x}\right]\{\bar{u}_i\}$$

which leads to the stiffness matrix for the axial rod:

$$[\bar{K}_A] = A\int_0^L [B_A]^{\mathrm{T}} E[B_A]\, d\bar{x}$$

For constant E, the integrals are easily evaluated to obtain $[\bar{K}_A]$ in terms of A, E, and L:

$$[\bar{K}_A] = \left[\begin{array}{c|c} \frac{AE}{L} & -\frac{AE}{L} \\ \hline -\frac{AE}{L} & \frac{AE}{L} \end{array}\right] \tag{4.61}$$

Next consider the *bending* effect of the frame element. There are four nodal DOFs associated with bending (a lateral translation and a rotation at each node), so the displacement variation is written as a cubic polynomial with four coefficients:

$$\bar{u}(x) = a_0 + a_1\bar{x} + a_2\bar{x}^2 + a_3\bar{x}^3$$

Coefficients a_0 through a_3 are evaluated based on the boundary conditions $\bar{v}(0) = \bar{v}_1$, $\bar{\theta}(0) = \bar{\theta}_1$, $\bar{v}(L) = \bar{v}_2$, and $\bar{\theta}(L) = \bar{\theta}_2$, in which $\bar{\theta} = d\bar{v}/d\bar{x}$. In terms of the nodal displacements, the displacement function is

$$\bar{v}(\bar{x}) = [N]\{\bar{v}_i\} \tag{4.62}$$

where $\{\bar{v}_i\} = [\bar{v}_1 \quad \bar{\theta}_1 \quad \bar{v}_2 \quad \bar{\theta}_2]^{\mathrm{T}}$, and the shape function matrix $[N]$ is

$$[N] = [N_1 \quad N_2 \quad N_3 \quad N_4] \tag{4.63a}$$

for which the individual shape functions are

$$\begin{aligned} N_1 &= 1 - 3\frac{\bar{x}^2}{L^2} + 2\frac{\bar{x}^3}{L^3} \\ N_2 &= \bar{x} - 2\frac{\bar{x}^2}{L} - \frac{\bar{x}^3}{L^2} \\ N_3 &= 3\frac{\bar{x}^2}{L^2} - 2\frac{\bar{x}^3}{L^3} \\ N_4 &= -\frac{\bar{x}^2}{L} + \frac{\bar{x}^3}{L^2} \end{aligned} \tag{4.63b}$$

These shape functions are illustrated in Figure 4.13.

The strain energy in a beam subjected to bending is given by

$$U = \int_0^L \frac{M^2}{2EI}\, d\bar{x} \tag{4.63c}$$

If the curvature $\bar{v}''$ is taken as a *generalized strain* quantity, the strain–nodal displacement relation is

$$\bar{v}''(\bar{x}) = [B_B]\{\bar{v}_i\} \tag{4.64a}$$

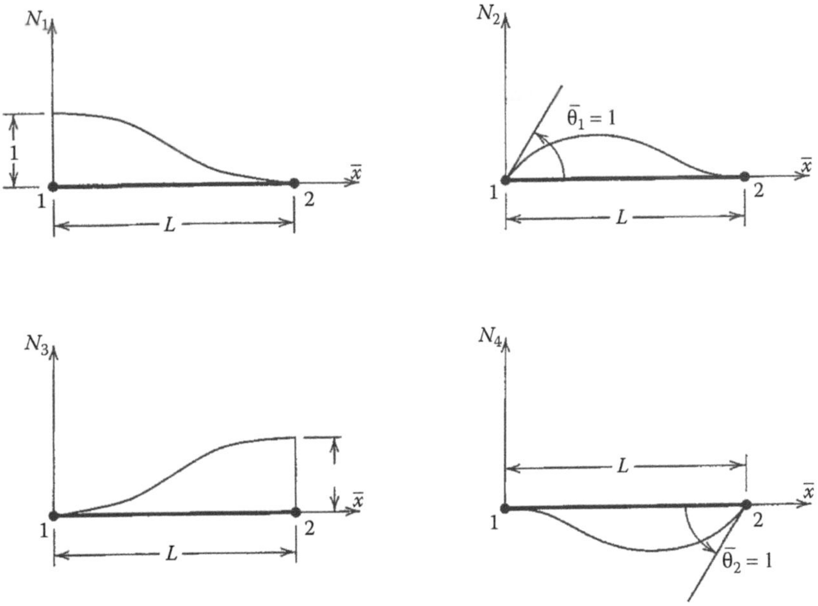

Figure 4.13 Beam element shape functions.

where the subscript B represents *bending* response and where

$$[B_B] = \left[\frac{d^2N_1}{d\bar{x}^2} \quad \frac{d^2N_2}{d\bar{x}^2} \quad \frac{d^2N_3}{d\bar{x}^2} \quad \frac{d^2N_4}{d\bar{x}^2} \right] \tag{4.64b}$$

Substitution of $M = EI\bar{v}''$ into Equation 4.63c gives

$$U = \int_0^L \frac{EI(\bar{v}'')^2}{2}\, d\bar{x} \tag{4.65}$$

from which the first variation of the strain energy is

$$\delta U = \int_0^L (\delta\bar{v})'' EI\bar{v}''\, d\bar{x}$$

In terms of nodal DOFs, from Equation 4.64a, δU is

$$\delta U = \int_0^L \{\delta\bar{v}_i\}^{\mathrm{T}} [B_B]^{\mathrm{T}} EI[B_B]\{\bar{v}_i\}\, d\bar{x} \tag{4.66}$$

In the manner followed with other elements, only nodal loads are assumed, Equation 4.66 is substituted into Equation 4.1, $\{\delta v_i\}$ is eliminated, and the bending stiffness matrix is found to be

$$[\bar{K}_B] = \int_0^L [B_B]^{\mathrm{T}} EI[B_B]\, d\bar{x}$$

Since EI is constant, integration yields the bending stiffness matrix in terms of E, I, and L as

$$[\bar{K}_B]=\left[\begin{array}{cc|cc} \frac{12EI}{L^3} & \frac{6EI}{L^2} & \frac{-12EI}{L^3} & \frac{6EI}{L^2} \\ \frac{6EI}{L^2} & \frac{4EI}{L} & \frac{-6EI}{L^2} & \frac{2EI}{L} \\ \hline \frac{-12EI}{L^3} & \frac{-6EI}{L^2} & \frac{12EI}{L^3} & \frac{6EI}{L^2} \\ \frac{6EI}{L^2} & \frac{2EI}{L} & \frac{-6EI}{L^2} & \frac{4EI}{L} \end{array}\right] \tag{4.67}$$

The stiffness matrix for the plane frame element (see Equation 4.68) is a combination of the axial stiffness matrix, Equation 4.61, and the bending stiffness matrix, Equation 4.67. Note that the ordering of the DOFs in the element first lists all three DOFs at node 1 and then the three DOFs at node 2.

$$[\bar{K}]=\left[\begin{array}{ccc|ccc} \frac{AE}{L} & 0 & 0 & \frac{-AE}{L} & 0 & 0 \\ 0 & \frac{12EI}{L^3} & \frac{6EI}{L^2} & 0 & \frac{-12EI}{L^3} & \frac{6EI}{L^2} \\ 0 & \frac{6EI}{L^2} & \frac{4EI}{L} & 0 & \frac{6EI}{L^2} & \frac{2EI}{L} \\ \hline \frac{-AE}{L} & 0 & 0 & \frac{AE}{L} & 0 & 0 \\ 0 & \frac{-12EI}{L^3} & \frac{-6EI}{L^2} & 0 & \frac{12EI}{L^3} & \frac{-6EI}{L^2} \\ 0 & \frac{6EI}{L^2} & \frac{2EI}{L} & 0 & \frac{-6EI}{L^2} & \frac{4EI}{L} \end{array}\right] \tag{4.68}$$

The displacement vector $\{\bar{u}_i\}$ for the element is

$$\{\bar{u}_i\}=[\,\bar{u}_1 \quad \bar{v}_1 \quad \bar{\theta}_1 \quad \bar{u}_2 \quad \bar{v}_2 \quad \bar{\theta}_2\,]^{\mathrm{T}} \tag{4.69}$$

and the element end action (load) vector $\{\bar{P}_i\}$ is

$$\{\bar{P}_i\}=[\,\bar{P}_{x1} \quad \bar{P}_{y1} \quad \bar{M}_1 \quad \bar{P}_{x2} \quad \bar{P}_{y2} \quad \bar{M}_2\,]^{\mathrm{T}} \tag{4.70}$$

Finally, the relationship between nodal loads and nodal displacements for an element in local coordinates is given by the familiar form

$$[\bar{K}]\{\bar{u}_i\} = \{\bar{P}_i\} \tag{4.71}$$

4.5.2 Equivalent nodal load vector

As for most other elements, actual loads that are applied over the element must be converted to equivalent nodal loads. We consider only element loads that affect beam behavior. Two cases are considered: a distributed load over a portion of the element and a transverse concentrated force. Equivalent nodal loads for axial behavior are derived in a similar fashion.

For a distributed load along the beam, not necessarily over the full length, the variation of work δW_D of the load is

$$\delta W_D = \int_{L_a}^{L_b} \delta\bar{v}\, \bar{q}(\bar{x})\, d\bar{x} \tag{4.72}$$

where $\bar{q}(\bar{x})$ is the load function that exists over the domain $L_a < \bar{x} < L_b$ (see Figure 4.14a) and the subscript D denotes a distributed load. Equation 4.62 is substituted into 4.72 and the equivalent nodal load vector is obtained as

$$\{\bar{P}_{Di}\} = \int_{L_a}^{L_b} [N]^{\mathrm{T}} \bar{q}(\bar{x})\, d\bar{x} \tag{4.73}$$

For a concentrated load $\bar{P}_C$ located at $\bar{x} = L_c$ along the beam (see Figure 4.14b), the variation of work δW_C of the load is

$$\delta W_C = \delta\bar{v}\Big|_{\bar{x}=L_c} \bar{P}_C \tag{4.74}$$

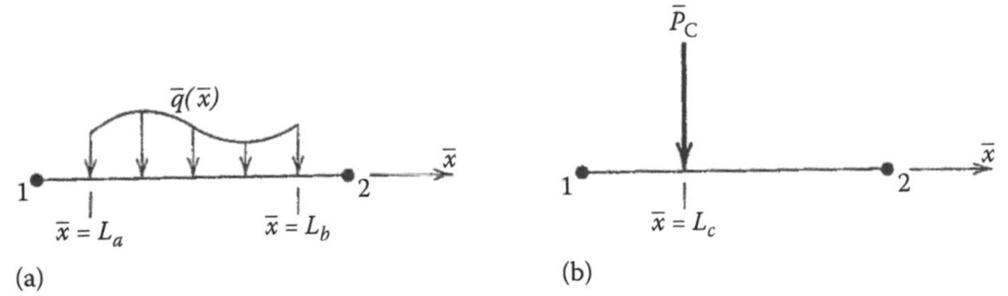

Figure 4.14 Element loads for beam element. (a) Distributed load and (b) concentrated load.

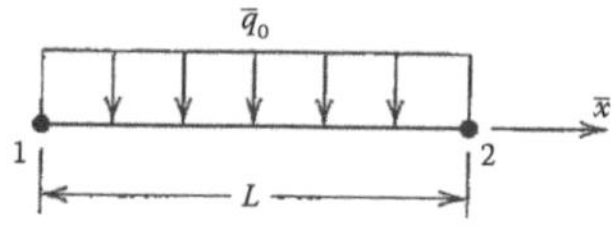

$$\{\bar{P}_{Di}\} = \left\{\begin{array}{c} -\dfrac{\bar{q}_0 L}{2} \\ -\dfrac{\bar{q}_0 L^2}{12} \\ \hline -\dfrac{\bar{q}_0 L}{2} \\ \dfrac{\bar{q}_0 L^2}{12} \end{array}\right\}$$

(a)

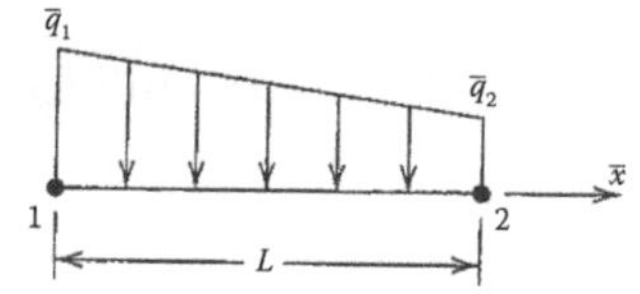

$$\{\bar{P}_{Di}\} = \left\{\begin{array}{c} -\dfrac{L}{20}(7\bar{q}_1 + 3\bar{q}_2) \\ -\dfrac{L^2}{60}(3\bar{q}_1 + 2\bar{q}_2) \\ \hline -\dfrac{L}{60}(3\bar{q}_1 + 7\bar{q}_2) \\ \dfrac{L^2}{60}(2\bar{q}_1 + 3\bar{q}_2) \end{array}\right\}$$

(b)

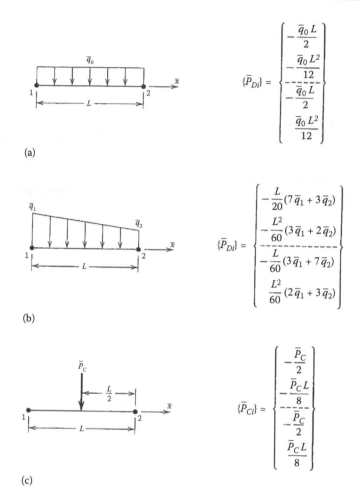

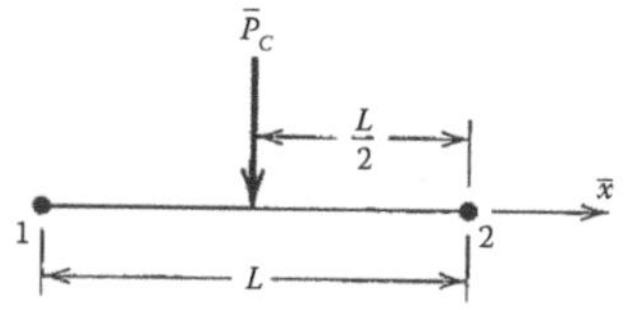

$$\{\bar{P}_{Ci}\} = \left\{\begin{array}{c} -\dfrac{\bar{P}_C}{2} \\ -\dfrac{\bar{P}_C L}{8} \\ \hline -\dfrac{\bar{P}_C}{2} \\ \dfrac{\bar{P}_C L}{8} \end{array}\right\}$$

(c)

Figure 4.15 Equivalent nodal loads for beam element. (a) Uniformly distributed load. (b) Linearly distributed load. (c) Concentrated load.

Variation of displacement $\delta\bar{v}$ at $\bar{x} = L_c$ is written in terms of the variation of nodal displacements by Equation 4.62 with the shape functions evaluated at $\bar{x} = L_c$. The equivalent nodal load vector is

$$\{\bar{P}_{Ci}\} = [N]\Big|_{\bar{x}=L_c}^{\mathrm{T}} \bar{P}_C \tag{4.75}$$

By Equations 4.73 and 4.75, equivalent nodal load vectors for several load patterns on a beam element were determined and are shown in Figure 4.15.

4.5.3 Coordinate rotations

Consider an element in a structure oriented at an angle ϕ with respect to the global x-axis (Figure 4.16). To assemble the stiffness matrix and load vector for this element with those

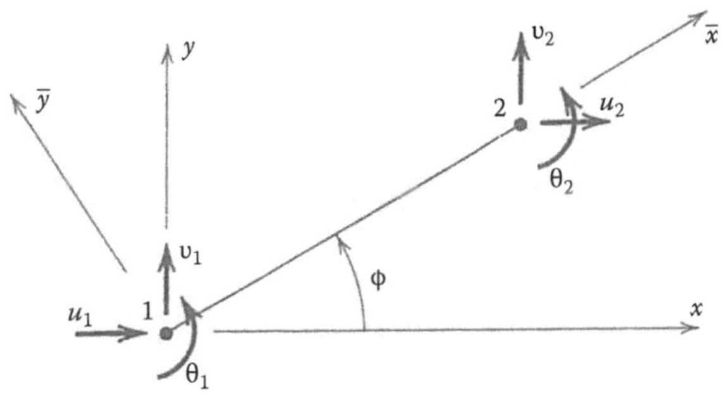

Figure 4.16 Frame element in global coordinates.

of other elements, all nodal DOFs must be defined in terms of the global coordinate system. For node i, the displacements in the two coordinate systems are related by

$$\begin{Bmatrix} \bar{u}_i \\ \bar{v}_i \\ \bar{\theta}_i \end{Bmatrix} = [\lambda] \begin{Bmatrix} u_i \\ v_i \\ \theta_i \end{Bmatrix} \tag{4.76}$$

where

$$[\lambda] = \begin{bmatrix} \cos\phi & \sin\phi & 0 \\ -\sin\phi & \cos\phi & 0 \\ 0 & 0 & 1 \end{bmatrix}$$

For a plane frame element, with two nodes, the displacements are related by

$$\{\bar{u}_i\} = [T]\{u_i\} \tag{4.77}$$

where the rotation (transformation) matrix $[T]$ is

$$[T] = \left[\begin{array}{c|c} \lambda & 0 \\ \hline 0 & \lambda \end{array} \right]$$

In a similar manner, element end actions (loads) are rotated by

$$\{\bar{P}_i\} = [T]\{P_i\} \tag{4.78}$$

Substitution of Equations 4.77 and 4.78 into 4.71 yields

$$[\bar{K}][T]\{u_i\} = [T]\{P_i\} \tag{4.79}$$

Table 4.4 Element stiffness matrix for plane frame element in global coordinates

$[K] =$

$j \rightarrow$ (Column index) / Row index $i \downarrow$	1	2	3	4	5	6
1	$c^2\frac{AE}{L}+s^2\frac{12EI}{L^3}$	$sc\left(\frac{AE}{L}-\frac{12EI}{L^3}\right)$	$-s\left(\frac{6EI}{L^2}\right)$	$-c^2\frac{AE}{L}-s^2\frac{12EI}{L^3}$	$-sc\left(\frac{AE}{L}-\frac{12EI}{L^3}\right)$	$-s\left(\frac{6EI}{L^2}\right)$
2	$sc\left(\frac{AE}{L}-\frac{12EI}{L^3}\right)$	$s^2\frac{AE}{L}+c^2\frac{12EI}{L^3}$	$c\left(\frac{6EI}{L^2}\right)$	$-sc\left(\frac{AE}{L}-\frac{12EI}{L^3}\right)$	$-s^2\frac{AE}{L}-c^2\frac{12EI}{L^3}$	$c\left(\frac{6EI}{L^2}\right)$
3	$-s\left(\frac{6EI}{L^2}\right)$	$c\left(\frac{6EI}{L^2}\right)$	$\frac{4EI}{L}$	$s\left(\frac{6EI}{L^2}\right)$	$-c\left(\frac{6EI}{L^2}\right)$	$\frac{2EI}{L}$
4	$-c^2\frac{AE}{L}-s^2\frac{12EI}{L^3}$	$-sc\left(\frac{AE}{L}-\frac{12EI}{L^3}\right)$	$-s\left(\frac{6EI}{L^2}\right)$	$c^2\frac{AE}{L}+s^2\frac{12EI}{L^3}$	$sc\left(\frac{AE}{L}-\frac{12EI}{L^3}\right)$	$s\left(\frac{6EI}{L^2}\right)$
5	$-sc\left(\frac{AE}{L}-\frac{12EI}{L^3}\right)$	$-s^2\frac{AE}{L}-c^2\frac{12EI}{L^3}$	$-c\left(\frac{6EI}{L^2}\right)$	$sc\left(\frac{AE}{L}-\frac{12EI}{L^3}\right)$	$s^2\frac{AE}{L}+c^2\frac{12EI}{L^3}$	$-c\left(\frac{6EI}{L^2}\right)$
6	$-s\left(\frac{6EI}{L^2}\right)$	$c\left(\frac{6EI}{L^2}\right)$	$\frac{2EI}{L}$	$s\left(\frac{6EI}{L^2}\right)$	$-c\left(\frac{6EI}{L^2}\right)$	$\frac{4EI}{L}$

$c = \cos\phi; \quad s = \sin\phi$

Premultipling both sides of Equation 4.79 by $[T]^{-1}$ and observing that $[T]^{-1} = [T]^T$, since $[T]$ is an orthogonal matrix, we obtain

$$[T]^{\mathrm{T}}[\bar{K}][T]\{u_i\} = \{P_i\}$$

Thus, since $\{u_i\}$ and $\{P_i\}$ are in global coordinates, the stiffness matrix for the plane frame element, in global coordinates, is

$$[K] = [T]^{\mathrm{T}}[\bar{K}][T] \tag{4.80}$$

The final form of $[K]$ is given in Table 4.4. The load vector for the element, in global coordinates, is obtained from Equation 4.78 as

$$\{P_i\} = [T]^{\mathrm{T}}\{\bar{P}_i\} \tag{4.81}$$

4.6 GENERALIZATIONS

Most of the problems in mechanics of materials, even in all areas of engineering science, are expressed in partial differential equations (PDEs). Finite element method can be employed to convert PDE to ordinary differential equations in time. Therefore, finite element method can be considered as the most powerful numerical method in engineering science. In this chapter, we provide the essential background and knowledge in finite element method to avoid blunders and blind acceptance of computer results. However, our approach does not prevent the readers to model advanced material systems and analyze complex physical phenomena. In this section, we will demonstrate that the knowledge provided in previous sections can be easily generalized from two dimensional space to three dimensional space, from static to dynamic, from elastic material to viscoelastic material or elastic-plastic material, from small strain theory to finite strain theory, and from local theory to nonlocal theory.

4.6.1 Three-dimensional isoparametric elements

Similar to two-dimensional isoparametric elements, the most frequently used three-dimensional solid element is the eight-node linear element, which has eight shape functions:

$$\begin{aligned}
N_1 &= (1-\xi)(1-\eta)(1+\varsigma)/8, & N_2 &= (1+\xi)(1-\eta)(1+\varsigma)/8 \\
N_3 &= (1+\xi)(1+\eta)(1+\varsigma)/8, & N_4 &= (1-\xi)(1+\eta)(1+\varsigma)/8 \\
N_5 &= (1-\xi)(1-\eta)(1-\varsigma)/8, & N_6 &= (1+\xi)(1-\eta)(1-\varsigma)/8 \\
N_7 &= (1+\xi)(1+\eta)(1-\varsigma)/8, & N_8 &= (1-\xi)(1+\eta)(1-\varsigma)/8
\end{aligned} \tag{4.82}$$

where $\{\xi, \eta, \varsigma\}$ are dimensionless natural coordinates (cf. Equation 4.51).

Then the displacements of a generic point, u_i (i = 1,2,3), can be linked to the nodal displacement values, U_α [α = 1,2,3,...,3 × 8], of the element in which the generic point resides as follows (cf. Equation 4.24):

$$\begin{Bmatrix} u_1 \\ u_2 \\ u_3 \end{Bmatrix} = \begin{bmatrix} N_1 & 0 & 0 & N_2 & 0 & 0 & \cdots & N_8 & 0 & 0 \\ 0 & N_1 & 0 & 0 & N_2 & 0 & \cdots & 0 & N_8 & 0 \\ 0 & 0 & N_1 & 0 & 0 & N_2 & \cdots & 0 & 0 & N_8 \end{bmatrix} \begin{Bmatrix} U_1 \\ U_2 \\ U_3 \\ U_4 \\ U_5 \\ U_6 \\ \cdot \\ \cdot \\ \cdot \\ U_{22} \\ U_{23} \\ U_{24} \end{Bmatrix} \tag{4.83}$$

which can be expressed in tensorial form as

$$u_i = N_{i\alpha} U_\alpha \quad \{i \in [1,2,3], \quad \alpha \in [1,2,3,....,24]\} \tag{4.84}$$

In a similar way, one may express the strain/nodal–displacement relation as (cf. Equations 4.25 and 4.26)

$$\varepsilon_{ij} = B_{ij\alpha} U_\alpha \tag{4.85}$$

4.6.2 Equilibrium equation

The balance law of linear momentum can be written as (Boresi et al., 2011; Chen et al., 2006; Eringen, 1980)

$$\sigma_{ji,j} + \rho f_i - \rho \ddot{u}_i = 0 \tag{4.86}$$

where σ, f, ü, ρ are Cauchy stress tensor, body force per unit mass, acceleration, mass density, respectively.

It is emphasized that this balance law is valid for any kind of materials as long as we are willing to consider that the materials in question can be modeled by classical continuum mechanics. It is noticed that, in classical continuum mechanics, the balance law of angular momentum implies that the Cauchy stress tensor, σ, is symmetric, i.e., $\sigma_{ij} = \sigma_{ji}$. Also, if we focus our attention to small-strain theory, then $\rho \approx \rho^o$ (the initial mass density).

The equilibrium equation, Equation 4.86, is usually referred to as the strong form. Now, we multiply Equation 4.86 by the virtual displacements δu_i and integrate over an volume v; then it results in

$$\int_v (\sigma_{ji,j} + \rho f_i - \rho\ddot{u}_i)\delta u_i \, dv = 0 \tag{4.87}$$

which is usually referred to as the weak form, because Equation 4.86 is a PDE and demands a solution for all points in the domain, while Equation 4.87 only demands a solution that makes the integral vanish. Term by term, Equation 4.87 yields

$$\int_v \rho\ddot{u}_i \, \delta u_i \, dv = \delta U_\alpha \left\{ \int_v \rho N_{i\beta} N_{i\alpha} \, dv \right\} \ddot{U}_\beta \equiv \delta U_\alpha M_{\alpha\beta} \ddot{U}_\beta \tag{4.88}$$

$$\int_v \rho f_i \, \delta u_i \, dv = \delta U_\alpha \left\{ \int_v \rho f_i N_{i\alpha} \, dv \right\} \equiv \delta U_\alpha F_\alpha^1 \tag{4.89}$$

and

$$\begin{aligned}
\int_v \sigma_{ji,j} \, \delta u_i \, dv &= \int_v \left\{ [\sigma_{ji} \delta u_i]_{,j} - \sigma_{ji} \delta u_{i,j} \right\} dv \\
&= \oint_s \sigma_{ji} n_j \, \delta u_i \, ds - \int_v \sigma_{ji} \, \delta\varepsilon_{ij} \, dv \\
&= \int_{s_\sigma} \tilde{\sigma}_i \, \delta u_i \, ds - \int_v \sigma_{ji} \, \delta\varepsilon_{ij} \, dv \\
&= \delta U_\alpha \left\{ \int_{s_\sigma} \tilde{\sigma}_i N_{i\alpha} \, ds - \int_v \sigma_{ji} B_{ij\alpha} \, dv \right\} \\
&\equiv \delta U_\alpha \left\{ F_\alpha^2 - F_\alpha \right\}
\end{aligned} \tag{4.90}$$

During the process of deriving Equation 4.90, it is noticed that we have utilized (1) the Green-Gauss theorem to convert volume integral to surface integral, (2) the symmetry of the Cauchy stress tensor so that $\sigma_{ji}\delta u_{i,j} = \sigma_{ji}\delta\varepsilon_{ij}$, (3) the requirement that the virtual displacement δu_i should be vanishing at s_u where displacement-specified boundary condition is imposed, and (4) the requirement that the surface traction $\sigma_i \equiv \sigma_{ji} n_j$ equals to the specified value $\tilde{\sigma}_i$ at s_σ, where the applied force per unit area is imposed. Note that $s = s_u + s_\sigma$.

Because the nodal value of the virtual displacement, δU_α is arbitrary, Equation 4.87 leads to

$$M_{\alpha\beta} \ddot{U}_\beta + F_\alpha = F_\alpha^1 + F_\alpha^2 \tag{4.91}$$

It is noticed that F_α^1 and F_α^2 are the nodal forces due to body force f_i and surface traction $\tilde{\sigma}_i$, respectively; F_α is the nodal force due to the Cauchy stress. If the constitutive equation can be expressed as

$$\sigma_{ij} = A_{ijkl} e_{kl} \tag{4.92}$$

then one may obtain

$$F_\alpha \equiv \int_v \sigma_{ji} B_{ij\alpha}\, dv = \left\{ \int_v A_{ijkl} B_{kl\beta} B_{ij\alpha}\, dv \right\} U_\beta \equiv K_{\alpha\beta} U_\beta \tag{4.93}$$

Then the governing equation, Equation 4.91, is reduced to

$$M_{\alpha\beta} \ddot{U}_\beta + K_{\alpha\beta} U_\beta = F_\alpha^1 + F_\alpha^2 \tag{4.94}$$

To be precise, Equation 4.94 is the finite element equation for linear anisotropic elastic solid. If the linear elastic solid is isotropic, then

$$A_{ijkl} = \lambda \delta_{ij} \delta_{kl} + \mu(\delta_{ik} \delta_{jl} + \delta_{il} \delta_{jk}) \tag{4.95}$$

where λ and μ are the two Lame constants related to Young's modulus and Poisson's ratio as

$$\begin{aligned} E &= \frac{\mu(3\lambda + 2\mu)}{\lambda + \mu}, \qquad \upsilon = \frac{\lambda}{2(\lambda + \mu)} \\ \lambda &= \frac{\upsilon E}{(1+\upsilon)(1-2\upsilon)}, \qquad \mu = \frac{E}{2(1+\upsilon)} \end{aligned} \tag{4.96}$$

For linear isotropic elastic solid, the stress–strain relation reduces to

$$\sigma_{ij} = \lambda \varepsilon_{kk} \delta_{ij} + 2\mu \varepsilon_{ij} \tag{4.97}$$

In Voigt's convention, Equation 4.97 can be rewritten as

$$\begin{Bmatrix} \sigma_{11} \\ \sigma_{22} \\ \sigma_{33} \\ \sigma_{23} = \sigma_{32} \\ \sigma_{31} = \sigma_{13} \\ \sigma_{12} = \sigma_{21} \end{Bmatrix} = \begin{bmatrix} \lambda + 2\mu & \lambda & \lambda & 0 & 0 & 0 \\ \lambda & \lambda + 2\mu & \lambda & 0 & 0 & 0 \\ \lambda & \lambda & \lambda + 2\mu & 0 & 0 & 0 \\ 0 & 0 & 0 & \mu & 0 & 0 \\ 0 & 0 & 0 & 0 & \mu & 0 \\ 0 & 0 & 0 & 0 & 0 & \mu \end{bmatrix} \begin{Bmatrix} \varepsilon_{11} \\ \varepsilon_{22} \\ \varepsilon_{33} \\ \gamma_{23} = \varepsilon_{23} + \varepsilon_{32} \\ \gamma_{31} = \varepsilon_{31} + \varepsilon_{13} \\ \gamma_{12} = \varepsilon_{12} + \varepsilon_{21} \end{Bmatrix} \tag{4.98}$$

If we further reduce the stress–strain relation to two-dimensional case (either plane strain or plane stress. We will obtain Equations 4.10–4.128.

4.6.3 Viscoelastic solid

We should recognize that

$$F_\alpha \equiv \int_v \sigma_{ji} B_{ij\alpha}\, dv \tag{4.99}$$

is the general formula for the calculation of the nodal force due to Cauchy stress (cf. Equation 4.93). In other words, for other kind of materials or other kind of theories, if one has the stress σ no matter from what kind of constitutive theory, one still can calculate the forcing term $\mathbf{F}$ using Equation 4.99 and proceed to solve Equation 4.91, which is now an ordinary differential equation in time.

To illustrate this point, let us express the constitutive equation for linear anisotropic viscoelastic solid (Kelvin-Voigt model) as

$$\sigma_{ij} = A_{ijkl}\varepsilon_{kl} + a_{ijkl}\dot{\varepsilon}_{kl} \tag{4.100}$$

where A_{ijkl} and a_{ijkl} maybe called the elastic and damping constants, respectively. Then Equation 4.99 becomes

$$\begin{aligned} F_\alpha &\equiv \int_v \sigma_{ji} B_{ij\alpha}\, dv = \left\{ \int_v A_{ijkl} B_{kl\beta} B_{ij\alpha}\, dv \right\} U_\beta + \left\{ \int_v a_{ijkl} B_{kl\beta} B_{ij\alpha}\, dv \right\} \dot{U}_\beta \\ &\equiv K_{\alpha\beta} U_\beta + C_{\alpha\beta} \dot{U}_\beta \end{aligned} \tag{4.101}$$

and finite element equation is reduced to the following familiar form

$$M_{\alpha\beta}\ddot{U}_\beta + C_{\alpha\beta}\dot{U}_\beta + K_{\alpha\beta}U_\beta = F_\alpha^1 + F_\alpha^2 \tag{4.102}$$

Of course, after the mass matrix $\mathbf{M}$, damping matrix $\mathbf{C}$, and the stiffness matrix $\mathbf{K}$ for all elements are obtained, one has to go through the assembly process as indicated in Table 4.2.

4.6.4 Nonlocal elasticity

For nonlocal elasticity, we generalize the stress–strain relation as (Chen et al., 2002, 2006; Eringen, 2002)

$$\sigma_{ij}(\mathbf{x}) = \frac{\left\{ \lambda\delta_{ij}\delta_{mn} + \mu(\delta_{im}\delta_{jn} + \delta_{in}\delta_{jm}) \right\} \int_{\Omega^*} f(\tilde{r}) e_{mn}(\mathbf{x}^*)\, d\Omega(\mathbf{x}^*)}{\int_{\Omega^*} f(\tilde{r})\, d\Omega(\mathbf{x}^*)} \tag{4.103}$$

where

$$f(\tilde{r}) = \frac{35}{4\pi R^3}\begin{cases} 1-6\tilde{r}^2+8\tilde{r}^3-3\tilde{r}^4, & \text{if } \tilde{r}\le 1 \\ 0, & \text{if } \tilde{r}\ge 1 \end{cases}$$
$$r = \|\mathbf{x}-\mathbf{x}^*\|, \quad \tilde{r}\equiv r/R \tag{4.104}$$

This simply means the stress tensor at position **x** depends on strain tensors at all points x^* within Ω, not just at x only. It is seen that R is a newly introduced length scale and

$$\lim_{R\to 0} f(\tilde{r}) = \delta(r) = \delta\left(\|\mathbf{x}-\mathbf{x}^*\|\right) \tag{4.105}$$

As $R \to 0$

$$\begin{aligned}\lim_{R\to 0}\sigma_{ij}(\mathbf{x}) &= \lim_{R\to 0}\frac{\left\{\lambda\delta_{ij}\delta_{mn}+\mu(\delta_{im}\delta_{jn}+\delta_{in}\delta_{jm})\right\}\int_{\Omega^*} f(\tilde{r})e_{mn}(\mathbf{x}^*)\,d\Omega(\mathbf{x}^*)}{\int_{\Omega^*} f(\tilde{r})\,d\Omega(\mathbf{x}^*)} \\ &= \frac{\left\{\lambda\delta_{ij}\delta_{mn}+\mu(\delta_{im}\delta_{jn}+\delta_{in}\delta_{jm})\right\}\int_{\Omega^*} \delta\left(\|\mathbf{x}-\mathbf{x}^*\|\right)e_{mn}(\mathbf{x}^*)\,d\Omega(\mathbf{x}^*)}{\int_{\Omega^*} \delta\left(\|\mathbf{x}-\mathbf{x}^*\|\right)d\Omega(\mathbf{x}^*)} \\ &= \left\{\lambda\delta_{ij}\delta_{mn}+\mu(\delta_{im}\delta_{jn}+\delta_{in}\delta_{jm})\right\}e_{mn}(\mathbf{x})\end{aligned} \tag{4.106}$$

In other words, Equation 4.103 in nonlocal elasticity is reduced to the stress–strain relation in classical elasticity when the length scale R is approaching zero.

Now it is seen that, for nonlocal elasticity, F_α can be obtained as

$$\begin{aligned}F_\alpha &\equiv \int_{\Omega(\mathbf{x})} \sigma_{ij}(\mathbf{x})B_{ij\alpha}(\mathbf{x})\,d\Omega(\mathbf{x}) = \int_{\Omega(x)} \frac{A_{ijmn}\int_{\Omega^*} f(\tilde{r})e_{mn}(\mathbf{x}^*)\,d\Omega(\mathbf{x}^*)}{H(\mathbf{x})}B_{ij\alpha}(\mathbf{x})d\Omega(\mathbf{x}) \\ &= \left(\int_{\Omega(\mathbf{x})} \frac{A_{ijmn}}{H(\mathbf{x})}B_{ij\alpha}(\mathbf{x})\left\{\int_{\Omega(\mathbf{x}^*)} f(\tilde{r})B_{mn\beta}(\mathbf{x}^*)\,d\Omega(\mathbf{x}^*)\right\}d\Omega(\mathbf{x})\right)U_\beta \\ &\equiv \tilde{K}_{\alpha\beta}U_\beta\end{aligned} \tag{4.107}$$

where

$$H(\mathbf{x}) \equiv \int_{\Omega^*} f(\tilde{r})\,d\Omega(\mathbf{x}^*) \tag{4.108}$$

The finite element equation for nonlocal elasticity becomes

$$M_{\alpha\beta}\ddot{U}_\beta + \tilde{K}_{\alpha\beta}U_\beta = F_\alpha^1 + F_\alpha^2 \tag{4.109}$$

It has the same form as that in elasticity except that the stiffness matrix now has a larger band width.

4.6.5 Plasticity

The fundamental concept of computational plasticity with return mapping algorithm has been introduced by Casey (1998), Simo and Hughes (1998), Lee and Chen (2001), and Chen et al. (2006). Here, we present a simple theory of plasticity.

In continuum mechanics, the dependent and the independent constitutive variables are denoted by $\mathbf{Z} = \{\mathbf{T}, \mathbf{Q}, \psi, \eta\}$ and $\mathbf{U} = \{\mathbf{E}, \dot{\mathbf{E}}, \ldots, \theta, \nabla\theta, \ldots\}$, respectively. $\mathbf{T}$, $\mathbf{Q}$, ψ, η are the stress, heat flux, Helmholtz free energy, entropy, respectively; $\mathbf{E}$, $\dot{\mathbf{E}}$, θ, $\nabla\theta$ are the strain, strain rate, absolute temperature, temperature gradient, respectively. In this work, let $\mathbf{U} = \{\mathbf{E}, \theta\}$. Then the constitutive equations are expressed as

$$\mathbf{Z} = \mathbf{Z}(\mathbf{U}) \tag{4.110}$$

In small strain theory, it does not matter whether $\mathbf{T}$ is the second-order Piola–Kirchhoff stress or the Cauchy stress and whether $\mathbf{E}$ is the Lagrangian strain or the infinitesimal strain either. This seemingly general theoretical framework turns out to be insufficient for the construction of the constitutive theory of plasticity. For plasticity, one needs to introduce a new set of variables, called internal variables, as

$$\mathbf{W} = \{\mathbf{E}^p, \mathbf{R}\} \tag{4.111}$$

where $\mathbf{E}^p$ is the plastic strain tensor and $\mathbf{R}$, named as the hardening parameters, is a generalized vector of internal variables. Also, one may define the elastic strain tensor $\mathbf{E}^e$ as

$$\mathbf{E}^e \equiv \mathbf{E} - \mathbf{E}^p \tag{4.112}$$

It should be emphasized that the (total) strain $\mathbf{E}$ is derivable from the displacement field; if the plastic strain $\mathbf{E}^p$ is obtained, then the elastic strain $\mathbf{E}^e$ follows the simple rule, Equation 4.112.

To separate the material behavior into two distinct parts—elastic state and elastic-plastic state—a scalar-valued yield function is introduced as

$$f = f(\mathbf{U}, \mathbf{W}) \tag{4.113}$$

and, for a fixed set of values for $\mathbf{W}$, a hyper surface, named yield surface, is determined by the equation

$$f(\mathbf{U}, \mathbf{W}) = 0 \tag{4.114}$$

The yield function can be chosen in such a way that the elastic region corresponds to $f < 0$. The states of $f > 0$ are nonadmissible and ruled out in plasticity. We now define the loading rate as the inner product between the outward normal to the yield surface and the tangent vector to the trajectory in the **U** space, i.e.,

$$\xi \equiv \frac{\partial f}{\partial \mathbf{U}} \cdot \dot{\mathbf{U}} = \frac{\partial f}{\partial \mathbf{E}} : \dot{\mathbf{E}} + \frac{\partial f}{\partial \theta} \dot{\theta} \tag{4.115}$$

Three distinct cases, unloading, neutral loading, and loading, can be defined by (a) $f < 0$, (b) $f = \xi = 0$, and (c) $f = 0$, $\xi > 0$, respectively. The internal variables **W** will remain constants in cases of unloading and neutral loading. The evolution equations for the internal variables are postulated to be

$$\dot{\mathbf{W}} = \hat{\xi} \pi \varphi(\mathbf{U}, \mathbf{W}) \tag{4.116}$$

where $\hat{\xi} \equiv 0$ if $\xi \le 0$ and $\hat{\xi} \equiv \xi$ if $\xi > 0$. One should impose the consistency condition of plasticity: $f = 0$ and $\xi \ge 0$ lead to $\dot{f} = 0$; in other words, an elastic-plastic state leads to another elastic-plastic state. To enforce this consistency condition in the case of loading, one must have

$$\dot{f} = 0 = \frac{\partial f}{\partial \mathbf{U}} \cdot \dot{\mathbf{U}} + \frac{\partial f}{\partial \mathbf{W}} \cdot \xi \, \pi \boldsymbol{\varphi} = \xi + \frac{\partial f}{\partial \mathbf{W}} \cdot \xi \, \pi \boldsymbol{\varphi} \tag{4.117}$$

which implies

$$1 + \pi \frac{\partial f}{\partial \mathbf{W}} \cdot \boldsymbol{\varphi} = 0 \quad \Rightarrow \quad \pi = -\frac{1}{\dfrac{\partial f}{\partial \mathbf{W}} \cdot \boldsymbol{\varphi}} \tag{4.118}$$

Now, to begin with, let

$$\mathbf{Z} = \mathbf{Z}(\mathbf{U}, \mathbf{W}) \tag{4.119}$$

Recall the Clausius-Duhem inequality as (Eringen, 1980)

$$-\rho^{o}(\dot{\psi} + \eta \dot{\theta}) + \mathbf{T} : \dot{\mathbf{E}} - \theta^{-1} \mathbf{Q} \cdot \nabla \theta \ge 0 \tag{4.120}$$

Substituting Equations 4.116 and 4.119 into the Clausius-Duhem inequality, Equation 4.120, results in

$$-\rho^{o} \left\{ \frac{\partial \psi}{\partial \mathbf{E}} \cdot \dot{\mathbf{E}} + \frac{\partial \psi}{\partial \theta} \dot{\theta} + \frac{\partial \psi}{\partial \mathbf{W}} \cdot \hat{\xi} \, \pi \varphi + \eta \, \dot{\theta} \right\} + \mathbf{T} \cdot \dot{\mathbf{E}} - \theta^{-1} \mathbf{Q} \cdot \nabla \theta \ge 0 \tag{4.121}$$

which implies

$$\eta = -\frac{\partial \psi}{\partial \theta}, \qquad \mathbf{T} = \rho^o \frac{\partial \psi}{\partial \mathbf{E}}, \quad \mathbf{Q} = 0$$
$$\frac{\partial \psi}{\partial \mathbf{W}} \cdot \boldsymbol{\varphi} \le 0, \qquad \frac{\partial f}{\partial \mathbf{W}} \cdot \boldsymbol{\varphi} < 0 \tag{4.122}$$

Green and Naghdi [1965], in their pioneer work on theory of plasticity, proposed that the stresses and entropy for thermo-elastic-plastic continuum are only functions of temperature and elastic Lagrangian strains. Following this idea, we further assume that the Helmholtz free energy density can be expressed as a polynomial of $\mathbf{E}^e = \mathbf{E} - \mathbf{E}^p$ and θ up to second order as

$$\psi = \psi(\mathbf{E} - \mathbf{E}^p, \theta) = \psi(\mathbf{E}^e, \theta) = \left\{ S^o - \rho^o \eta^o T - \frac{1}{2} \rho^o c T^2 / T^o + \frac{1}{2} A_{KLMN} E^e_{KL} E^e_{MN} \right\} \Big/ \rho^o \tag{4.123}$$

It results in

$$\eta = \eta^o + cT/T^o, \qquad T_{KL} = A_{KLMN} E^e_{MN} \tag{4.124}$$

where T^o is the temperature of the natural state, which may also be referred to as reference temperature.

It is noticed that, from Equation 4.124, if one knows the plastic strain tensor $\mathbf{E}^p$ and calculates the total strain $\mathbf{E}$ and the elastic strain $\mathbf{E}^e = \mathbf{E} - \mathbf{E}^p$, the stress tensor can be obtained. In other words, we need an algorithm, a return mapping algorithm, to calculate the plastic strain tensor $\mathbf{E}^p$ (Simo and Hughes, 1988).

Now we propose the yield function and the hardening parameters as

$$f(\mathbf{T}, \mathbf{R}) = \|\xi\| - \sqrt{2/3}(\sigma_Y + cH\bar{\beta}) \tag{4.125}$$

$$R = \{\boldsymbol{\beta}, \bar{\beta}\} \tag{4.126}$$

where

$$S_{KL} \equiv T_{KL} - \frac{1}{3} T_{MM} \delta_{KL} \qquad \text{or} \qquad \mathbf{S} \equiv \mathbf{T} - \frac{1}{3} tr(\mathbf{T})\mathbf{I} \tag{4.127}$$

$$\xi \equiv S - \boldsymbol{\beta}, \qquad \xi \equiv \sqrt{\xi : \xi} \tag{4.128}$$

Notice that β is a second order symmetric tensor with $\mathrm{tr}(\beta) = 0$ and defines the center of the von Mises yield surface in the stress space and $\bar{\beta}$ defines the isotropic hardening of the von

Mises yield surface. Then the evolution of $\mathbf{E}^p$, $\boldsymbol{\beta}$, and $\bar{\beta}$ can be obtained as follows (Chen et al., 2006):

$$\begin{aligned} &\dot{\mathbf{E}}^p = \gamma\boldsymbol{\xi}/\xi \\ &\dot{\boldsymbol{\beta}} = \gamma\frac{2}{3}(1-c)H\boldsymbol{\xi}/\xi \\ &\dot{\bar{\beta}} = \gamma\sqrt{2/3} \end{aligned} \tag{4.129}$$

In Equation 4.125, σ_Y is the von Mises strength; H is a constant representing the slope of the stress–strain relation in plastic loading; c is a constant with $0 \le c \le 1$. Also, it is noticed that (1) $c = 1$ implies $\dot{\boldsymbol{\beta}} = 0$, which is the case of isotropic hardening and (2) $c = 0$ implies that $\bar{\beta}$ does not affect the size of the yield surface–it is the case of kinematic hardening.

Now we are going to describe the numerical procedures to implement the return mapping algorithm, which splits the problem into two parts: (1) elastic-predictor and (2) plastic-corrector. First, the elastic-predictor creates a trial elastic state as

$$\begin{aligned} &\tilde{\mathbf{E}}^p_{n+1} = \mathbf{E}^p_n \\ &\tilde{\mathbf{R}}_{n+1} = \mathbf{R}_n \\ &\tilde{\mathbf{T}}_{n+1} = \mathbf{A} : \left(\mathbf{E}_{n+1} - \tilde{\mathbf{E}}^p_{n+1}\right) \\ &\tilde{f}_{n+1} = f(\tilde{\mathbf{T}}_{n+1}, \tilde{\mathbf{R}}_{n+1}) \end{aligned} \tag{4.130}$$

where the subscripts n and $n + 1$ indicate the values of the variables at $t_n = n\ \Delta t$ and $t_{n+1} = (n + 1)\ \Delta t$, respectively; $\tilde{\mathbf{E}}^p_{n+1}$, $\tilde{\mathbf{R}}_{n+1}$, $\tilde{\mathbf{T}}_{n+1}$, and $\tilde{f}_{n+1}$ indicate that they are the trial values. If the trial value $\tilde{f}_{n+1} \le 0$, it means the prediction is correct and the nothing needs to be done. On the other hand, if the trial value $\tilde{f}_{n+1} > 0$, then the plastic corrector demands the following updates:

$$\begin{aligned} &E^p_{n+1} = E^p_n + \Delta\gamma(\tilde{S}_{n+1} - \boldsymbol{\beta}_n)/\left\|\tilde{S}_{n+1} - \boldsymbol{\beta}_n\right\| \\ &\boldsymbol{\beta}_{n+1} = \boldsymbol{\beta}_n + \frac{2}{3}(1-c)H\Delta\gamma(\tilde{\mathbf{S}}_{n+1} - \boldsymbol{\beta}_n)/\left\|\tilde{\mathbf{S}}_{n+1} - \boldsymbol{\beta}_n\right\| \\ &\bar{\beta}_{n+1} = \bar{\beta}_n + \sqrt{2/3}\Delta\gamma \end{aligned} \tag{4.131}$$

where

$$\Delta\gamma = \frac{\tilde{f}_{n+1}}{2\mu + (2/3)H} \tag{4.132}$$

One may readily prove that, with the updated values of E^p_{n+1}, β_{n+1}, and $\bar{\beta}_{n+1}$, from Equation 4.131, and the updated values of

$$\begin{aligned} E^e_{n+1} &= E_{n+1} - E^p_{n+1} \\ T_{n+1} &= \lambda\, tr\left(E^e_{n+1}\right) I + 2\mu E^e_{n+1} \\ S_{n+1} &= T_{n+1} - \frac{1}{3}\, tr(T_{n+1}) I \end{aligned} \tag{4.133}$$

the value of the yield function

$$f_{n+1} = \left\| \mathbf{S}_{n+1} - \boldsymbol{\beta}_{n+1} \right\| - \sqrt{2/3}(\sigma_Y + cH\bar{\beta}_{n+1}) \tag{4.134}$$

indeed returns to zero.

4.6.6 Finite strain theory

Most of the engineering mechanics problems are treated by small strain theory, in which the magnitude of any independent constitutive variables, especially the strains, are small. In its counterpart, finite strain theory, no restrictive assumption has been made to the magnitude of any independent constitutive variables. Relatively speaking, it is still easy to perform finite element analysis with finite strain. We will show the derivation in the following.

First, we recall the relation between the Cauchy stress tensor and the second-order Piola–Kirchhoff stress tensor

$$T_{KL} = j\sigma_{kl} X_{K,k} X_{L,l}, \qquad \sigma_{kl} = j^{-1} T_{KL} x_{k,K} x_{l,L} \tag{4.135}$$

where the motion and the inverse motion are expressed as

$$x_k = x_k(\mathbf{X}, t), \qquad X_K = X_K(\mathbf{x}, t) \tag{4.136}$$

the displacement gradient and its inverse are $x_{k,K} = \dfrac{\partial x_k}{\partial X_K}$, $X_{K,k} = \dfrac{\partial X_K}{\partial x_k}$; j is the Jacobian, defined as

$$j \equiv \det(x_{k,K}) \tag{4.137}$$

Also, we mention the well-known relation between the volume of deformed state and the volume of the undeformed state:

$$dv = j dV \tag{4.138}$$

Now, we recall the nodal force and further derive it as follows:

$$
\begin{aligned}
\int_v \sigma_{ij}\,\delta u_{i,j}\,dv &= \delta U_\alpha \int_v \sigma_{ij} N_{i\alpha,j}\,dv \\
&= \delta U_\alpha \int_v \sigma_{ij} N_{i\alpha,j}\,dv \\
&= \delta U_\alpha \int_v j^{-1} x_{i,K} x_{j,L} T_{KL} N_{i\alpha,j}\,dv \\
&= \delta U_\alpha \int_V x_{i,K} x_{j,L} T_{KL} N_{i\alpha,j}\,dV \\
&= \delta U_\alpha \int_V x_{i,K} T_{KL} N_{i\alpha,L}\,dV \\
&= \delta U_\alpha \int_V (\delta_{iK} + N_{i\beta,K} U_\beta) T_{KL} N_{i\alpha,L}\,dV \\
&= \delta U_\alpha F_\alpha
\end{aligned}
\tag{4.139}
$$

where δ_{iK} is called the shifter and defined as

$$
\delta_{iK} \equiv \mathbf{i}_k \cdot \mathbf{I}_K \tag{4.140}
$$

$\mathbf{i}_k$ is the unit vector along the $k - th$ axis of the Eulerian coordinate system (for the deformed state); $\mathbf{I}_K$ is the unit vector along the $K - th$ axis of the Lagrangian coordinate system (for the undeformed state).

Equation 4.139 says again, if one has the stress σ, no matter from what kind of constitutive theory, no matter it is based on small strain theory or finite strain theory, one still can calculate the forcing term **F** and proceed to solve the ordinary differential equation, Equation 4.91, in time. The numerical method to solve ordinary differential equations in time will be discussed in Chapter 7.

4.7 CLOSING REMARKS

4.7.1 Requirements for accuracy

The accuracy of a finite element solution depends strongly on two conditions. First, it is important that the equations of equilibrium be satisfied throughout the model. Second, it is important that compatibility (continuity of displacements) be maintained. In certain circumstances, these conditions are violated, as noted in the following paragraphs.

Equilibrium at the structure nodes is satisfied since the basic system of equations, Equation 4.45, is fundamentally a system of nodal equilibrium equations. Thus, within the accuracy of the equation-solving process (numerical error), the structure nodes are in equilibrium.

For elements with only displacement DOFs, equilibrium along element edges is generally not satisfied. This is due to the fact that while displacements might be continuous across element boundaries, their derivatives are not, and thus stresses are not continuous. For instance, consider two CST elements, such as those shown in Figure 4.17. Nodes 1, 2, and 3

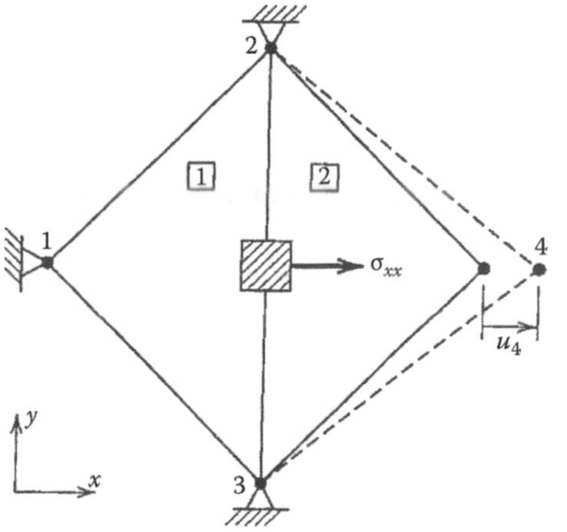

Figure 4.17 Equilibrium along element edges.

are fully constrained while node 4 has an imposed displacement in the x direction. Hence, element 1 is unstressed while element 2 has nonzero σ_{xx}. Due to the stress discontinuity, a differential element located at the boundary between the two elements does not satisfy equilibrium in the x direction.

Equilibrium within an element is not satisfied unless body forces are of relatively low order or are entirely absent. For a CST, the stress state is constant throughout the element. Thus, equilibrium of a differential element is satisfied only when body forces are absent. Similarly, for elements that can represent linear stress variation, body forces must be, at most, constant in magnitude for equilibrium.

Compatibility at the nodes is ensured due to the assembly process. That is, the displacements of adjacent elements are the same at their common nodes. However, to ensure that compatibility is maintained along the common edge between two adjacent elements, the displacements along that edge, viewed from either element, must be expressed entirely in terms of the displacement of nodes on that edge. Elements that maintain compatibility along common edges are known as *conforming* elements. Generally, this condition is satisfied for elements that possess only translational DOFs. However, certain plate-bending and shell elements, for instance, are nonconforming. Compatibility within an element is ensured as long as the displacement interpolation polynormals is continuous.

4.7.2 Requirements for convergence

As discussed at the beginning of this chapter, a major source of error in a finite element solution is due to the use of approximation functions to describe element response (formulation error). To reduce formulation error, we successively refine our finite element models with the expectation that the numerical solution will converge to the *exact* solution. Under certain conditions, convergence can be guaranteed. These conditions are as follows:

1. *The elements must be complete.* That is, the shape function must be a complete polynomial. For instance, a complete quadratic contains all possible quadratic terms and omits no linear or constant terms. Inclusion of a few cubic terms, such as for the quadratic serendipity and Lagrange elements, does not destroy completeness of the quadratic polynomial.

2. *The elements must be compatible.* Hence, continuity of displacements must be ensured throughout the entire structural model.
3. *The elements must be capable of representing rigid-body motion and constant strain.* For two- and three-dimensional elasticity problems, these are ensured if the displacement field contains at least a complete linear polynomial. For shell elements, constant strain implies constant curvature w_{xx} and w_{yy} and constant twist w_{xy}. Some shell elements have difficulty representing rigid-body motion.

Generally, a finite element model is too stiff. That is, displacements converge from below. A qualitative explanation is as follows. The elements are *constrained*, by the shape functions, to deform in a specific (unnatural) manner. This constraint adds stiffness, relative to the physical system, that results in smaller displacements when the external influences on the system are loads. If all external loads are zero and the only external influences on the system are imposed (nonzero) displacements, additional energy is required to force the model into the imposed deformed shape.

In the case of isoparametric elements, reduced integration can be used effectively to *soften* the element such that its response improves relative to full integration. Problem 4.1 demonstrates how the use of approximation functions to represent displacements results in a model that is stiff relative to the actual system.

4.7.3 Modeling recommendations

As an aid to the application of the finite element method to analysis of practical problems in elasticity, the following recommendations are offered. The list is not exhaustive and the recommendations themselves are not rigid rules that cannot be violated.

1. Avoid abrupt transitions in element size and geometry. Limit the change in *element stiffness* (approximated by E/V_e, where V_e is the volume of the element) from one element to the next to roughly a factor of 3.
2. Avoid unnecessary element irregularity. Keep aspect ratios (the length ratio of the longest side to the shortest side) below 10:1. Interior angles of quadrilaterals should be as regular as possible. They should not exceed 150° and they should not be less than 30°. Midside nodes on quadratic elements should be within the middle third of the edge.
3. Maintain compatibility between elements. For instance, it is not appropriate to attach one quadratic quadrilateral to two linear quadrilaterals simply because they have three nodes in common. Such an assembly would not maintain compatibility because of the difference in displacement interpolation on the two sides of the boundary (see Figure 4.18).
4. Use a fine mesh in regions of high stress gradient (stress concentration). Use a coarse mesh where gradients are low.

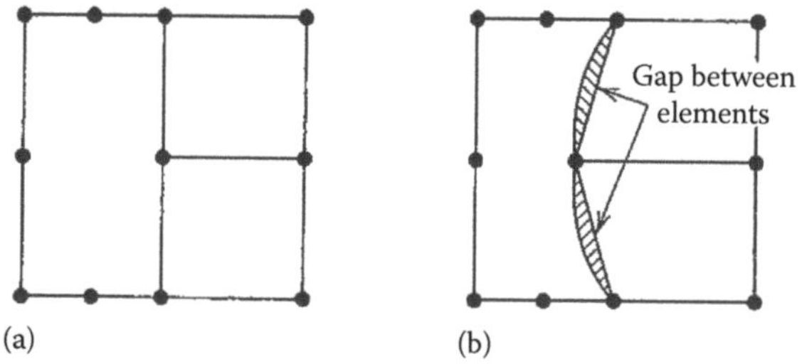

Figure 4.18 Assembly of incompatible elements. (a) Undistorted assembly. (b) Loss of compatibility under distortion.

5. When Choleski decomposition, or any other *band* solver, is used, minimize the bandwidth of the assembled structure stiffness matrix by proper node numbering. The nonzero entries in the structure stiffness matrix are clustered about the diagonal in a *band*. The *bandwidth* is the number of terms across a row (or down a column) of the band. The *half-bandwidth* is the number of terms from the diagonal out to the edge of the band. The nodal half-bandwidth is computed as $(n_{max} - n_{min} + 1)$, where n_{max} and n_{min} are the largest and smallest structure node numbers in the incidence list for an element. Hence, to minimize bandwidth, keep the range of node numbers that define the incidences for a single element as small as possible. Examples of poor and good node numbering schemes are illustrated in Figure 4.19.
6. Exploit symmetry in the geometry and loads of the physical system to build the smallest reasonable model.

The finite element method and its use in engineering practice are evolving continuously. For instance, not long ago, material and/or geometric nonlinear analyses were rarely attempted. Today, such analyses are not limited to research but are performed by practicing engineers as well. The popularity of the finite element method is due primarily to the greater availability, and affordability, of user-friendly software that integrates sophisticated analysis capabilities with solid modeling and CAD. Unfortunately, user training and experience are not always equal to the capabilities of the software. Hence, the danger exists that these powerful analytical tools will be used as *black boxes*, without proper understanding of the physical system or of the algorithms used in the analysis. There is no substitute for common sense and sound judgment, and one should remain skeptical of computer-generated results until they can be verified by other means.

An effective means for an engineer to gain experience in performing finite element analysis and to develop confidence in the finite element program is to solve a series of relatively simple *benchmark* problems. Such problems are specially designed to test the accuracy of the individual elements in the program. However, they can also be used as a training device for novice users. A reasonable set of benchmark problems has been proposed by MacNeil and Harder (1984, 1985). Additional problems can be found in AIAA (1985).

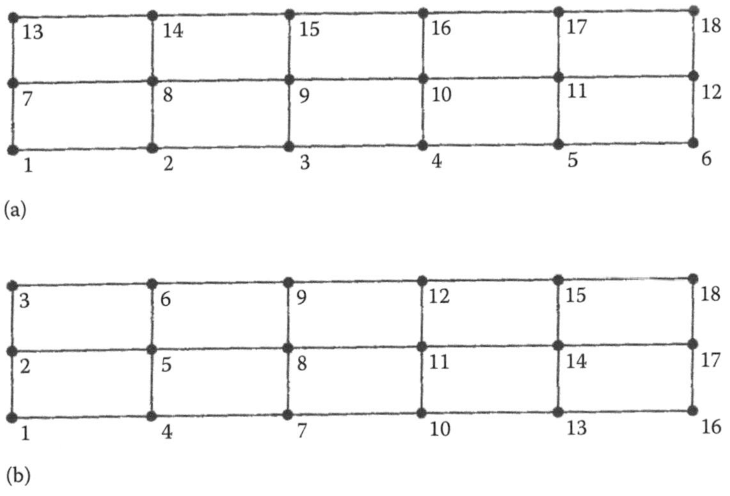

Figure 4.19 Node numbering to minimize bandwidth. (a) Poor numbering scheme, half-band-width = 8. (b) Good numbering scheme, half-bandwidth = 5.

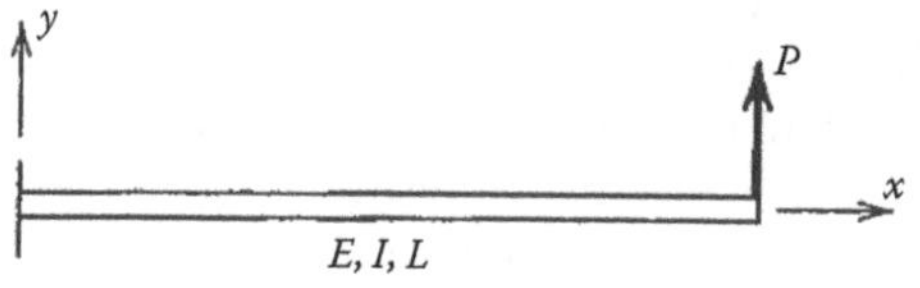

Figure P4.1 A transverse load P applied to the end of a cantilever beam.

4.8 PROBLEMS

4.8.1 Section 4.2

4.1. A transverse load P is applied to the end of a cantilever beam (Figure P4.1). The beam has length L, moment of inertia I, and modulus of elasticity E. The displaced shape of the beam is assumed to be of the following forms:

(1) $v(x) = a_0 + a_1x + a_2x^2$

(2) $v(x) = b_0\left(1 - \cos\dfrac{\pi x}{2l}\right)$

(3) $v(x) = c_0 + c_1x + c_2x^2 + c_3x^3$

Consider only strain energy due to bending as given by Equation 4.65 and the potential of the load [$\Omega = -Pv(L)$] with respect to the undeformed beam.

(a) To the extent possible, simplify each of the assumed displaced shapes to account for the boundary conditions.

(b) Calculate the elastic strain energy U and the potential Ω of the external load P for each of the assumed displaced shapes.

(c) Solve for the parameters (a_0,..., c_3) using the principle of stationary potential energy, where for equilibrium $\delta\Pi = \delta U + \delta\Omega = 0$. *Hint:* The virtual displacement δv is first written in terms of a variation in the parameters (δa_0,..., δc_3). Then simultaneous equations are written from

$$\delta\Pi = \frac{\partial\Pi}{\partial a_0}\partial a_0 + \cdots = 0$$

(d) Compute values of Π and $v(L)$ for each of the assumed displaced shapes. Compare the values of $v(L)$ to each other and to the elasticity solution of $v(L) = PL^3/3EI$.

(e) Discuss the results.

4.2. For the CST element shown in Figure P4.2:

(a) Write the shape function for each node.

(b) Evaluate each shape function at point P.

(c) Show, numerically for each shape function, that the value of the shape function for node i is equal to the ratio A_{Pjk}/A_{ijk}, where A_{Pjk} is the area of triangle Pjk and A_{ijk} is the area of the element.

4.3. For the mesh shown in Figure 4.5, construct the Boolean connectivity matrix $[M]$ for elements 2, 3, and 4. Refer to Example 4.2.

4.4. For the mesh shown in Figure 4.5, assemble the complete stiffness matrix for the structure. Use the notation k_{ij}^e to represent each stiffness coefficient, where the superscript identifies the element number. Refer to Example 4.2.

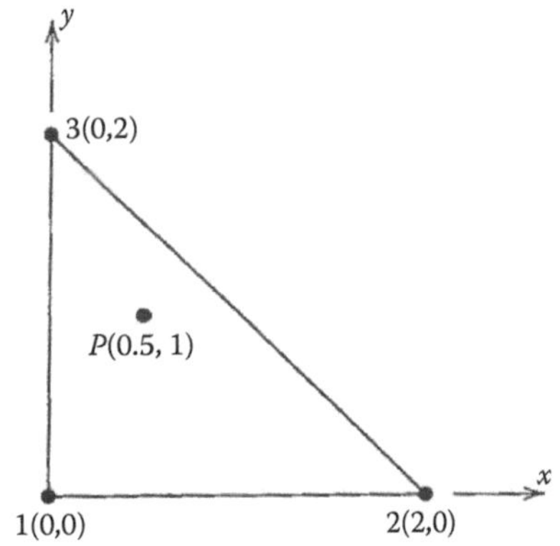

Figure P4.2 A CST element.

4.8.2 Section 4.4

4.5. A four-node isoparametric element has nodes at the following (x, y) coordinates: 1(0,0), 2(1,0), 3(2,2), 4(0,1).

(a) Sketch to scale the element and the lines for which $\xi = \pm\frac{1}{2}$, $\xi = \pm\frac{1}{4}$, $\eta = \pm\frac{1}{2}$, and = $\eta = \pm\frac{1}{4}$.

(b) Write the coordinate interpolation functions: $x(\xi,\eta) = \sum_{i=1}^{4} N_i(\xi,\eta)x_i$ and $y(\xi,\eta) = \sum_{i=1}^{4} N_i(\xi,\eta)y_i$.

(c) Compute the terms in the Jacobian matrix $[J]$ given by Equation 4.53.

(d) Evaluate $|J|$ at $\xi = 0$, $\eta = 0$. Compare this value to the ratio of the area of the element in (x, y) coordinates to that in (ξ, η) coordinates.

4.6. For the linear isoparametric element shown in Figure P4.6, compute $[B_1]$ at the point $\xi = 0$, $\eta = 0$.

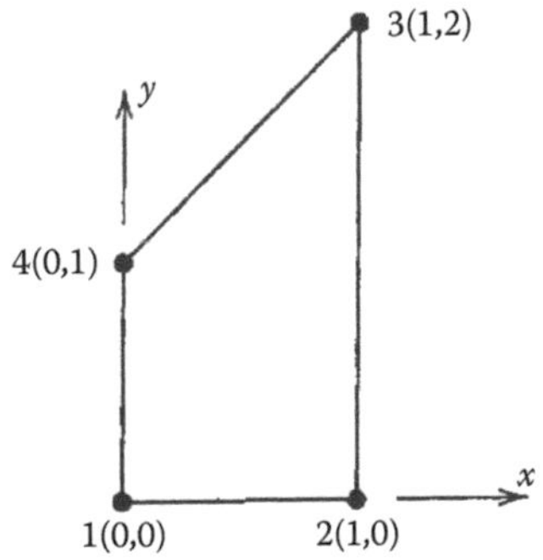

Figure P4.6 A linear isoparametric element.

4.7. Using the one-, two-, and three-point Gauss quadrature rules, evaluate the following integrals numerically. Compare the numerical results to the exact solutions.

(a) $I = \int_{-1}^{1} (6x^3 - 4x^2 + 3x - 2)\, dx$

(b) $I = \int_{-1}^{1} \cosh \xi \, d\xi$

(c) $I = \int_{-1}^{1} e^{\xi} \, d\xi$

4.8. Using the one-, two-, and three-point symmetric Gauss quadrature rules, evaluate the following integrals numerically. Compare the numerical results to the exact solutions.

(a) $I = \int_{-1}^{1}\int_{-1}^{1} \cos \xi \cos \eta \, d\xi \, d\eta$

(b) $I = \int_{-1}^{1}\int_{-1}^{1} \sin^2 \xi \cos \eta \, d\xi \, d\eta$

4.8.3 Section 4.5

4.9. Derive the equivalent nodal load vector for an axial rod element subjected to a concentrated axial force $\bar{P}_C$ acting at L_c from node 1 (see Figure P4.9).

4.10. Derive the equivalent nodal load vector for an axial rod subjected to a uniformly distributed axial force of magnitude $\bar{q}_0$ acting over the domain $L_a < \bar{x} < L_b$ (see Figure P4.10).

4.11. Derive the equivalent nodal load vector for a beam element subjected to a concentrated bending moment $\bar{M}_C$ acting at L_c from node 1 (see Figure P4.11).

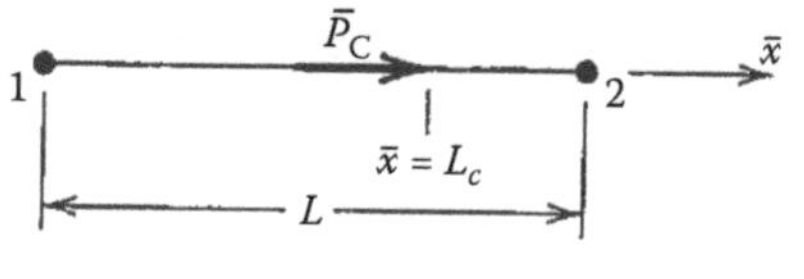

Figure P4.9 An axial rod element subjected to a concentrated axial force $\bar{P}_C$ acting at L_c from node 1.

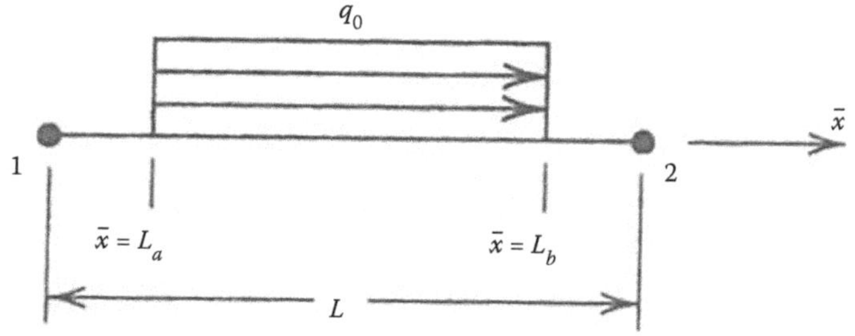

Figure P4.10 An axial rod subjected to a uniformly distributed axial force of magnitude $\bar{q}_0$ acting over the domain $L_a < \bar{x} < L_b$.

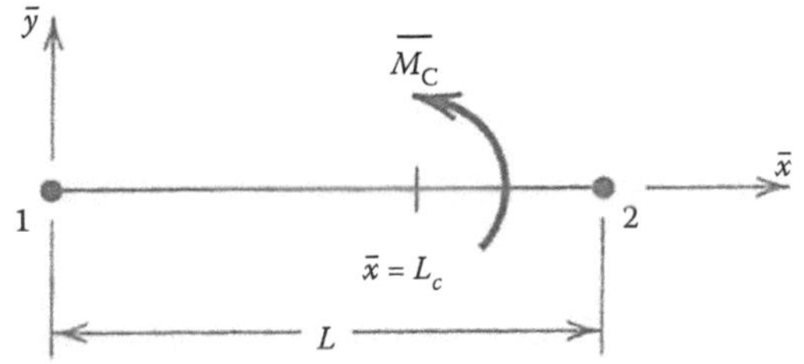

Figure P4.11 A beam element subjected to a concentrated bending moment $\overline{M}_C$ acting at L_c from node 1.

REFERENCES

AIAA. 1985. *Proc. Finite Element Standards Forum*, AIAA/ASME/ASCE/AHS 26th Structural Dynamics and Materials Conference, Apr. 15, Orlando, FL.

Boresi, A. P., Chong, K. P., and Lee, J. D. 2011. *Elasticity in Engineering Mechanics*, 3rd ed., Wiley, New York.

Boresi, A. P., and Schmidt, R. J., 2003. *Advanced Mechanics of Materials*, Wiley, New York.

Casey, J. 1998. On elastic-thermo-plastic materials at finite deformations, *Int. J. Plast.*, **14**, 173–191.

Chen, Y., Lee, J. D., and Eskandarian, A. 2002. Dynamic meshless method applied to nonlocal crack problems, *Theor. Appl. Fract. Mech*, **38**, 293–300.

Chen, Y., Lee, J. D., and Eskandarian, A. 2006. *Meshless Methods in Solid Mechanics*, Springer, New York.

Clough, R. W. 1960. The finite element method in plane stress analysis, *Proc. 2nd ASCE Conference on Electronic Computation*, Pittsburgh, PA, pp. 345–378.

Cook, R. D., Malkus, D. S., and Plesha, M. E. 2007. *Concepts and Applications of Finite Element Analysis*, 4th ed., Wiley, New York.

Courant, R. 1943. Variational methods for the solution of problems of equilibrium and vibrations, *Bull. Am. Math. Soc.*, **49**, 1–23.

Courant, R. 2011. *Differential and Integral Calculus*, Wiley, New York.

Eringen, A., C. 1980. *Mechanics of Continua*, Krieger, Malabar, FL.

Eringen, A., C. 2002. *Nonlocal Continuum Field Theories*, Springer, New York.

Green, A. E., and Naghdi, P. M. 1965. A general theory of an elastic-plastic continuum, *Arch. Ration. Mech. Anal*, **18**, 251–281.

Lee, J. D., and Chen, Y. 2001. A theory of thermo-visco-elastic-plastic materials: Thermomechanical coupling in simple shear, *Theor. Appl. Fract. Mech*, **35**, 187–209.

MacNeil, R. H., and Harder, R. L. 1984. A proposed standard set of problems to test finite element accuracy, *Proc. 25th Structural Dynamics and Materials Conference*, AIAA/ASME/ASCE/AHS, May 14, Palm Springs, CA.

MacNeil, R. H., and Harder, R. L. 1985. A proposed standard set of problems to test finite element accuracy, *Finite Elements Anal. Des.*, **1**(1), 3–20.

Simo, J. C, and Hughes, T. R. 1998. *Computational Inelasticity*, Springer, New York.

Turner, M. J., Clough, R. W, Martin, H. C., and Topp, L. J. 1956. Stiffness and deflection analysis of complex structures, *J. Aeronaut. Sci.* **25**(9), 805–823.

Zienkiewicz, O. C., and Taylor, R. L. 2005. *The Finite Element Method for Solid and Structural Mechanics*, Butterworth-Heinemann, Oxford, U.K.

BIBLIOGRAPHY

Baker, A. J., and Pepper, D. W. 1991. *Finite Elements 1–2–3*, McGraw-Hill, New York.

Bathe, K.-J. 2006. *Finite Element Procedures*, Klaus-Jurgen Bathe, Watertown, MA.

Belytschko, T., Liu, W. K., Moran, B., and Elkhodary, K. 2013. *Nonlinear Finite Elements for Continua and Structures*, Wiley, New York.

Bickford, W. B. 1990. *A First Course in the Finite Element Method*, Richard D. Irwin, Homewood, IL.

Burnett, D. S. 1987. *Finite Element Analysis, from Concepts to Applications*, Prentice Hall, Upper Saddle River, NJ.

Caendish, J. C, Field, D. A., and Frey, W. H. 1985. An approach to automatic three-dimensional finite element mesh generation, *Int. J. Numer. Methods Eng.*, **21**(2), 329–347.

Cheung, Y. K., Lo, S. H., and Leung, A. Y. T. 1996. *Finite Element Implementation*, Blackwell Science, Maiden, MA.

Courtin, S., Gardin, C., Bezine, G., and Hamouda, H. B. H. 2005. Advantages of the J-integral approach for calculating stress intensity factors when using the commercial finite element software ABAQUS, *Eng. Fract. Mech.*, **72**(14), 2174–2185.

De Borst, R., Crisfield, M. A., Remmers, J. J., and Verhoosel, C. V. 2012. *Nonlinear Finite Element Analysis of Solids and Structures*, John Wiley & Sons, Ltd., Chichester, U.K.

Dolbow, J. O. H. N., and Belytschko, T. 1999. A finite element method for crack growth without remeshing, *Int. J. Numer. Methods Eng.*, **46**(1), 131–150.

Finlayson, B. A. 1992. *Numerical Methods for Problems with Moving Fronts*, Ravenna Park Publishing, Seattle, WA.

Ghali, A., Neville, A. M., and Brown, T. G. 2003. *Structural Analysis: A Unified Classical and Matrix Approach*, 3rd ed., CRC Press, Boca Raton, FL.

Grandin, H., Jr. 1991. *Fundamentals of the Finite Element Method*, Waveland Press, Prospect Heights, IL.

Huebner, K. H. Dewhirst, D. L., Smith, D. E., and Byrom, T. G. 2008. *The Finite Element Method for Engineers*, 3rd ed., Wiley, New York.

Hughes, T. J. R. 2012. *The Finite Element Method: Linear Static and Dynamic Finite Element Analysis*, Courier Corp, North Chelmsford, MA.

Madenci, E., and Guven, I. 2015. *The Finite Element Method and Applications in Engineering Using ANSYS®*. Springer, New York.

Melenk, J. M., and Babuška, I. 1996. The partition of unity finite element method: Basic theory and applications, *Comput. Methods Appl. Mech. Eng.*, **139**(1), 289–314.

Melosh, R. J. 1990. *Structural Engineering Analysis by Finite Elements*, Prentice Hall, Upper Saddle River, NJ.

Potts, J. F., and Oler, J. W. 1989. *Finite Element Applications with Microcomputers*, Prentice Hall, Upper Saddle River, NJ.

Potts, D. M., Zdravkovic, L., and Zdravković, L. 2001. *Finite Element Analysis in Geotechnical Engineering: Application* (Vol. 2), Thomas Telford Publishing, London, U.K.

Przemieniecki, J. S. 1985. *Theory of Matrix Structural Analysis*, Dover Publications, Mineola, NY.

Rao, S. S. 2013. *The Finite Element Method in Engineering*, Pergamon Press, Oxford.

Reddy, J. N. 2014. *An Introduction to Nonlinear Finite Element Method*, Oxford University Press, Oxford, U.K.

Rybicki, E. F., and Kanninen, M. F. 1977. A finite element calculation of stress intensity factors by a modified crack closure integral, *Eng. Fract. Mech.*, **9**(4), 931–938.

Sack, R. L. 1994. *Matrix Structural Analysis*, Waveland Press, Prospect Heights, IL.

Segerlind, L. J. 1984. *Applied Finite Element Analysis*, Wiley, New York.

Stasa, F. L. 1995. *Applied Finite Element Analysis for Engineers*, HBJ College & School Division, Holt, Rinehart and Winston, Austin, TX.

Weaver, W., Jr., and Gere, J. M. 1990. *Matrix Analysis of Framed Structures*, 3rd ed., Van Nostrand Reinhold, New York.

Whirley, R. G., and Engelmann, B. E. 1993. *DYNA3D: A Nonlinear, Explicit, Three-Dimensional Finite Element Code for Solid and Structural Mechanics, User Manual. Revision 1* (No. UCRL-MA–107254-Rev. 1), Lawrence Livermore National Lab., Livermore, CA.

Yang, T. Y. 1986. *Finite Element Structural Analysis*, Prentice Hall, Upper Saddle River, NJ.

Chapter 5

Specialized methods

5.1 INTRODUCTION

The finite strip method (FSM), pioneered in 1968 by Y. K. Cheung (1968a, 1968b), is an efficient tool for analyzing structures with regular geometric planform and simple boundary conditions. Basically, the FSM reduces a two-dimensional problem to a one-dimensional problem. In some cases, computational savings by a factor of 10 or more are possible compared to the finite element method (Cheung, 2013).

Originally, the FSM was designed for rectangular plate problems (similar to Levy's solution; Timoshenko and Woinowsky-Krieger, 1971). Later, the FSM was extended to treat curved plates (Cheung, 1969b), skewed (quadrilateral) plates, folded plates, and box girders. Formulated as an eigenvalue problem, the FSM can be applied to vibration and stability problems of plates and shells with relative ease. Finite prism and finite layer methods (FLMs) were also introduced in the 1970's by Cheung (2013). These methods reduce three-dimensional problems to two- and one-dimensional problems, respectively, by choosing an appropriate choice of displacement functions. In addition, composite structures, such as sandwich panels with cold-formed facings, can be analyzed efficiently by coupling the FSM with the finite prism or FLMs (Cheung et al., 1982b; Chong, 1986; Chong et al., 1982a, 1982b; Tham et al., 1982).

The FSM has been modified through the use of spline functions to analyze plates of arbitrary shape (Cheung et al., 1982a; Chong and Chen, 1986; Li et al., 1986; Tham et al., 1986; Yang and Chong, 1982, 1984). Complicated boundary conditions can be accommodated. In general, FSMs based on spline functions are more involved than the FSM based on trigonometric series. However, they are still more efficient than finite element methods, and they require less input and less computational effort.

5.2 FINITE STRIP METHOD

The FSM, first proposed by Y. K. Cheung (1968a, 1968b), is approaching a state of maturity as a structural analysis technique (Puckett et al., 1987; Wiseman et al., 1987). Two comprehensive books on the method and its applications are available (Cheung, 2013; Loo and Cusens, 1978), as are papers on advances in the field (Cheung, 1981; and others). The paper by Wiseman et al. (1987) summarizes the developments in the methods of finite strips, finite layers, and finite prisms and includes a review of 114 references.

To examine the FSM, consider a rectangular plate with x and y axes in the plane of the plate and axis z in the thickness direction (Figure 5.1). Let the corresponding displacement components be denoted by (u, v, w). Then, similar to Levy's solution (Timoshenko and

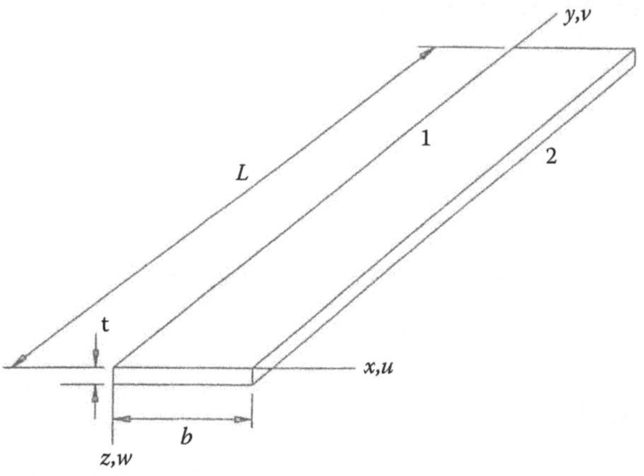

Figure 5.1 Rectangular bending strips.

Woinowsky-Krieger, 1971), for a typical strip (Figure 5.1), the w displacement component is (Cheung, 2013)

$$w(x,y) = \sum_{m=1}^{r} f_m(x) Y_m(y) \tag{5.1}$$

in which the functions $f_m(x)$ are polynomials and the functions $Y_m(y)$ are trigonometric terms that satisfy the end conditions in the y direction. The functions Y_m can be taken as basic functions (mode shapes) of the beam vibration equation

$$\frac{d^4Y}{dx^4} - \frac{\mu^4}{a^4} Y = 0 \tag{5.2}$$

where a is the beam (strip) length and μ is a parameter related to frequency, material, and geometric properties.

The general solution of Equation 5.2 is

$$Y(y) = C_1 \sin\frac{\mu y}{a} + C_2 \cos\frac{\mu y}{a} + C_3 \sinh\frac{\mu y}{a} + C_4 \cosh\frac{\mu y}{a} \tag{5.3}$$

Four boundary conditions are needed to determine the coefficients C_1 to C_4. For example, for both ends simply supported,

$$Y(0) = Y''(0) = Y(a) = Y''(a) = 0 \tag{5.4}$$

Equations 5.3 and 5.4 yield the mode shape functions

$$Y_m(y) = \sin\frac{\mu_m y}{a} \qquad m = 1,2,3,\ldots,r \tag{5.5}$$

where $\mu_m = m\pi$; $m = 1, 2, 3,\ldots, r$.

Since the functions Y_m are mode shapes, they are orthogonal (Meirovitch, 1986); that is, they satisfy the relations

$$\int_0^a Y_m Y_n \, dy = 0 \qquad \text{for } m \neq n \tag{5.6}$$

Also, it can be shown that (Cheung, 2013)

$$\int_0^a Y_m'' Y_n'' dy = 0 \qquad \text{for } m \neq n \tag{5.7}$$

The orthogonal properties of Y_m result in structural matrices with very narrow bandwidths, thus minimizing computational storage and computational time.

Similarly to the finite element method, the functions $f_m(x)$ in Equation 5.1 can be expressed as

$$f_m(x) = [[C_1] \quad [C_2]]\begin{Bmatrix} \{\delta_1\} \\ \{\delta_2\} \end{Bmatrix}_m \tag{5.8}$$

where the subscripts 1 and 2 denote sides 1 and 2 of the plate (strip), respectively; $[C_1]$ and $[C_2]$ are interpolating functions, equivalent to shape functions in one-dimensional finite elements; and $\{\delta_1\}$ and $\{\delta_2\}$ are nodal parameters.

Let b be the width of a strip in the plate and let $\bar{x} = x/b$. Taking the functions $f_m(x)$ as linear functions of x, and considering nodal displacements only, we have

$$\delta_1 = w_1, \qquad \delta_2 = w_2 \qquad C_1 = 1 - \bar{x} \qquad C_2 = \bar{x}$$

Hence, by Equation 5.8, we have

$$f_m(x) = [(1-\bar{x}) \quad \bar{x}]\begin{Bmatrix} w_1 \\ w_2 \end{Bmatrix} = (1-\bar{x})w_1 + \bar{x}w_2 \tag{5.9}$$

Equation 5.9 is equivalent to a one-dimensional linear finite element (Chapter 4), which employs nodal displacements only.

For higher-order functions, with nodal displacements w_i and first derivatives (nodal slopes) $\theta_i = w_i'$, we have

$$\{\delta_1\} = \begin{Bmatrix} w_1 \\ \theta_1 \end{Bmatrix} \qquad \{\delta_2\} = \begin{Bmatrix} w_2 \\ \theta_2 \end{Bmatrix}$$
$$[C_1] = [(1 - 3\bar{x}^2 + 2\bar{x}^3) \qquad x(1 - 2\bar{x} + \bar{x}^2)]$$
$$[C_2] = [(3\bar{x}^2 - 2\bar{x}^3) \qquad x(\bar{x}^2 - \bar{x})] \tag{5.10}$$

By Equations 5.8 and 5.10, we obtain

$$f_m(x) = [(1 - 3\bar{x}^2 + 2\bar{x}^3) \qquad x(1 - 2\bar{x} + \bar{x}^2) \qquad (3\bar{x}^2 - 2\bar{x}^3) \qquad x(\bar{x}^2 - \bar{x})] \times \begin{Bmatrix} w_1 \\ \theta_1 \\ w_2 \\ \theta_2 \end{Bmatrix} \tag{5.11}$$

Equation 5.11 is equivalent to a one-dimensional beam element, which employs both nodal displacements and nodal slopes. Other higher-order functions can be derived in a similar manner (Cheung, 2013).

5.3 FORMULATION OF THE FSM

In general (see Equations 5.1 and 5.8), the displacement function may be written as

$$w = \sum_{m=1}^{r} Y_m \sum_{k=1}^{s} [C_k]\{\delta_k\}_m \tag{5.12}$$

in which r is the number of mode shape functions (Equation 5.5) and s is the number of nodal line parameters.

Let

$$[N_k]_m = Y_m[C_k] \tag{5.13}$$

Then, by Equation 5.12,

$$w = \sum_{m=1}^{r} \sum_{k=1}^{s} [N_k]_m \{\delta_k\}_m = [N]\{\delta\} \tag{5.14}$$

where $[N]$ denotes the shape functions and $\{\delta\}$ denotes the nodal parameters.

The formulation of the FSM is similar to that of the finite element method (Section 4.2). For example, for a strip subjected to bending, the strain matrix (vector), {ε}, is given by

$$\{\varepsilon\} = \begin{Bmatrix} M_x \\ M_y \\ M_{xy} \end{Bmatrix} = \begin{Bmatrix} -\partial^2 w/\partial x^2 \\ -\partial^2 w/\partial y^2 \\ 2\partial^2 w/\partial x\,\partial y \end{Bmatrix} \tag{5.15}$$

where M_x, M_y, and M_{xy} are moments per unit length.

Differentiating w, Equation 5.14, and substituting into Equation 5.15, we obtain the result

$$\{\varepsilon\} = [B]\{\delta\} = \sum_{m=1}^{r} [B]_m \{\delta\}_m \tag{5.16}$$

in which

$$[B]_m = \begin{Bmatrix} -\partial^2 [N]_m/\partial x^2 \\ -\partial^2 [N]_m/\partial y^2 \\ 2\partial^2 [N]_m/\partial x\,\partial y \end{Bmatrix} \tag{5.17}$$

Equation 5.16 is similar to Equation 4.25.

By Hooke's law (Boresi et al., 2011) and Equation 5.16, the stress matrix (vector), {σ}, is

$$\begin{aligned}\{\sigma\} &= [D]\{\varepsilon\} \\ &= [D]\sum_{m=1}^{r} [B]_m \{\delta\}_m\end{aligned} \tag{5.18}$$

where $[D]$ is the elasticity matrix, defined in Section 4.2 (Equations 4.10–4.12) for isotropic plane strain and plane stress problems.

Minimization of the total potential energy gives (Cheung, 2013)

$$[K]\{\delta\} = \{F\} \tag{5.19}$$

where

$$[K] = \int_{\text{vol}} [B]^{\mathrm{T}}[D][B]\,dv \tag{5.20}$$

is the stiffness matrix and $\{F\}$ is the load vector.

For a distributed load $\{q\}$, the load vector is

$$\{F\} = \int_A [N]^{\mathrm{T}}\{q\}\, dA \tag{5.21}$$

Expanding Equation 5.20, we obtain the stiffness matrix in the form

$$\begin{aligned}
[K] &= \int_{\mathrm{vol}} [[B]_1^{\mathrm{T}}[B]_2^{\mathrm{T}} \cdots [B]_r^{\mathrm{T}}][D][[B]_1[B]_2 \cdots [B]_r]dv \\
&= \int_{\mathrm{vol}} \begin{bmatrix} [B]_1^{\mathrm{T}}[D][B]_1 & [B]_1^{\mathrm{T}}[D][B]_2 & \cdots [B]_1^{\mathrm{T}}[D][B]_r \\ \vdots & \vdots & \vdots \\ [B]_r^{\mathrm{T}}[D][B]_1 & [B]_r^{\mathrm{T}}[D][B]_2 & \cdots [B]_r^{\mathrm{T}}[D][B]_r \end{bmatrix} dv \\
&= \begin{bmatrix} [k]_{11} & [k]_{12} \cdots & [k]_{1r} \\ \vdots & \vdots & \vdots \\ [k]_{r1} & [k]_{r2} \cdots & [k]_{rr} \end{bmatrix}
\end{aligned} \tag{5.22}$$

where

$$[k]_{mn} = \int_{\mathrm{vol}} [B]_m^{\mathrm{T}}[D][B]_n\, dv \tag{5.23}$$

For each strip (with s nodal line parameters),

$$[k]_{mn} = \begin{bmatrix} [k_{11}] & [k_{12}] & \cdots & [k_{1s}] \\ \vdots & \vdots & & \vdots \\ [k_{s1}] & [k_{s2}] & \cdots & [k_{ss}] \end{bmatrix}_{mn}$$

In general, the individual elements of matrix $[k]_{mn}$ are given by

$$[k_{ij}]_{mn} = \int_{\mathrm{vol}} [B_i]_m^{\mathrm{T}}[D][B_j]_n\, dv \tag{5.24}$$

Similarly, the basic element of the load vector (for distributed loads) is

$$\{F_i\} = \int_A [N_i]_m^{\mathrm{T}}\{q\}dA \tag{5.25}$$

For simple functions Equations 5.24 and 5.25 can be evaluated in closed form (Cheung, 2013). Alternatively, they can be integrated numerically using Gaussian quadrature or other numerical integration methods.

5.4 EXAMPLE OF THE FSM

In this section, the FSM is applied to the analysis of a uniformly loaded plate simply supported on two parallel edges. Each nodal line is free to move in the z direction and rotate about the y axis (Figure 5.1). Thus, Equations 5.1, 5.5, and 5.10 are to be used. The matrix $[B]_m$, Equations 5.17 and 5.23, is (Cheung, 2013)

$$[B]_m = \begin{bmatrix} \frac{6}{b^2}(1-2\bar{x})Y_m & \frac{2}{b}(2-3\bar{x})Y_m & \frac{6}{b^2}(-1+2\bar{x})Y_m & \frac{2}{b}(-3\bar{x}+1)Y_m \\ -(1-3\bar{x}^2+2\bar{x}^3)Y''_m & -x(1-2\bar{x}+\bar{x}^2)Y''_m & -(3\bar{x}^2-2\bar{x}^3)Y''_m & -x(\bar{x}^2-\bar{x})Y''_m \\ \frac{2}{b}(-6\bar{x}+6\bar{x}^2)Y'_m & 2(1-4\bar{x}+3\bar{x}^2)Y'_m & \frac{2}{b}(6\bar{x}-6\bar{x}^2)Y'_m & 2(3\bar{x}^2-2\bar{x})Y'_m \end{bmatrix}$$

For orthotopic plates, the elasticity matrix is (Boresi et al., 2011; Cheung, 2013)

$$[D] = \begin{bmatrix} D_x & D_1 & 0 \\ D_1 & D_y & 0 \\ 0 & 0 & D_{xy} \end{bmatrix} \tag{5.26}$$

where

$$D_x = \frac{E_x t^3}{12(1-\nu_x\nu_y)} \qquad D_y = \frac{E_x t^3}{12(1-\nu_x\nu_y)} \qquad D_{xy} = \frac{Gt^3}{12}$$
$$D_1 = \frac{\nu_x E_y t^3}{12(1-\nu_x\nu_y)} = \frac{\nu_y E_x t^3}{12(1-\nu_x\nu_y)} \tag{5.27}$$

For isotropic materials,

$$E_x = E_y = E, \quad \nu_x = \nu_y = \nu \qquad \text{and} \qquad G = \frac{E}{2(1+\nu)} \tag{5.28}$$

By the orthogonal properties, $[k]_{mn} = 0$ if $m \neq n$. Thus, Equation 5.19 reduces to

$$\begin{bmatrix} [k]_{11} & & & \\ & [k]_{22} & & \\ & & \ddots & \\ & & & [k]_{rr} \end{bmatrix} \begin{Bmatrix} w_1 \\ \theta_1 \\ w_2 \\ \theta_2 \\ \vdots \\ w_{r+1} \\ \theta_{r+1} \end{Bmatrix} = \begin{Bmatrix} \{F_1\} \\ \\ \{F_2\} \\ \vdots \\ \{F_r\} \end{Bmatrix} \tag{5.29}$$

where the matrices $[k]_{mm}$ are 4 × 4 symmetrical matrices (Cheung, 1998). The load vector $\{F_m\}$ can be computed by integrating Equation 5.25. It is a 4 × 1 matrix given by

$$\{F\} = q \begin{Bmatrix} b/2 \\ b^2/12 \\ b/2 \\ -b^2/12 \end{Bmatrix} \int_0^a \sin\frac{m\pi y}{a}\,dy \tag{5.30}$$

The load matrices $\{F\}_m$ for eccentric and concentric concentrated loads have been derived around 1970 by Cheung (2013). With two degrees of freedom (w, θ) along the nodal line, the moments (Equation 5.15) are linear functions of x. Thus, to obtain the maximum moment at a known location (e.g., at the center for uniform load and under load point for concentrated load), a nodal line must pass through this location for good accuracy. In a convergence test applied to a uniformly loaded simply supported square plate, Cheung (2013) has shown that with four strips (10 equations), greater accuracy was obtained than that by the finite element method with 25 nodes (75 equations). In addition, considerable saving in computational labor is achieved (Yang and Chong, 1984). Higher-order strips have also been applied to plates (Cheung, 2013). Finite strip formulations for curved and skewed plates have also been derived (Cheung, 2013). These techniques are applicable to the analysis of curved bridges, stiffened plates, and so on.

5.5 FINITE LAYER METHOD

By selecting functions satisfying the boundary conditions in two directions, the philosophy of the FSM can be extended to layered systems. The resulting method is called the *finite layer method* (FLM). The method was first proposed by Cheung and Chakrabarti (1971). The FLM is useful for layered materials, rectangular in planform. To illustrate the method, consider Figure 5.2. Let $\bar{z} = z/c$. Then the lateral displacement component, w, can be expressed as

$$w = \sum_{m=1}^{r}\sum_{n=1}^{t}[(1-\bar{z})w_{1mn} + \bar{z}w_{2mn}]X_m(x)Y_n(y) \tag{5.31}$$

in which, w_{1mn} and w_{2mn} are displacement parameters for side 1 (top) and side 2 (bottom), respectively. The displacement component, w, is assumed to vary linearly in the z direction.

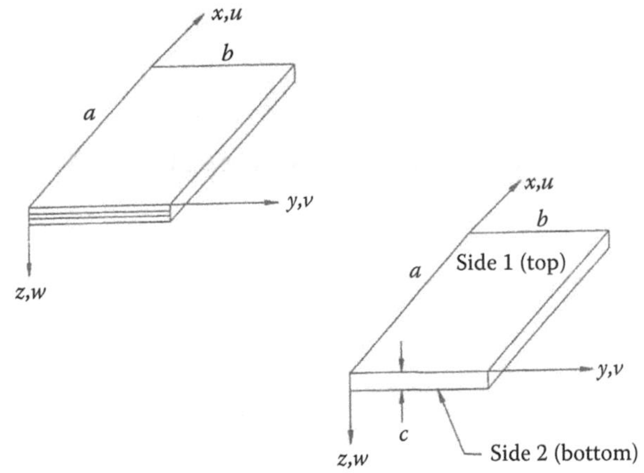

Figure 5.2 Finite layers.

The functions X_m and Y_n are taken as terms in a trigonometric series, satisfying the boundary conditions. In this manner a three-dimensional problem is reduced to a one-dimensional problem with considerable saving in computer storage and computational time (Cheung, 2013; Cheung et al., 1982b).

By linear theory, the (x, y) displacement components (u, v) are linearly related to the derivatives of w; that is,

$$u = A\frac{\partial w}{\partial x} \tag{5.32}$$

and

$$v = B\frac{\partial w}{\partial y} \tag{5.33}$$

Thus, as with Equation 5.31, we may write

$$\begin{aligned} v &= \sum_{m=1}^{r}\sum_{n=1}^{t}[(1-\bar{z})v_{1mn} + \bar{z}v_{2mn}]X_m(x)Y_n'(y) \\ u &= \sum_{m=1}^{r}\sum_{n=1}^{t}[(1-\bar{z})u_{1mn} + \bar{z}u_{2mn}]X_m'(x)Y_n(y) \end{aligned} \tag{5.34}$$

in which u_{1mn}, v_{1mn}, u_{2mn} and v_{2mn} are displacement parameters of sides 1 and 2.

The displacement is

$$\{f\} = \begin{Bmatrix} u \\ v \\ w \end{Bmatrix} = \sum_{m=1}^{r}\sum_{n=1}^{t}[N]_{mn}\{\delta\}_{mn} = [N]\{\delta\} \tag{5.35}$$

where

$$\{\delta\}_{mn} = [u_{1mn}, v_{1mn}, w_{1mn}, u_{2mn}, v_{2mn}, w_{2mn}]^T \tag{5.36}$$

The strain–displacement relationship is (Boresi et al., 2011)

$$\{\varepsilon\} = \begin{Bmatrix} \varepsilon_{xx} \\ \varepsilon_{yy} \\ \varepsilon_{zz} \\ \varepsilon_{xy} \\ \varepsilon_{yz} \\ \varepsilon_{zx} \end{Bmatrix} = \begin{Bmatrix} \partial u/\partial x \\ \partial v/\partial y \\ \partial w/\partial z \\ \frac{1}{2}(\partial u/\partial y + \partial v/\partial x) \\ \frac{1}{2}(\partial v/\partial z + \partial w/\partial y) \\ \frac{1}{2}(\partial w/\partial x + \partial u/\partial z) \end{Bmatrix} = \sum_{m=1}^{r}\sum_{n=1}^{t} [B]_{mn}\{\delta\}_{mn} = [B]\{\delta\} \tag{5.37}$$

The stress–strain relationship is

$$\{\sigma\} = [\sigma_{xx}\ \sigma_{yy}\ \sigma_{zz}\ \sigma_{xy}\ \sigma_{yz}\ \sigma_{zx}]^T = [D][B]\{\delta\} \tag{5.38}$$

Proceeding as in Section 5.4, we obtain the stiffness matrix (Cheung, 2013)

$$[S] = \int_0^a \int_0^b \int_0^c [B]^{\mathrm{T}}[D][B]\,dx\,dy\,dz \tag{5.39}$$

For a simply supported rectangular plate, we have the conditions

$$\int_0^a \int_0^b \int_0^c [B]_{mn}^T [D][B]_{rs}\,dx\,dy\,dz = 0 \quad \text{for} \quad mn \neq rs \tag{5.40}$$

Hence, the off-diagonal terms are zero, and the stiffness matrix reduces to

$$[K] = \int_0^a \int_0^b \int_0^c \begin{bmatrix} [B]_{11}^{\mathrm{T}}[D][B]_{11} & & & & & & \\ & [B]_{12}^{\mathrm{T}}[D][B]_{12} & & & & & \\ & & \ddots & & & & \\ & & & [B]_{1t}^{\mathrm{T}}[D][B]_{1t} & & & \\ & & & & [B]_{21}^{\mathrm{T}}[D][B]_{21} & & \\ & & & & & \ddots & \\ & & & & & & [B]_{rt}^{\mathrm{T}}[D][B]_{rt} \end{bmatrix} dx\,dy\,dz \tag{5.41}$$

The remaining formulation is similar to that of Sections 5.3 and 5.4.

5.6 FINITE PRISM METHOD

Similarly to Equation 5.1, for prismatic members with rectangular planform, the three-dimensional problem can be reduced to a two-dimensional problem by expressing the displacement function f as (Cheung, 2013)

$$f = \sum_{m=1}^{r} f_m(x,z)Y_m \tag{5.42}$$

where $f_m(x, z)$ is a function of (x, z) only and Y_m is a trigonometric function of y, which satisfy the boundary conditions in the y direction. In formulating the finite prism method (FPM), including nodal displacements and shape functions as in the finite element method, it is convenient to use nodal coordinates. Therefore, let ξ, η be the local coordinates of an element, ξ_k, η_k the nodal coordinates of the element, δ the displacement of a point in the element, δ_k the nodal displacements, and (ϕ_k, ψ_k) functions associated with a particular coordinate system (such as Cartesian, skew, or curvilinear). Then we can represent the local coordinates in terms of nodal coordinates in the form

$$\xi = \sum_{k=1}^{s} \phi_k \xi_k \tag{5.43}$$

where s refers to the number of nodes of the element.

Similarly, the displacement δ of a point within the element can be expressed in terms of the nodal displacements δ_k as

$$\delta = \sum_{k=1}^{s} \psi_k \delta_k \tag{5.44}$$

In general, $\phi_k \neq \psi_k$; however, if $\phi_k = \psi_k$, the element is termed *isoparametric*. Using Equation 5.42, for a prismatic member with two ends simply supported, the lateral (out-of-plane) displacement component is

$$w = \sum_{m=1}^{r} w_m(x,z)\sin k_m y \tag{5.45}$$

in which

$$w_m(x,z) = \sum_{k=1}^{s} C_k w_{km} \tag{5.46}$$

and w_{km} are the mth term nodal displacements at the kth node, the C_k are the shape functions for the two-dimensional element (in the ξ–η plane). For an isoparametric six-node (ISW'6) model (Figure 5.3), the shape functions are given as follows:

Corner nodes:

$$C_k = \frac{1}{4}\eta_k\eta(1+\eta_k\eta)(1+\xi_k\xi) \tag{5.47}$$

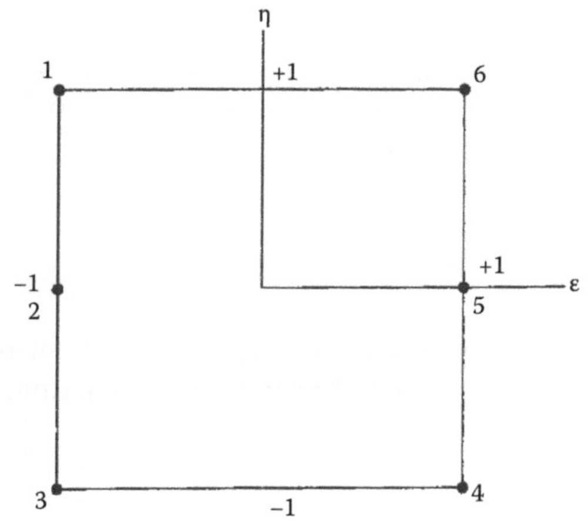

Figure 5.3 ISW'6 model.

Midside nodes:

$$C_k = \frac{1}{2}(1+\xi_k\xi)(1-\eta^2) \tag{5.48}$$

Therefore, the lateral (out-of-plane) displacement component for the isoparametric six-node model is

$$w = \sum_{m=1}^{r}\sum_{k=1}^{6} C_k w_{km} \sin k_m y \tag{5.49}$$

The corresponding in-plane displacement components (u, v) are

$$u = \sum_{m=1}^{r}\sum_{k=1}^{6} C_k u_{km} \sin k_m y \tag{5.50}$$

and

$$v = \sum_{m=1}^{r}\sum_{k=1}^{6} C_k v_{km} \cos k_m y \tag{5.51}$$

The stiffness matrix is developed in a manner similar to the development in Sections 5.3 and 5.4. It is (Chong et al., 1982b)

$$^{\mathrm{P}}K_{ijmn} = \int {}^{\mathrm{P}}B_{im}^{\mathrm{T}}\,{}^{\mathrm{P}}D\,{}^{\mathrm{P}}B_{jn}\,d(\mathrm{vol}) \tag{5.52}$$

where

$$
{}^{P}B_{im} = \begin{bmatrix}
\frac{\partial C_i}{\partial x}\sin\frac{m\pi y}{L} & 0 & 0 \\
0 & -C_i\frac{m\pi}{L}\sin\frac{m\pi y}{L} & 0 \\
0 & 0 & \frac{\partial C_i}{\partial z}\sin\frac{m\pi y}{L} \\
C_i\frac{m\pi}{L}\cos\frac{m\pi y}{L} & \frac{\partial C_i}{\partial x}\cos\frac{m\pi y}{L} & 0 \\
0 & \frac{\partial C_i}{\partial x}\cos\frac{m\pi y}{L} & C_i\frac{m\pi}{L}\cos\frac{m\pi y}{L} \\
\frac{\partial C_i}{\partial x}\sin\frac{m\pi y}{L} & 0 & \frac{\partial C_i}{\partial z}\sin\frac{m\pi y}{L}
\end{bmatrix} \tag{5.53}
$$

${}^{P}D$ is the elasticity matrix for isotropic materials. In Equation 5.52, the superscript P indicates the finite prism model.

The geometric transformation from the natural coordinate (x, z) to the local coordinate (ξ, η) can be carried out as in the standard finite element method, and the stiffness matrix can be obtained accordingly (Cheung, 2013).

5.7 APPLICATIONS AND DEVELOPMENTS OF FSM, FLM, AND FPM

Owing to their narrow bandwidth and reduction in dimensions, the finite strip, finite layer, and finite prism methods (FSM, FLM, and FPM) are especially adaptable for personal computers (Rhodes, 1987) and minicomputers, with significant savings in computational labor. Developments and applications are presented by Wiseman et al. (1987) and Graves-Smith (1987).

One of the first reported applications of the method was to orthotropic right box girder bridge decks (Cheung, 1969a). Cheung suggested that the FSM could be used in a composite slab-beam system, where the longitudinal beam stiffness could be derived separately and added to the structure stiffness. This procedure is a refinement to systems in which the action of longitudinal beams is approximated by assuming orthotropic properties for the slab. Cheung further asserted that the beam would have to coincide with a strip nodal line. This condition was verified by comparison with theoretical and model test results for simply supported slab-beam models with varying slab thicknesses (Cheung et al., 1970). A force method for introducing rigid column supports in the strips was also described.

An analytical method using a separation-of-variables procedure and series solution for the resulting differential equations was developed by Harik and his associates (1984, 1985, 1986) for orthotropic sectorial plates and for rectangular plates under various loadings. The plates were divided into strips where point and patch loads were applied, and the accuracy of the solutions was demonstrated for various boundary conditions.

Two refinements aimed at increasing the accuracy of finite strips in the in-plane transverse direction were developed by Loo and Cusens (1970). One refinement ensued curvature compatibility at the nodal lines. Although this requirement limited the analysis to plates with uniform properties across the section, convergence was shown to be faster for appropriate plate systems. The other refinement introduced an auxiliary nodal line within the strips. A drawback of these refinements is that they both increase the size of the stiffness matrix by 50%.

Other formulations for increasing the accuracy or the range of applicability of the FSM have been published. Brown and Ghali (1978) used a subparametric finite strip for the analysis of quadrilateral plates. Bucco et al. (1979) proposed a deflection contour method which allows the FSM to be used for any plate shape and loading for which the deflection contour is known.

Other applications of the FSM have used spline functions for the boundary component of the displacement function. Yang and Chong (1982, 1984) used *X*-spline functions, allowing extension of the FSM to plate bending problems with irregular boundaries. The solution was shown to converge correctly for a trapezoidal plate, and techniques for approximating a plate with more irregularly shaped sides and ends were discussed. The buckling of irregular plates has been investigated by Chong and Chen (1986). Extensive development of the FSM using the cubic *B*-spline as the boundary component of the displacement function was done by Cheung et al. (1982a). After initial application to flat plates and box girder bridges by Cheung and Fan (1983), the cubic *B*-spline method was extended to skewed plates by Tham et al. (1986), to curved slabs by Cheung et al. (1986), and to arbitrarily shaped general plates by Li et al. (1986). Shallow shells were analyzed by Fan and Cheung (1983) using a spline finite strip formulation based on the theory of Vlasov (1984).

The application of the FSM to folded plates by Cheung (1969c) extended the method to three-dimensional plate structures. The strips used were simply supported with respect to out-of-plane displacement and had similar in-plane longitudinal displacement functions. These end conditions suggest a real plate system having end diaphragms or supports with infinite in-plane stiffness. This system is an acceptable approximation for many structures, including roof systems resting on walls and box girder bridges with plate diaphragms.

A study of folded plates, continuous over rigid supports, has been presented by Delcourt and Cheung (1978). They represented longitudinal displacement functions as eigenfunctions of a continuous beam with the same span and relative span stiffnesses as the plate structure. For comparison, the longitudinal displacement function for a structure with pinned ends and no internal supports is a sine function.

The FSM has become popular for the analysis of slab girder and box girder bridges. Cheung and Chan (1978) used folded plate elements to study 392 theoretical bridge models in order to suggest refinements to American and Canadian design codes. Cusens and Loo (1974) have done extensive work in the application of the FSM to the analysis of prestressed concrete box girder bridges, and their book (Loo and Cusens, 1978) contains an extensive treatment of the subject with many examples.

The analysis of cylindrically orthotropic curved slabs using the FSM was described by Cheung (1969c). The primary difference between this technique and the FSM for right bridge decks is the use of a polar coordinate system for the displacement function. It was noted that a model in this coordinate system could be used to approximate a rectangular slab by specifying a very large radius for the plate. A radial shell element was subsequently developed (Cheung and Cheung, 1971a), which allowed the analysis of curved box girder bridges supported by rigid end diaphragms.

Because of its efficiency and ease of input, the FSM has been readily adapted to free vibration and buckling problems. The formulation for frequency analysis was outlined by Cheung et al. (1971). Characteristic functions for strips with varying end conditions have also been developed (Cheung and Cheung, 1971b). Finite strips and prisms were combined by Chong et al. (1982a, 1982b) to study free vibrations and to calculate the effects of different face

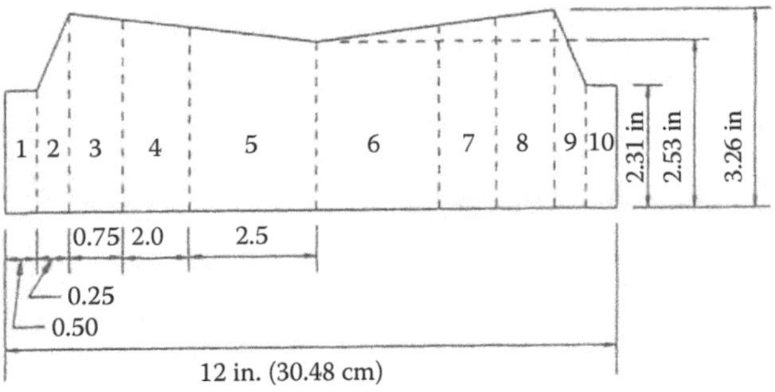

Figure 5.4 **Finite strips and prisms in a sandwich panel.**

temperatures on foam-core architectural sandwich panels (Figure 5.4). The panel facings were modeled by finite strips and the panel core by finite prisms. The eigenvalue problem for free vibrations is characterized by the equation (Chong et al., 1982a)

$$-\bar{\omega}^2\, {}^{S}M_m^{P}\, {}^{S}\bar{\delta}_m^{P} - \bar{\omega}^2\, {}^{S}M_m^{B}\, {}^{S}\bar{\delta}_m^{B} - \bar{\omega}^2\, {}^{P}M_m\, {}^{P}\bar{\delta}_m + {}^{S}K_m^{P}\, {}^{S}\bar{\delta}_m^{P} + {}^{S}K_m^{B}\, {}^{S}\bar{\delta}_m^{B} + {}^{P}K_m\, {}^{P}\bar{\delta}_m = 0 \qquad (5.54)$$

where $\bar{\omega}$ is the frequency, ${}^{S}M_m^{P}$ is the in-plane mass matrix of the bending strip, ${}^{S}M_m^{B}$ is the bending mass matrix of the bending strip, ${}^{P}M_m$ is the mass matrix of the finite prism, ${}^{S}K_m^{P}$ is the in-plane stiffness matrix of the bending strip, ${}^{S}K_m^{B}$ is the bending stiffness matrix of the bending strip, ${}^{P}K_m$ is the stiffness matrix of the finite prism, ${}^{S}\bar{\delta}_m^{P}$ and ${}^{S}\bar{\delta}_m^{B}$ are the in-plane and bending interpolation parameters of the bending strip, respectively, and ${}^{P}\bar{\delta}_m$ is the interpolating parameter of the finite prism.

Yoshida (1971) was among the first to develop the geometric stiffness matrix for buckling of rectangular plates. Przemieniecki (1972) used the same displacement function as in the FSM (Cheung, 2013) for the analysis of local stability in plates and stiffened panels. Chong and Chen (1986) investigated the buckling of nonrectangular plates by spline finite strips. Cheung et al. (1982b) used the FLM to derive buckling loads of sandwich plates. Mahendran and Murray (1986) modified the standard finite strip displacement function to include out-of-phase displacement parameters, allowing shearing loads to be applied to the strips in the buckling analysis.

The FSM has been used as the basis for parametric and theoretical studies of column buckling. Both local and overall buckling of H-columns with residual stress under axial load was studied by Yoshida and Maegawa (1978). Buckling of columns with I-sections was studied by Hancock (1981). The interaction of the two modes of buckling for other prismatic columns was also investigated by Sridharan and Benito (1984) using the FSM. Hasegawa and Maeno (1979) determined the effects of stiffeners on cold-formed steel shapes using finite strips. The postbuckling behavior of columns has been the subject of papers by Graves-Smith and Sridharan (1979), Sridharan and Graves-Smith (1981), and Graves-Smith and Gierlinski (1982).

The compound strip method, an enhancement of the flexibility and applicability of the FSM, was developed by Puckett and Gutkowski (1986). This formulation added both longitudinal and transverse beams at any location in a strip and included the effect of column supports. A stiffness formulation, which included the added stiffness of all elements associated with a beam, eliminated the repetitive solution passes required for each redundant in a flexibility analysis. The strain energies for flexure and torsion of the beams as well as for axial and flexural deformation of the columns were derived in terms consistent with the strip displacement function, allowing the total strip stiffness to be calculated by adding the element stiffnesses. With the compound strip method, the analysis of continuous beams can be performed using simple displacement functions previously suitable only for single spans. The compound strip method was extended to curved plate systems and was used for dynamic analysis by Puckett and Lang (1986).

Gutkowski et al. (1991) developed a cubic *B*-spline FSM for the analysis of thin plates. Spline series with unequal spacing allow local discretization refinement near patch and concentrated loads. Oscillatory convergence (Gibb's phenomenon) is avoided. Spline finite strips have also been used to analyze non-prismatic space structures (Tham, 1990).

Transient responses of noncircular cylindrical shells subject to shock waves were studied by Cheung et al. (1991) using finite strips and a coordinate transformation. Arbitrarily shaped sections were mapped approximately into circular cylindrical shells in which finite strips can be applied for discretization. The FSM was used for the free vibration and buckling analysis of plates with abrupt thickness changes and a nonhomogeneous winkler elastic foundation (Cheung et al., 2000).

REFERENCES

Boresi, A. P., Chong, K. P., and Lee, J. D. 2011. *Elasticity in Engineering Mechanics*, Wiley, New York.

Brown, T. G., and Ghali, A. 1978. Finite strip analysis of quadrilateral plates in bending, *J. Eng. Mech. Div., Proc. ASCE*, **104**(EM2), 481–484.

Bucco, D., Mazumdar, J., and Sved, G. 1979. Application of the finite strip method combined with the deflection contour method to plate bending problems, *Comput. Struct.*, **1**, 827–830.

Cheung, Y. K. 1968a. The finite strip method in the analysis of elastic plates with two opposite simply supported ends, *Proc. Inst. Civ. Eng.*, **40**, 1–7.

Cheung, Y. K. 1968b. Finite strip method analysis of elastic slabs, *J. Eng. Mech. Div., Proc. ASCE*, **94**(EM6), 1365–1378.

Cheung, Y. K. 1969a. Orthotropic right bridges by the finite strip method, Publ. SP-26, American Concrete Institute, Detroit, MI, pp. 182–203.

Cheung, Y. K. 1969b. The analysis of cylindrical orthotropic curved bridge decks, *Proc. Int. Assoc. Bridge Struct. Eng.*, **29**, 41–51.

Cheung, Y. K. 1969c. Folded plate structures by the finite strip method, *J. Struct. Div., Proc. ASCE*, **95**(ST2), 2963–2979.

Cheung, Y. K. 1981. Finite strip method in structural and continuum mechanics, *Int. J. Struct.*, **1**, 19–37.

Cheung, Y. K. 2013. *The Finite Strip Method*, Elsevier.

Cheung, Y. K., Au, F. T. K., and Zheng, D. Y. 2000. Finite strip method for the free vibration and buckling analysis of plates with abrupt changes in thickness and complex support conditions, *Thin-Walled Struct.*, **36**, 89–110.

Cheung, Y. K., and Chakrabarti, S. 1971. Analysis of simply supported thick, layered plates, *J. Eng. Mech. Div., Proc. ASCE*, **97**(EM3), 1039–1044.

Cheung, M. S., and Chan, M. Y. T. 1978. Finite strip evaluation of effective flange width of bridge girders, *Can. J. Civ. Eng.*, **5**, 174–185.

Cheung, M. S., and Cheung, Y. K. 1971a. Analysis of curved box girder bridges by finite strip method, *Proc. Int. Assoc. Bridge Struct. Eng.*, **31**(I), 1–19.

Cheung, M. S., and Cheung, Y. K. 1971b. Natural vibrations of thin, flatwalled structures with different boundary conditions, *J. Sound Vib.*, **18**, 325–337.

Cheung, M. S., Cheung, Y. K., and Ghali, A. 1970. Analysis of slab and girder bridges by the finite strip method, *Build. Sci.*, **5**, 95–104.

Cheung, M. S., Cheung, Y. K., and Reddy, D. V. 1971. Frequency analysis of certain single and continuous span bridges, in *Developments in Bridge Design and Construction*, Rocky, K. C., Bannister, J. L., and Evans, H. R. (eds.), Crosby Lockwood, London, pp. 188–199.

Cheung, Y. K., and Fan, S. C. 1983. Static analysis of right box girder bridges by spline finite strip method, *Proc. Inst. Civ. Eng.*, **75**(Pt. 2), 311–323.

Cheung, Y. K., Fan, S. C., and Wu, C. Q. 1982a. Spline finite strip in structure analysis, in *Proc. International Conference on Finite Element Methods*, Shanghai, Gordon & Breach, London, pp. 704–709.

Cheung, Y. K., Tham, L. G., and Chong, K. P. 1982b. Buckling of sandwich plate by finite layer method, *Comput. Struct.*, **15**(2), 131–134.

Cheung, Y. K., Tham, L. G., and Li, W. Y. 1986. Application of spline-finite strip method in the analysis of curved slab bridge, *Proc. Inst. Civ. Eng.*, Pt. 2, **81**, 111–124.

Cheung, Y. K., Yuan, C. Z., and Xiong, Z. J. 1991. Transient response of cylindrical shells with arbitrary shaped sections, *J. Thin-Walled Struct.*, **11**(4), 305–318.

Chong, K. P. 1986. Sandwich panels with cold-formed thin facings, Keynote paper, *Proc. IABSE International Colloquium on Thin-Walled Metal Structures in Buildings*, Stockholm, Sweden, Vol. 49, pp. 339–348.

Chong, K. P., and Chen, J. L. 1986. Buckling of irregular plates by splined finite strips, *AIAAJ.*, **24**(3), 534–536.

Chong, K. P., Cheung, Y. K., and Tham, L. G. 1982a. Free vibration of formed sandwich panel by finite-prism-strip method, *J. Sound Vib.*, **81**(4), 575–582.

Chong, K. P., Tham, L. G., and Cheung, Y. K. 1982b. Thermal behavior of formed sandwich plate by finite-prism-method, *Comput. Struct.*, **15**, 321–324.

Cusens, A. R., and Loo, Y. C. 1974. Applications of the finite strip method in the analysis of concrete box bridges, *Proc. Inst. Civ. Eng.*, Pt. 2, **57**, 251–273.

Delcourt, C., and Cheung, Y. K. 1978. Finite strip analysis of continuous folded plates, in *Proc. International Association Bridge and Structural Engineers*, May, pp. 1–16.

Fan, S. C., and Cheung, Y. K. 1983. Analysis of shallow shells by spline finite strip method, *Eng. Struct.*, **5**, 255–263.

Graves-Smith, T. R. 1987. The finite strip analysis of structures, in *Developments in Thin-Walled Structures* (Vol. 3), Rhodes, J., and Walker, A. C. (eds.), pp. 205–235, Elsevier Applied Science, Barking, Essex, England.

Graves-Smith, T. R., and Gierlinski, J. T. 1982. Buckling of stiffened webs by local edge loads, *J. Struct. Div., Proc. ASCE*, **108**(ST6), 1357–1366.

Graves-Smith, T. R., and Sridharan, S. 1979. Elastic collapse of thin-walled columns, in *Recent Technical Advances and Trends in Design, Research, and Construction*, International Conference at the University of Strathclyde, Glasgow, Rhodes, J., and Walker, A. C. (eds.), pp. 718–729.

Gutkowski, R. M., Chen, C. J., and Puckett, J. A. 1991. Plate bending analysis by unequally spaced splines, *J. Thin-Walled Struct.*, **11**(4), 413–435.

Hancock, G. J. 1981. Interactive buckling in I-section columns, *J. Struct. Div., Proc. ASCE*, **107**, 165–180.

Harik, I. E. 1984. Analytical solution to orthotopic sector, *J. Eng. Mech.*, **110**, 554–568.

Harik, I. E., and Pashanasangi, S. 1985. Curved bridge desks: Analytical strip solution, *J. Struct. Eng.*, **111**, 1517–1532.

Harik, I. E., and Salamoun, G. L. 1986. Analytical strip solution to rectangular plates, *J. Eng. Mech.*, **112**, 105–118.

Hasegawa, A., and Maeno, H. 1979. Design of edge stiffened plates, in *Recent Technical Advances and Trends in Design, Research, and Construction*, International conference at the University of Strathclyde, Glasgow, Rhodes, J., and Walker, A. C. (eds.), pp. 679–680.

Li, W. Y., Cheung, Y. K., and Tham, L. G. 1986. Spline finite strip analysis of general plates, *J. Eng. Mech.*, **112**(1), 43–54.

Loo, Y. C., and Cusens, A. R. 1970. A refined finite strip method for the analysis of orthotropic plates, *Proc. Inst. Civ. Eng.*, **48**, 85–91.

Loo, Y. C., and Cusens, A. R. 1978. *The Finite-Strip Method in Bridge Engineering*, Viewpoint Press, Tehachapi, CA.
Mahendran, M., and Murray, N. W. 1986. Elastic buckling analysis of ideal thin walled structures under combined loading using a finite strip method, *J. Thin-Walled Struct.*, **4**, 329–362.
Meirovitch, L. 1986. *Elements of Vibration Analysis*, 2nd ed., McGraw-Hill, New York.
Przemieniecki, J. S. 1972. Matrix analysis of local instability in plates, stiffened panels and columns, *Int. J. Numer. Methods Eng.*, **5**, 209–216.
Puckett, J. A., and Gutkowski, R. M. 1986. Compound strip method for analysis of plate systems, *J. Struct. Eng.*, **112**(1), 121–138.
Puckett, J. A., and Lang, G. J. 1986. Compound strip method for continuous sector plates, *J. Eng. Mech.*, **112**(5), 1375–1389.
Puckett, J. A., and Lang, G. J. 1989. Compound strip method for the frequency analysis of continuous sector plates, *J. Thin-Walled Struct.*, **8**(3), 165–182.
Puckett, J. A., Wiseman, D. L., and Chong, K. P. 1987. Compound strip method for the buckling analysis of continuous plates, *Int. J. Thin-Walled Struct.*, **5**(5), 385–402.
Rhodes, J. 1987. A simple microcomputer finite strip analysis, in *Dynamics of Structures*, Roesset, J. M. (ed.), American Society of Civil Engineers, Reston, VA, pp. 276–291.
Sridharan, S., and Benito, R. 1984. Columns: Static and dynamic interactive buckling, *J. Eng. Mech.*, **110**, 49–65.
Sridharan, S., and Graves-Smith, T. R. 1981. Postbuckling analyses with finite strips, *J. Eng. Mech. Div., Proc. ASCE*, **107**(EM5), 869–888.
Tham, L. G. 1990. Application of spline finite strip method in the analysis of space structures, *J. Thin-Walled Struct.*, **10**(3), 235–246.
Tham, L. G., Chong, K. P., and Cheung, Y. K. 1982. Flexural bending and axial compression of architectural sandwich panels by finite-prism-strip methods, *J. Reinf. Plast. Compos. Mater.*, **1**, 16–28.
Tham, L. G., Li, W. Y., Cheung, Y. K., and Cheng, M. J. 1986. Bending of skew plates by spline-finite-strip method, *Comput. Struct.*, **22**(1), 31–38.
Timoshenko, S. P., and Woinowsky-Krieger, S. 1971. *Theory of Plates and Shells*, 2nd ed., McGraw-Hill, New York.
Vlasov, V. Z. 1984. *Thin-Walled Elastic Beams* (trans. Y. Schechtman), National Tech. Information Service, Alexandria, VA.
Wiseman, D. L., Puckett, J. A., and Chong, K. P. 1987. Recent developments of the finite strip method, in *Dynamics of Structures*, Roesset, J. M. (ed.), American Society of Civil Engineers, Reston, VA, pp. 292–309.
Yang, H. Y., and Chong, K. P. 1982. On finite strip method, *Proc. International Conference on Finite Element Methods*, Shanghai, China, Department of Civil Engineering, University of Hong Kong, Hong Kong, pp. 824–829.
Yang, H. Y., and Chong, K. P. 1984. Finite strip method with X-spline functions, *Comput. Struct.*, **18**(1), 127–132.
Yoshida, H. 1971. Buckling analysis of plate structures by finite strip method, *Proc. Jpn. Soc. Nav. Arch.*, **130**, 161–171.
Yoshida, H., and Maegawa, K. 1978. Local and member buckling of H-columns, *J. Struct. Mech.*, **6**, 1–27.

BIBLIOGRAPHY

Akhras, G., Cheung, M. S., and Li, W. 1993. Static and vibration analysis of anisotropic composite laminates by finite strip method, *Int. J. Solids Struct.*, **30**(22), 3129–3137.
Akhras, G., and Li, W. C. 2007. Three-dimensional static, vibration and stability analysis of piezoelectric composite plates using a finite layer method, *Smart Mater. Struct.*, **16**(3), 561.
Bradford, M. A., and Azhari, M. 1995. Buckling of plates with different end conditions using the finite strip method, *Comput. Struct.*, **56**(1), 75–83.

Chau, K. T. 2018. *Theory of Differential Equations in Engineering and Mechanics*, CRC Press, Boca Raton, FL.

Cheung, Y. K., Lo, S. H., and Leung, A. Y. T. 1996. *Finite Element Implementation*, Blackwell Science, Maiden, MA.

Guingand, M., De Vaujany, J. P., and Icard, Y. 2004. Fast three-dimensional quasi-static analysis of helical gears using the finite prism method, *J. Mech. Des.*, **126**(6), 1082–1088.

Lau, S. C. W., and Hancock, G. J. 1986. Buckling of thin flat-walled structures by a spline finite strip method, *Thin-Walled Struct.*, 4(4), 269–294.

Olakorede, A. A., and Play, D. 1991. Load sharing, load distribution and stress analysis of cylindrical gears by finite prism method, In *Design Productivity International Conference, Honolulu, Hawaii*, pp. 921–927.

Reddy, J. N. 2006. *Theory and Analysis of Elastic Plates and Shells*, CRC Press, Boca Raton, FL.

Wood, R. D., and Zienkiewicz, O. C. 1977. Geometrically nonlinear finite element analysis of beams, frames, arches and axisymmetric shells, *Comput. Struct.*, 7(6), 725–735.

Chapter 6

The boundary element method

6.1 INTRODUCTION

Boundary integral equations, which form the basis of the boundary element method, have been in existence for a long time. The boundary element method as it is known today, however, has been developed largely in the last four decades. The rapid development of the method may have drawn its motivation from the limitations of the finite element method. A major step in performing the finite element analysis is the discretization of the domain into finite elements, called *meshing*. For arbitrarily shaped three-dimensional objects, meshing is an extremely tedious job. Often, this step may take weeks, even months, to accomplish, whereas the rest of the analysis may require only a few days. There are situations where the mesh for the object may need to be defined anew several times for the same analysis. In metal-forming operations, the metal workpiece undergoes very large deformations, including large rotations. The individual elements in the mesh may become severely deformed and possess unduly large aspect ratios. Unless the mesh for the deformed workpiece is redefined several times during the analysis, the original mesh may lead to erroneous results. In the simulation of crack propagation in a solid object, the mesh is first defined with respect to the initial geometry of the crack. If the crack does not propagate along element boundaries, a rare occurrence, the crack will intersect one or more elements. To account for the new location of the crack, the mesh may need to be redefined for each advance of the crack front. In shape optimization problems (Saigal and Kane, 1990), the geometry of the object is revised continually in each calculation step to proceed toward its optimal configuration. For each revision of the geometry, a new mesh needs to be defined. The applications mentioned earlier and several others make use of the finite element method undesirable. There have been, and continue to be, several attempts in the literature to obviate the problems associated with meshing. Through the use of appropriate mathematical theorems, the boundary element method reduces the dimensions of the problem by one degree. Thus, for a three-dimensional object, a two-dimensional discretization—that of the surfaces bounding the object—is required. Similarly, for two-dimensional analyses, a one-dimensional discretization of the lines enclosing the object is required. This reduction in the dimensions of the problem by one degree leads to significant advantages in terms of ease of discretization of the domain and of reducing the overall time to perform the analysis of objects with complex geometries. Consider the plate with centrally located hole under uniaxial loading shown in Figure 6.1a. For this problem, the boundary element meshes using quadratic boundary elements are shown in Figure 6.1b and c, respectively. No boundary elements are shown on the axes of symmetry of the plate in the discretized model shown in Figure 6.1c. The effects due to symmetry can be included within the analytical formulation of the boundary elements (Kaljevic and Saigal, 1995b; Saigal et al., 1990a). This eliminates the need for providing boundary elements on the axes of symmetry (Kaljevic and Saigal, 1995a). In view of the meshes shown

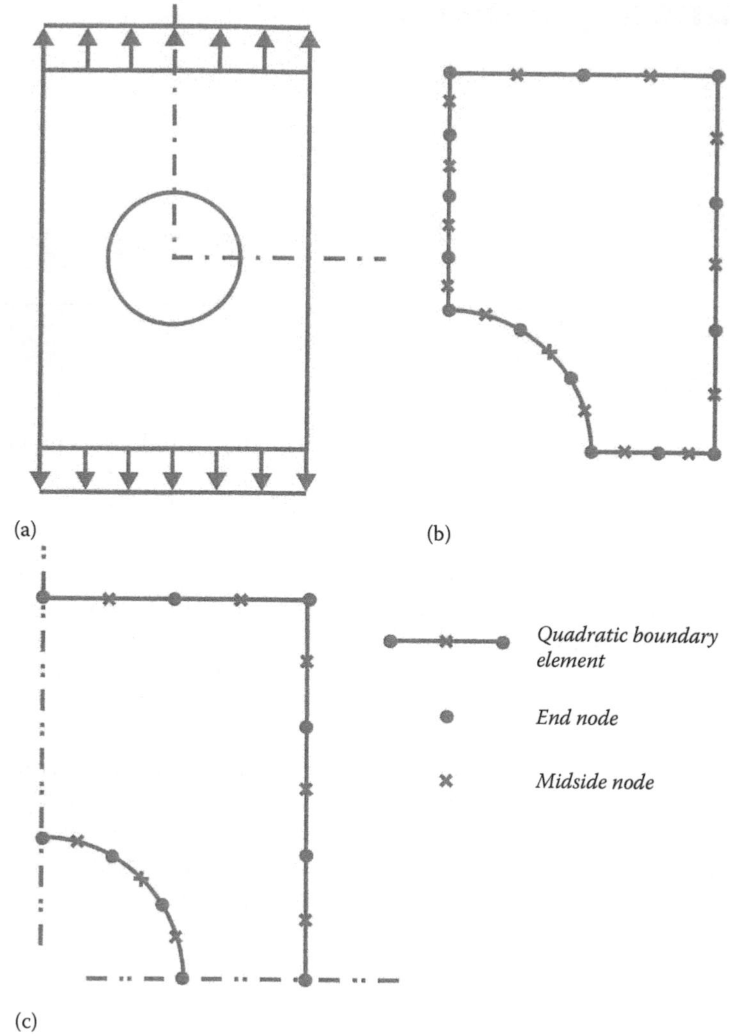

Figure 6.1 Boundary element models of a plate with centrally located hole. (a) Plate geometry and loading. (b) Boundary element mesh. (c) Boundary element mesh with symmetry conditions included in formulation.

in Figure 6.1b and c, respectively, the advantage offered by boundary elements in terms of meshing requirements is quite apparent.

There are, however, several limitations of the boundary element method. A true boundary-only formulation is typically obtained only for linear elastic problems and in the absence of body forces. Several developments, including the particular integral approach (Ahmad and Banerjee, 1986; Henry and Banerjee, 1988a; Henry et al., 1987), the dual reciprocity method (Niku and Brebbia, 1988; Partridge and Brebbia, 1990; Partridge et al., 1991; Wrobel and Brebbia, 1987), and the multiple reciprocity method (Neves and Brebbia, 1991; Nowak and Neves, 1994; Sladek and Sladek, 1996), have been reported that reduce the volume integrals due to body forces into equivalent surface integrals. Similar developments that account for nonlinear effects through boundary-only formulations have also been reported (Henry and Banerjee, 1988b).

A finite element formulation leads to sparse, banded symmetric matrices, for which highly efficient solution procedures are available in the numerical analysis literature. Boundary element formulations, however, lead to dense, fully populated, unsymmetric matrices (Banerjee, 1994). This aspect of boundary element method seriously compromises the advantages due to the reduction of the problem by one degree. There have also been recent developments in the literature toward the development of symmetric boundary element formulations (Balakrishna et al., 1994; Gray and Paulino, 1997; Hartmann et al., 1985; Layton et al., 1997; Maier et al., 1990).

The boundary element method is used today in several industries for the solution of complex analysis problems (Banerjee and Wilson, 1989). It does not, however, appear to have gained the popularity enjoyed by the finite element method. While numerous commercial finite element codes are available to the analyst, only a handful of commercial boundary element codes are in existence. Despite the lack of its popularity, the boundary element method is a powerful tool and, for certain classes of problem, the tool of choice for their analysis.

6.2 INTEGRALS IN THE BOUNDARY ELEMENT METHOD

A brief introduction of certain integrals is provided to facilitate an understanding of the nature of integrals resulting from a boundary element formulation. Consider the one-dimensional integral

$$I = \int_{L_1}^{L_2} \frac{f(r)}{|r|^{\alpha}} dr \tag{6.1}$$

where $f(r)$ is a regular, integrable function between the limits L_1 and L_2, and α is a constant.

1. For $\alpha = 0$, the denominator in Equation 6.1 is unity, the integrand, $f(r)$, is regular, and the integral, I, is easily evaluated.
2. For $\alpha < 1$, the integrand, $f(r)/|r|^{\alpha}$, is singular at $r = 0$ but presents an integrable singularity. The integral, I, in this case is said to be *weakly singular*. For example, for $f(r) = 1$ and $\alpha = 0.5$,

$$I = \int_{L_1}^{L_2} \frac{1}{|r|^{0.5}} dr = \frac{1}{2} |r|^{0.5} \Big|_{L_1}^{L_2} \tag{6.2}$$

The right-hand side expression in Equation 6.2 is easily evaluated.

3. For $\alpha = 1$, the integrand, $f(r)/|r|$, is singular at $r = 0$ and the integral, I, may only be evaluated in the Cauchy principal value (CPV) sense (Churchill and Brown, 1987). The integral, I, is said to be *strongly singular*. From Figure 6.2,

$$I = \int_{L_1}^{L_2} \frac{f(r)}{|r|} dr = \lim_{\varepsilon \to 0} \left[\int_{L_1}^{-\varepsilon} \frac{f(r)}{|r|} dr + \int_{C} \frac{f(r)}{\varepsilon} ds + \int_{\varepsilon}^{L_2} \frac{f(r)}{|r|} dr \right] \tag{6.3}$$

where C is the semicircular contour around the singularity at $r = 0$. Noting that $ds = \varepsilon\, d\theta$ and $0 \le \theta \le \pi$, the singularity in the second integral is removed as ε cancels out in

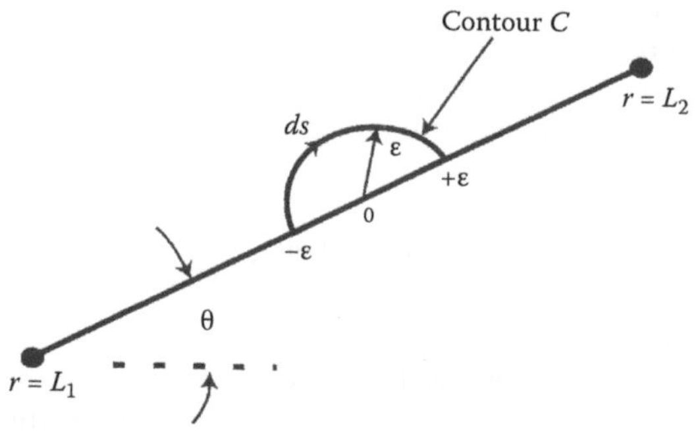

Figure 6.2 Singular integral domain in one dimension.

both the denominator and the numerator. Further, if the effects due to $-\varepsilon$ in the upper limit and $+\varepsilon$ in the lower limit in Equation 6.3 cancel out, I is said to be integrable in the CPV sense.

4. For $\alpha > 1$, the integrand, $f(r)/|r|^\alpha$, is singular at $r = 0$ and cannot be evaluated in the CPV sense. The integral, I, in this case is said to be *hypersingular.*

Consider next a two-dimensional integral of the form

$$I = \int_A \frac{f(x,y)}{|r|^\alpha} dA \tag{6.4}$$

where $f(x, y)$ is a regular function, $r = [(x - x_0)^2 + (y - y_0)^2]^{0.5}$ is the radial distance from an origin (x_0, y_0), A is the domain under consideration containing the origin, and α is a constant.

1. For $0 < \alpha < 2$, a transformation to polar coordinates leads to

$$I = \int_r \int_\theta \frac{F(\theta)}{r^\alpha} r\, dr\, d\theta = \int_r \int_\theta \frac{F(\theta)}{r^{\alpha-1}} dr\, d\theta \tag{6.5}$$

where $F(\theta)$ is obtained from $f(x, y)$ after performing the transformation substitutions. The integration in Equation 6.5 for the variable θ can be carried out in a straightforward manner. From the discussion earlier on one-dimensional integrals, Equation 6.4 for variable r is integrable for $\alpha - 1 < 1$ or $\alpha < 2$.

2. For $\alpha = 2$, the integral in Equation 6.4, using Figure 6.3, can be written as

$$I = \int_A \frac{f(x,y)}{r^2} dA = \lim_{A_\varepsilon \to 0} \left[\int_{A-A_\varepsilon} \frac{f(x,y)}{r^2} dA + \int_{A_\varepsilon} \frac{f(x,y)}{r^2} dA \right] \tag{6.6}$$

where A_ε is an infinitesimal area around the singularity at $r = 0$. The second integral on the right-hand side in Equation 6.6 contains the singular term. If $f(x, y)$ satisfies certain properties, the integral, I, would still remain finite in the CPV sense.

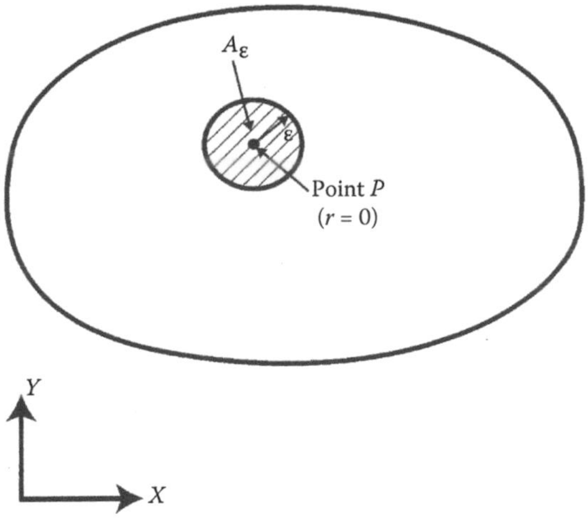

Figure 6.3 Singular integral domain in two dimensions.

3. For $\alpha > 2$, following the discussion on singular integrals in one dimension, the integral, I, in Equation 6.4 cannot be evaluated in the CPV sense, and it is termed as *hypersingular.*

In three dimensions, we consider the integral

$$I = \int_V \frac{f(x,y,z)}{r^\alpha} dV \tag{6.7}$$

where $r = [\{x - x_0)^2 + (y - y_0)^2 + (z - z_0)^2)]^{0.5}$, α is a constant, and V is the volume domain under consideration, including the origin at which $r = 0$. Following arguments similar to those presented previously for the one- and two-dimensional integrals and using the transformation $dV = r^2 \sin\phi \, dr \, d\phi \, d\theta$, where θ and ϕ are the spherical coordinate angles, respectively, we can deduce that

1. For $0 < \alpha < 3$, the singularity is integrable, the integral, I, in Equation 6.7 is finite and is termed *weakly singular.*
2. For $\alpha = 3$, the integral, I, in Equation 6.7 is finite provided that $f(x, y, z)$ satisfies certain conditions and it can be evaluated in the CPV sense. The integral, I, is termed *strongly singular.*
3. For $\alpha > 3$, the integral, I, in Equation 6.7 *is hypersingular* and cannot be evaluated in the CPV sense.

6.3 EQUATIONS OF ELASTICITY

The boundary element method is developed in this chapter with reference to elastic behavior of objects. It is advantageous, then, to present a brief outline of the equations of elasticity (Boresi et al., 2011). The stress components, σ_{ij}, must satisfy the differential equations of equilibrium in a domain, Ω, under consideration:

$$\sigma_{ij,j} + F_i = 0 \tag{6.8}$$

where F_i is the component of the body force acting on the object. The previous field equations are to be solved given certain boundary conditions as

$$\begin{aligned} u_i &= \bar{u}_i \qquad \text{on}\,\Gamma_u \\ t_i &= \bar{t}_i \qquad \text{on}\,\Gamma_t \end{aligned} \tag{6.9}$$

where the tractions on a surface with normal n_j are given as $t_i = \sigma_{ij}n_j$ and $\bar{u}_i$ and $\bar{t}_i$ are the prescribed values of displacements and tractions on boundaries Γ_u and Γ_t, respectively. $\Gamma_u \cup \Gamma_t = \Gamma$ is the boundary of the subject domain Ω. The stress components, σ_{ij}, in Equation 6.8 are symmetric (i.e., $\sigma_{ij} = \sigma_{ji}$). The kinematic equations, relating the strain components, ε_{ij}, to the displacements, u_i, are

$$\varepsilon_{ij} = \frac{1}{2}(u_{i,j} + u_{j,i}) \tag{6.10}$$

where a comma (,) denotes differentiation, thus, $u_{i,j} = \partial u_i/\partial x_j$. The constitutive relations for a linear elastic material are given as

$$\sigma_{ij} = \lambda\varepsilon_{kk}\delta_{ij} + 2\mu\varepsilon_{ij} = E_{ijkl}\varepsilon_{kl} \tag{6.11}$$

where λ and μ are the Lamé constants and E_{ijkl} is the constitutive tensor. Equations 6.8–6.11 summarize the equations of elastostatics for a linear, isotropic, homogenous elastic object.

6.4 FUNDAMENTAL OR KELVIN'S SOLUTION

The *fundamental solution*, also called the *Kelvin solution* (Love, 2013; Sokolnikoff, 1956), is now introduced. This solution is employed in the boundary element formulation to extract the solution of displacement components from the integral equations of the elastic body. The fundamental solution refers to the solution of the response of a linear, elastic solid of infinite extent due to the application of a point load. Consider a domain of infinite extent shown in Figure 6.4. A unit load is applied at the source point P, and the response at the field point Q is sought. Denoting all quantities related to the fundamental solution with an asterisk (*), the body force

$$F_i^* = \Delta(P,Q)\delta_{ij}e_j \tag{6.12}$$

is substituted in Equation 6.8 to obtain the fundamental solution. Here, $\Delta(P, Q)$ is the Dirac delta function, δ_{ij} is the Kronecker delta, and e_j is the unit vector in direction j. The fundamental (Kelvin) solution for displacement is then expressed symbolically as

$$u_i^* = U_{ij}^*(P,Q)e_j \tag{6.13}$$

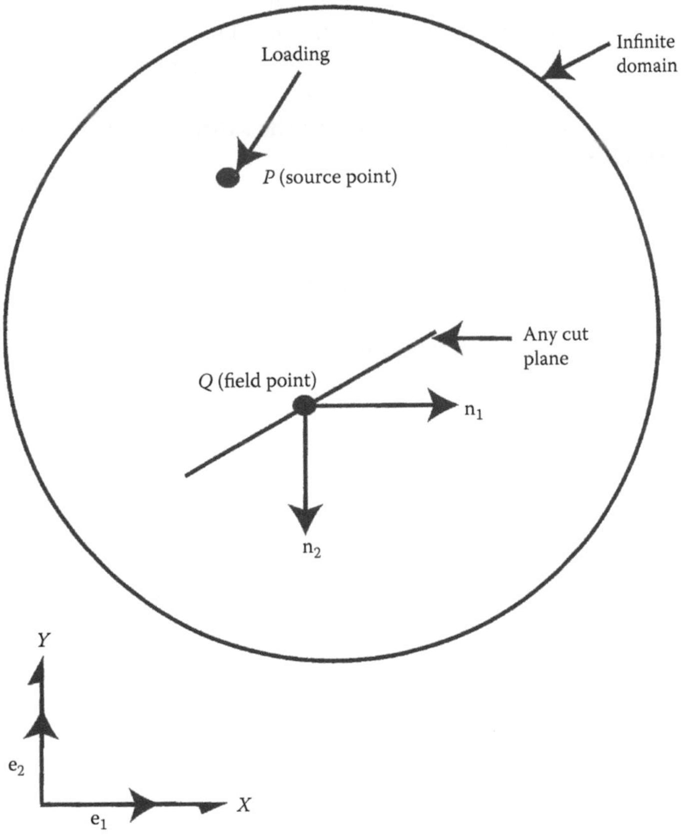

Figure 6.4 Domain of infinite extent under a concentrated point load.

Substituting Equation 6.13 into Equation 6.10 followed by Equation 6.11 and using the relation $t_i = \sigma_{ij}n_j$, the fundamental (Kelvin) solution for traction is obtained and is expressed here as

$$t_i^* = T_{ij}^*(P,Q)e_j \tag{6.14}$$

For a two-dimensional body, these solutions are given as

$$U_{ij}^*(P,Q) = \frac{-1}{8\pi\mu(1-\upsilon)}\left[(3-4\upsilon)\delta_{ij}\ln r - \frac{y_iy_j}{r^2}\right] \tag{6.15}$$

and

$$T_{ij}^*(P,Q) = \frac{1}{4\pi(1-\upsilon)r^2}\left\{(1-2\upsilon)(n_jy_i - n_iy_j) + \left[(1-2\upsilon)\delta_{ij} + \frac{2y_iy_j}{r^2}\right]y_kn_k\right\} \tag{6.16}$$

where $r = y_iy_i$; $y_i = x_{pi} - x_{qi}$; x_{pi} is the x_i coordinate of the source point P and, similarly, for x_{qi}.

It is important here to understand the notation for the subscripts and for the arguments of the displacement and traction fundamental solutions. Thus, $U^*_{ij}(P,Q)$ denotes displacement at location Q in the j direction due to a unit load applied at location P in the i direction. Similarly, $T^*_{ij}(P,Q)$ denotes the traction at location Q on an orientation defined by the normal n_j due to a unit load applied at location P in the i direction. It is easy, then, to see that $U^*_{ij}(P,Q)$ and $T^*_{ij}(P,Q)$ can each be written as 3 × 3 matrices for three-dimensional analyses and 2 × 2 matrices for two-dimensional analyses.

6.5 BOUNDARY ELEMENT FORMULATION

A starting point for the boundary element formulation is to multiply the equations of equilibrium in Equation 6.8 by an arbitrary weighing function, $\bar{w}_i$, and integrate over the region considered:

$$\int_\Omega (\sigma_{ij,j} + F_i)\bar{w}_i \, d\Omega = 0 \tag{6.17}$$

Noting that $(\sigma_{ij}\bar{w}_i),_j = \sigma_{ij,j}\bar{w}_i + \sigma_{ij}\bar{w}_{i,j}$ from the product rule of differentiation (Stewart, 2010), the first term in the integral in Equation 6.17 can be rewritten as

$$\int_\Omega [(\sigma_{ij}\bar{w}_i),_j - \sigma_{ij}\bar{w}_{i,j}] d\Omega + \int_\Omega F_i\bar{w}_i \, d\Omega = 0 \tag{6.18}$$

The first term in Equation 6.18 is now rewritten using the divergence theorem of calculus (Schey, 2005). The *divergence theorem* for a vector field with components, R_j, is written as

$$\int_\Omega R_{j,k} \, d\Omega = \int_\Gamma R_j n_k \, d\Gamma \tag{6.19}$$

where Γ is the bounding surface for the volume Ω and n_k is the outward normal to the surface $d\Gamma$. Substituting $R_j = \sigma_{ij}\bar{w}_i$, and replacing the index k by j in Equation 6.19, the first term under the integral in Equation 6.18 can be written as

$$\int_\Omega (\sigma_{ij}\bar{w}_i),_j \, d\Omega = \int_\Gamma \sigma_{ij}\bar{w}_i n_j \, d\Gamma \tag{6.20}$$

Since $\sigma_{ij}n_j = t_i$, this term is further simplified as

$$\int_\Omega (\sigma_{ij}\bar{w}_i),_j \, d\Omega = \int_\Gamma t_i\bar{w}_i \, d\Gamma \tag{6.21}$$

We now turn our attention to the second integral in Equation 6.18. Since the stress tensor σ_{ij} is symmetrical ($\sigma_{ij} = \sigma_{ji}$), the integrand for this term may be written as

$$\sigma_{ij}\bar{w}_{i,j} = \frac{1}{2}\sigma_{ij}(\bar{w}_{i,j} + \bar{w}_{j,i}) \tag{6.22}$$

Using the constitutive relation $\sigma_{ij} = E_{ijkl}\varepsilon_{kl}$ from Equation 6.11 and denoting $\frac{1}{2}(\bar{w}_{i,j} + \bar{w}_{j,i})$ as $\bar{\varepsilon}_{ij}$, by considering the weighing function $\bar{w}_i$ as a displacement field and employing the strain–displacement relationships of Equation 6.10, we obtain

$$\sigma_{ij}\bar{w}_{i,j} = E_{ijkl}\varepsilon_{kl}\bar{\varepsilon}_{ij} \tag{6.23}$$

The strain field ε_{kl} is next expressed in terms of its displacement components as $\varepsilon_{kl} = \frac{1}{2}(u_{k,l} + u_{l,k})$. Also, using the symmetry of the constitutive tensor $E_{ijkl} = E_{klij}$ (Chung, 1988), it is combined with the strain field ε_{ij} to give $\bar{\sigma}_{kl} = E_{klij}\bar{\varepsilon}_{ij}$. The integrand under consideration is then obtained as

$$\sigma_{ij}\bar{w}_{i,j} = \frac{1}{2}(u_{k,l} + u_{l,k})\bar{\sigma}_{kl} \tag{6.24}$$

Finally, noting the symmetry of $\bar{\sigma}_{kl}$, Equation 6.24 is written as

$$\sigma_{ij}\bar{w}_{i,j} = u_{k,l}\bar{\sigma}_{kl} \tag{6.25}$$

To allow use of the divergence theorem on this term, it is first rewritten using the product rule of differentiation. The second integral in Equation 6.18 can then be written as

$$\int_\Omega \sigma_{ij}\bar{w}_{i,j}\, d\Omega = \int_\Omega [(\bar{\sigma}_{kl}u_k)_{,l} - \bar{\sigma}_{kl,l}u_k]d\Omega \tag{6.26}$$

Applying the divergence theorem to the first term on the right-hand side of Equation 6.26, we obtain

$$\int_\Omega \sigma_{ij}\bar{w}_{i,j}\, d\Omega = \int_\Gamma \bar{\sigma}_{kl}u_k n_l\, d\Gamma - \int_\Omega \bar{\sigma}_{kl,l}u_k\, d\Omega \tag{6.27}$$

Expressing $\bar{\sigma}_{kl}n_l = \bar{t}_k$ and noting that the stress field $\bar{\sigma}_{kl}$ satisfies the equilibrium relations in Equation 6.8 for a body force field $\bar{F}_k$ (i.e., $\bar{\sigma}_{kl,l} + \bar{F}_k = 0$), Equation 6.27 is rewritten as

$$\int_\Omega \sigma_{ij}\bar{w}_{i,j}\, d\Omega = \int_\Gamma \bar{t}_k u_k\, d\Gamma + \int_\Omega \bar{F}_k u_k\, d\Omega \tag{6.28}$$

Substituting Equations 6.21 and 6.28 into 6.18 and upon rearranging terms, we get

$$\int_\Gamma t_i\bar{w}_i\, d\Gamma + \int_\Omega F_i\bar{w}_i\, d\Omega = \int_\Gamma \bar{t}_i u_i\, d\Gamma + \int_\Omega \bar{F}_i u_i\, d\Omega \tag{6.29}$$

Equation 6.29 is the familiar *Betti's reciprocal theorem* of structural analysis (Norris et al., 1976). Several boundary element formulations employ Equation 6.29 as their starting point.

The weighting function, $\bar{w}_i$, in our discussions earlier is an arbitrary function that has taken on the characteristics of a displacement field. The corresponding stress, strain, traction, and body force fields are denoted as $\bar{\sigma}_{ij}, \bar{\varepsilon}_{ij}, \bar{t}_i$, and $\bar{F}_i$, respectively. We now select the weighting function to be the fundamental displacement solution, the displacement field for an infinite elastic body under a unit point load, as outlined in Section 6.4. Thus,

$$\bar{F}_i = F_i^* = \Delta(P,Q)\delta_{ij}e_j \tag{6.30a}$$

$$\bar{w}_i = u_i^* = U_{ij}^*(P,Q)e_j \tag{6.30b}$$

and

$$\bar{t}_i = t_i^* = T_{ij}^*(P,Q)e_j \tag{6.30c}$$

Substituting these relations into Equation 6.29, and canceling e_j in each term of the resulting equation, we get

$$\int_\Gamma t_i(Q)U_{ij}^*(P,Q)\,d\Gamma + \int_\Omega F_i(Q)U_{ij}^*(P,Q)\,d\Omega = \int_\Gamma u_i(Q)T_{ij}^*(P,Q)\,d\Gamma + \int_\Omega u_i(Q)\Delta(P,Q)\delta_{ij}\,d\Omega \tag{6.31}$$

Equation 6.31 is the well-known *Somigliana's identity* (Somigliana, 1886) and has also been used as the starting point for several boundary element formulations.

To continue the boundary element formulation, we consider the second term on the right-hand side of Equation 6.31 in further detail. The quantity $\Delta(P, Q)$ in the integrand vanishes everywhere except at P. The integral, using Figure 6.5, may be written as

$$\int_\Omega u_i(Q)\Delta(P,Q)\delta_{ij}\,d\Omega = \int_{\Omega-\Omega_\varepsilon} u_i(Q)\Delta(P,Q)\delta_{ij}\,d\Omega + \int_{\Omega_\varepsilon} u_i(Q)\Delta(P,Q)\delta_{ij}\,d\Omega \tag{6.32}$$

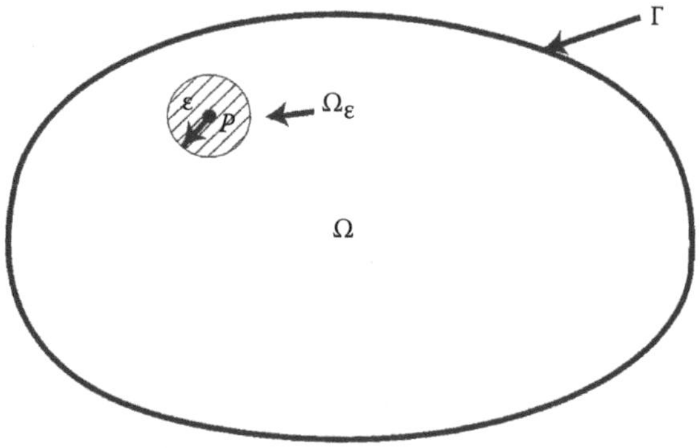

Figure 6.5 Domain of an elastic body and the source point.

where Ω_ε is a sphere of radius ε surrounding the singularity at $r = 0$. The first term on the right-hand side of Equation 6.32 vanishes as $\Delta(P, Q)$ is zero everywhere in $\Omega - \Omega_\varepsilon$. The second term is written as

$$\int_{\Omega_\varepsilon} u_i(Q)\Delta(P,Q)\delta_{ij}\, d\Omega = \varepsilon\delta_{ij}u_i(P) \tag{6.33}$$

where $\varepsilon = 0, \frac{1}{2}$, and 1, depending on whether P lies outside Γ, is within the volume Ω, or lies on a smooth boundary Γ, respectively (see Section 6.15). The term on the right-hand side of Equation 6.33 is called the *free term*. Our system, at this point, is represented by Equation 6.31. Considering the relation written in Equation 6.33, the only term containing a volume integration in Equation 6.31 is the one due to the body force $F_i(Q)$. To obtain a representation that involves only the surface integrals, we consider the cases for which the body forces, $F_i(Q)$, are absent in Equation 6.31. Using Equation 6.33 and $F_i(Q) = 0$, with a rearrangement of terms, Equation 6.31 reduces to

$$\varepsilon\delta_{ij}u_i(P) = \int_\Gamma [t_i(Q)U_{ij}^*(P,Q) - u_i(Q)T_{ij}^*(P,Q)]\,d\Gamma(Q) \tag{6.34}$$

Equation 6.34 is the basic boundary integral equation that forms the basis for the boundary element method. In the form given in Equation 6.34, no volume integrals are involved and only surface integrals need be evaluated. The order of the system in going from volume to surface integrals (and from surface to line integrals in two dimensions) has been reduced by one. This feature has constituted a major attraction of the boundary element method in the area of numerical analysis methods.

It is recognized here that the body forces have been neglected in obtaining the boundary-only representation of Equation 6.34. The inclusion of such forces at this point would lead to a volume integral and destroy the boundary-only nature of Equation 6.34. Various treatments of body forces that preserve the boundary-only nature of the boundary integral formulation are presented in Sections 6.12 and 6.13.

6.6 DISPLACEMENT AND TRACTION INTERPOLATION

The boundary element method is frequently referred to as a *mixed formulation*, as it involves both the displacements as well as the tractions as unknown variables. Interpolation functions, similar to those described in Chapter 4, are required to describe the variation of displacements and tractions, respectively, over an element in terms of their nodal values at the nodes of the element. A one-dimensional boundary element suitable for use in two-dimensional analysis is shown in Figure 6.6. A quadratic three-node element is considered here. The formulations for constant and linear elements follow along similar lines. The interpolations for the unknown variables, namely, displacements, u, and tractions, t, are written using the shape functions, N_i, as

$$u(\xi) = \sum_{i=1}^{3} N_i(\xi)u_i \tag{6.35a}$$

Figure 6.6 One-dimensional quadratic boundary element: ●, end nodes; ×, midside node.

$$t(\xi) = \sum_{i=1}^{3} N_i(\xi) t_i \tag{6.35b}$$

where u_i and t_i are the displacement and traction, respectively, at the node i of the boundary element.

The shape functions, $N_i(\xi)$, are given as

$$\begin{aligned} N_1(\xi) &= 2(\xi - 0.5)(\xi - 1) \\ N_2(\xi) &= -4\xi(\xi - 1) \\ N_3(\xi) &= 2\xi(\xi - 0.5) \end{aligned} \tag{6.36}$$

Nodes 1, 2, and 3 of the quadratic element lie at the nondimensional locations $\xi = 0$, 0.5, and 1.0, respectively. It may be verified that the shape functions satisfy their usual properties. The shape function $N_1 = 1$ at node 1 and vanishes at nodes 2 and 3, respectively. Shape functions N_2 and N_3 behave similarly. Additionally, by Equation 6.36, $\sum_{i=1}^{3} N_i(\xi) = 1$. The shape functions of Equation 6.36 are illustrated graphically in Figure 6.7. Equations 6.35a and 6.35b are written in the matrix notation as

$$\{u\} = [N]\{u_i\} \tag{6.37a}$$

and

$$\{t\} = [N]\{t_i\} \tag{6.37b}$$

where

$$\{u\} = \begin{Bmatrix} u_x \\ u_y \end{Bmatrix} \quad \{t\} = \begin{Bmatrix} t_x \\ t_y \end{Bmatrix} \tag{6.37c}$$

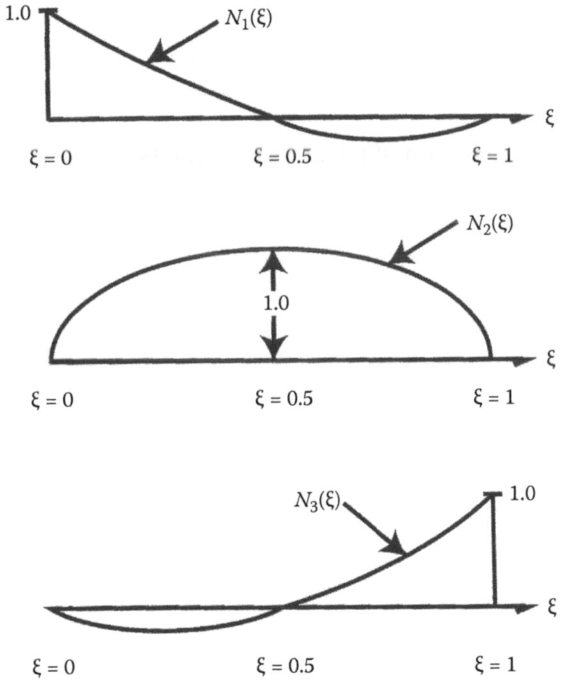

Figure 6.7 Shape functions of a one-dimensional quadratic boundary element.

$$[N]=\begin{bmatrix} N_1 & 0 & N_2 & 0 & N_3 & 0 \\ 0 & N_1 & 0 & N_2 & 0 & N_3 \end{bmatrix} \tag{6.37d}$$

$$\{u_i\}=\begin{Bmatrix} u_x^1 \\ u_y^1 \\ u_x^2 \\ u_y^2 \\ u_x^3 \\ u_y^3 \end{Bmatrix} \quad \{t_i\}=\begin{Bmatrix} t_x^1 \\ t_y^1 \\ t_x^2 \\ t_y^2 \\ t_x^3 \\ t_y^3 \end{Bmatrix} \tag{6.37e}$$

where, for example, u_x^1 in displacement vector $\{u_i\}$ is the displacement at element node 1 along the x direction and similarly for terms in the traction vector $\{t_i\}$.

6.7 ELEMENT CONTRIBUTIONS

We seek to evaluate the surface integral in Equation 6.34 numerically and develop a matrix form of that equation. The matrix relation so developed may be solved to obtain the unknown displacements and tractions at the nodes of the boundary mesh. To evaluate the surface integral, it is first expressed as the sum of integrals over the boundary of each element. Thus,

$$\int_{\Gamma} (\cdot)\, d\Gamma = \sum_{n=1}^{NEL} \int_{\Gamma_n} (\cdot)\, d\Gamma \tag{6.38}$$

where NEL represents the number of elements used to discretize the boundary and Γ_n is the boundary of the boundary element n. The summation in this equation is performed by assembling the matrix contributions from each element. The assembly process is explained in Section 6.8. The procedure for evaluation of the contribution from each element is now outlined. The first term on the right-hand side of Equation 6.34 for a single element appears as

$$\alpha_j = \int_{\Gamma_n} t_i(Q)\, U_{ij}^*(P,Q)\, d\Gamma(Q) \tag{6.39}$$

As explained previously, for two-dimensional analysis, the fundamental solution $U_{ij}^*(P, Q)$ may be written as a 2 × 2 matrix. Equation 6.39 can then be expressed in matrix form as

$$\{\alpha\} = \int_{\Gamma_n} [U^*]\{t\}\, d\Gamma(Q) \tag{6.40a}$$

where

$$\{\alpha\} = \begin{Bmatrix} \alpha_1 \\ \alpha_2 \end{Bmatrix} \tag{6.40b}$$

and

$$[U^*] = \begin{bmatrix} U_{11}^* & U_{21}^* \\ U_{12}^* & U_{22}^* \end{bmatrix} \tag{6.40c}$$

Substituting from Equation 6.37b, Equation 6.40a can be written in terms of element nodal tractions as

$$\{\alpha\} = \int_{\Gamma_n} [U^*][N]\{t_i\}\, d\Gamma(Q) \tag{6.41}$$

Since $\{t_i\}$ are nodal tractions, they can be taken outside the integral and Equation 6.41 can be written in compact matrix form as

$$\{\alpha\} = [G_n]\{t_i\} \tag{6.42}$$

with

$$[G_n] = \int_{\Gamma_n} [U^*][N]d\Gamma(Q) \tag{6.43}$$

where $[G_n]$ is a (2×6) matrix and is the contribution of boundary element n to the first term on the right-hand side of Equation 6.34. The contribution of this boundary element to the second term on the right-hand side of Equation 6.34 may similarly be written as

$$\{\beta\} = [F_n]\{u_i\} \tag{6.44}$$

with

$$[F_n] = \int_{\Gamma_n} [T^*][N]d\Gamma(Q) \tag{6.45}$$

where $[F_n]$ is a (2×6) matrix contribution due to element n. The expression in Equation 6.34 can now be rewritten after these substitutions and that from Equation 6.38 as

$$\varepsilon\delta_{ij}u_i(P) = \sum_{n=1}^{NEL}\int_{\Gamma_n} t_i(Q)U_{ij}^*(P,Q)\,d\Gamma(Q) - \sum_{n=1}^{NEL}\int_{\Gamma_n} u_i(Q)T_{ij}^*(P,Q)\,d\Gamma(Q) \tag{6.46}$$

and then expressed in the discretized form as

$$\varepsilon\delta_{ij}u_i(P) = \sum_{n=1}^{NEL}\int_{\Gamma_n} ([U^*][N]d\Gamma(Q))\{t_i\} - \sum_{n=1}^{NEL}\left(\int_{\Gamma_n} [T^*][N]d\Gamma(Q)\right)\{u_i\} \tag{6.47}$$

Equation 6.47 now deserves a closer inspection. It is an expression for the displacement at a point P that may lie anywhere on the body either within its domain Ω or on its bounding surface Γ. The displacement $u_i(P)$ is expressed in terms of the displacements $\{u_i\}$ and the tractions $\{t_i\}$ at nodes that lie on the boundary Γ. In the case of two-dimensional analysis, a discretization of the boundary using N nodes will lead to $2N$ displacement and $2N$ traction quantities from which the displacement $u_i(P)$ at point P can be found. It is realized further that, in general, a total of $2N$ of these quantities, either tractions or displacements, are known from the boundary conditions prescribed. Thus, there are a total of $2N$ unknowns on the right-hand side of Equation 6.47 for which we require $2N$ independent equations. Recall again that the source point P in Equation 6.47 may lie anywhere on the object, including its boundary. Taking advantage of this fact, the source point P is located successively at each nodal point on the boundary Γ. For each location of P on a boundary node, two independent equations, one corresponding to each coordinate direction, are obtained. Thus, the required $2N$ independent equations are obtained by considering the application of Equation 6.47 successively at each nodal point.

Consider now the load point P placed at node j on the boundary. The displacements $u_1(P)$ and $u_2(P)$ obtained from Equation 6.47 correspond to the displacements of node j in the x- and y-directions, respectively. These unknown displacements at node j also appear on the right-hand side of Equation 6.47 when considering integration over element(s) for which node j is either an end node or a midside node. In the assembly process discussed in Section 6.8, the coefficients of the unknown displacements coming from the left- and right-hand sides of Equation 6.47 are all lumped together. It is noted further that an interesting case arises when the load point P is at the node point j and the contributions to the surface integral on the right-hand side of Equation 6.47 from an element that contains node j are considered. In this case, the quantity r, the distance between the source point and the field point, approaches zero. Since r appears in the fundamental solutions, which are included in the integrands in Equation 6.47, the integrals are of a singular nature. For two-dimensional analysis, the singularities are due to an r term present in the displacement solution U_{ij}^* and a $1/r$ term present in the traction solution T_{ij}^*. Referring to the discussion on singular one-dimensional integrals in Section 6.2, the $\ln r$ term presents a weak singularity and the integrals can still be evaluated, albeit with care. The $1/r$ term, on the other hand, presents a strong singularity and cannot easily be evaluated. Fortunately, these strongly singular terms can be evaluated directly without having to calculate the singular integrals as shown in Sections 6.9 and 6.14.

6.8 ASSEMBLY OF BOUNDARY ELEMENT MATRICES

The contributions $[G_n]$ and $[F_n]$ of the element n with boundary Γ_n were obtained in Equations 6.43 and 6.45, respectively. Assuming that the boundary Γ of the object is smooth and without any sharp corners, the tractions at a node shared by two neighboring elements are continuous, otherwise they are not. Regardless of whether the boundary is smooth or not, the displacements at such shared nodes are always continuous. The continuity of displacements at shared nodes of two neighboring elements is exploited to perform the assembly of boundary element matrices. The assembly process for the $[G_n]$ and $[F_n]$ contributions arising from three neighboring elements $(n - 1)$, n, and $(n + 1)$ is shown in Figure 6.8 corresponding to the case when the load point, P, is located at node j. The double-hatched areas denote the assembly of two matrices where the contributions from these matrices are added to each other. These areas contain contributions corresponding to the node shared by two adjacent elements. A boundary discretization containing N nodes in a two-dimensional analysis will therefore lead to an assembled $[F]$ matrix of size $2N \times 2N$. The assembly of the matrices $[G_n]$ cannot be performed similarly, as the tractions at nodes shared by elements may not be continuous. One strategy for the assembly of these matrices is shown in Figure 6.9. The contributions arising from the neighboring elements $(n - 1)$, n, and $(n + 1)$ are simply placed next to each other. For a boundary element discretization consisting of N nodes and NEL elements in a two-dimensional analysis, the assembled $[G]$ matrix is of size $2N \times 6\text{NEL}$. The resulting [G] matrix is rectangular and therefore cannot be inverted.

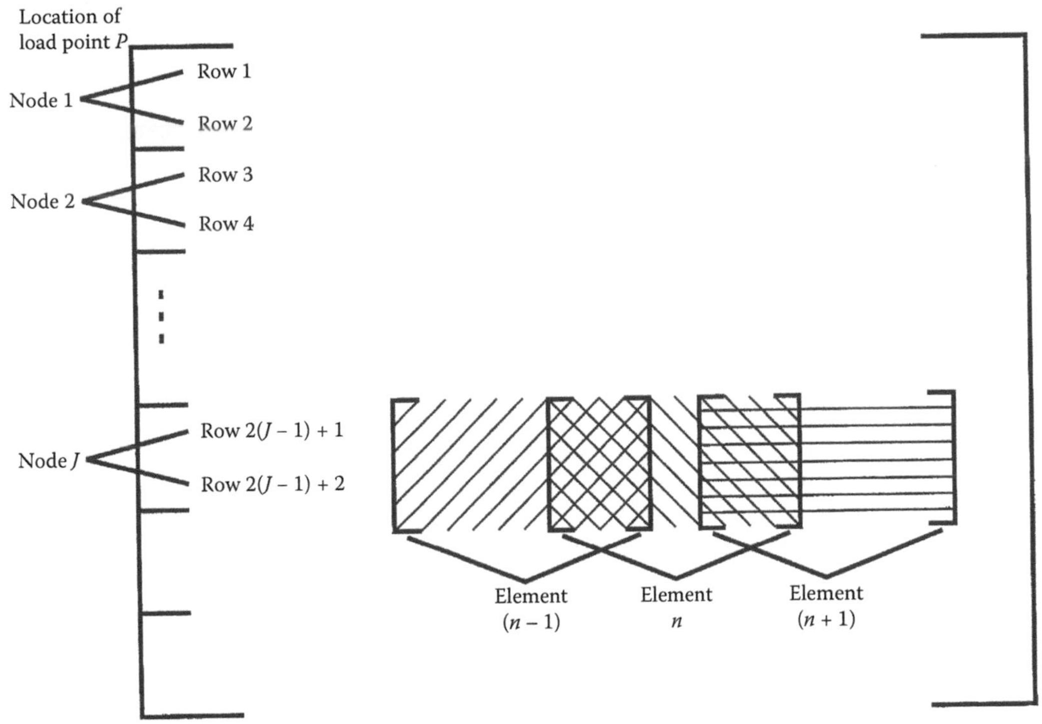

Figure 6.8 Assembly of two-dimensional boundary element matrices $[F_n]$.

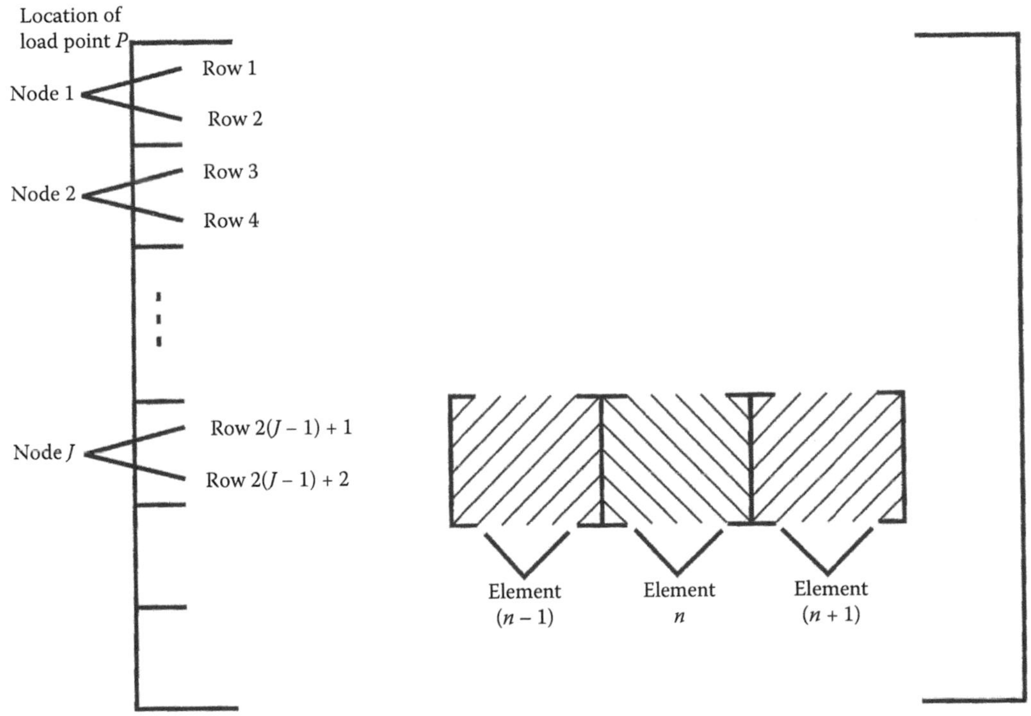

Figure 6.9 Assembly of two-dimensional boundary element matrices $[G_n]$.

Upon completing the assembly process and absorbing the $\varepsilon\delta_{ij}$ coefficient term with the corresponding coefficients for $u_i(P)$ on the right-hand side, Equation 6.47 is written in matrix form as

$$[F]\{u\}=[G]\{t\} \tag{6.48}$$

A solution of Equation 6.48 must be performed to obtain the unknown displacements and tractions at the boundary.

6.9 RIGID-BODY MOTION

Before performing the solution of the unknowns in Equation 6.48, attention is focused again on the elements of matrix $[F]$. For a two-dimensional analysis, the 2×2 diagonal matrix elements of this matrix need to be obtained by computing strongly singular integrations. As pointed out in Section 6.7, these terms arise when the load point P is at the node point j and the contributions to the surface integral on the right-hand side of Equation 6.47 from an element that contains node j are considered. In this case, the quantity r, the distance between the source point and field point, approaches zero. Since r appears in the fundamental solutions, which are included in the integrands in Equation 6.47, the integrals are of a singular nature. For two-dimensional analysis, the singularities are due to an ln r term present in the displacement solution U_{ij}^* and an $1/r$ term present in the traction solution T_{ij}^*. Referring to the discussion on singular one-dimensional integrals in Section 6.2, the ln r term presents a weak singularity and the integrals can still be evaluated, albeit with care. The $1/r$ term, on the other hand, presents a strong singularity and cannot be easily evaluated. A numerical procedure for such integrals is not desirable, as it may require excessive computational effort and still lead to erroneous results. The off-diagonal terms are, however, all easily computed using appropriate numerical integration schemes. Fortunately, the integrals for strongly singular 2×2 matrix contributions do not need to be computed and can be obtained using the rigid-body motion property of solid objects. This property states that a rigid-body motion of the object does not lead to the development of tractions on the body. Consider the two rigid-body motions given as

$$\{u\}=\begin{Bmatrix}1\\0\\1\\0\\\vdots\\\vdots\\1\\0\end{Bmatrix} \quad \{u\}=\begin{Bmatrix}0\\1\\0\\1\\\vdots\\\vdots\\0\\1\end{Bmatrix} \tag{6.49}$$

In Equation 6.49, the first $\{u\}$ vector denotes a rigid-body motion of magnitude unity in the x direction, while the second $\{u\}$ vector denotes a rigid-body motion of magnitude unity in the y direction. Under either of these rigid-body motions, the traction vector $\{t\} = \{0\}$, making the right-hand side of Equation 6.48 equal to zero. Substituting this result in Equation 6.48 first for the case of unit rigid-body motion in the x direction, we obtain

$$\begin{bmatrix} F_{11} & F_{12} & \cdots & \cdots & \cdots & \cdots & \cdots & F_{1,2N} \\ \cdots & \cdots & \cdots & \cdots & \cdots & \cdots & \cdots & \cdots \\ \cdots & \cdots & \cdots & \cdots & \cdots & \cdots & \cdots & \cdots \\ F_{k,1} & F_{k,2} & \cdots & F_{k,k} & F_{k,k+1} & \cdots & \cdots & F_{k,2n} \\ \cdots & \cdots & \cdots & \cdots & \cdots & \cdots & \cdots & \cdots \\ \cdots & \cdots & \cdots & \cdots & \cdots & \cdots & \cdots & \cdots \\ \cdots & \cdots & \cdots & \cdots & \cdots & \cdots & \cdots & \cdots \\ F_{2N,1} & F_{2N,2} & \cdots & \cdots & \cdots & \cdots & \cdots & F_{2N,2N} \end{bmatrix} \begin{Bmatrix} 1 \\ 0 \\ \vdots \\ 1 \\ 0 \\ \vdots \\ 1 \\ 0 \end{Bmatrix} = \begin{Bmatrix} 0 \\ 0 \\ 0 \\ 0 \\ 0 \\ 0 \\ 0 \\ 0 \end{Bmatrix} \tag{6.50}$$

where for $k = 2(j - 1) + 1$, F_{kk} and $F_{k,k+1}$ represent two of the four singular terms in the 2×2 diagonal submatrix arising from considering the load point P to lie on node j. The other two singular terms appear in the equation for the row $k + 1$. Expanding row k of Equation 6.50 and rearranging yields

$$F_{k,k} = -(F_{k,1} + F_{k,3} + \cdots + F_{k;2N-1}) \tag{6.51}$$

The term $F_{k,k+1}$ can be similarly obtained by considering the unit rigid-body motion in the y direction. The singular diagonal terms can thus be computed as the negative sum of the off-diagonal terms and do not need to be computed from the corresponding singular integrals. This property also leads to the diagonal dominance of the assembled $[F]$ matrix.

6.10 SOLUTION OF BOUNDARY ELEMENT EQUATIONS

The system of assembled boundary element equations in Equation 6.48 is solved to obtain the unknown displacements and tractions at the boundary. The matrices $\{u\}$ and $\{t\}$ in Equation 6.48 each contains both the known (through prescribed boundary conditions) and the unknown respective components. The system shown in Equation 6.48 is partitioned as

$$\left[\begin{array}{c|c} [F_{uu}] & [F_{uk}] \\ \hline [F_{ku}] & [F_{kk}] \end{array}\right] \begin{Bmatrix} \{u_u\} \\ \hline \{u_k\} \end{Bmatrix} = \left[\begin{array}{c|c} [G_{uu}] & [G_{uk}] \\ \hline [G_{ku}] & [G_{kk}] \end{array}\right] \begin{Bmatrix} \{t_u\} \\ \hline \{t_k\} \end{Bmatrix} \tag{6.52}$$

The subscripts u and k in Equation 6.52 denote unknown and known components, respectively. Thus, $\{u_k\}$ and $\{t_k\}$ are the known nodal displacements and tractions obtained from the boundary conditions prescribed. Similarly, $\{u_u\}$ and $\{t_u\}$ are the unknown nodal displacements and tractions. Expanding the matrix equations in Equation 6.52 and rearranging to have all unknowns on the left-hand side, we obtain

$$\left[\begin{array}{c|c} [F_{uu}] & [-G_{uu}] \\ \hline [F_{ku}] & [-G_{ku}] \end{array}\right]\left\{\begin{array}{c} \{u_u\} \\ \hline \{t_u\} \end{array}\right\}=\left[\begin{array}{c|c} [-F_{uk}] & [G_{uk}] \\ \hline [-F_{kk}] & [G_{kk}] \end{array}\right]\left\{\begin{array}{c} \{u_k\} \\ \hline \{t_k\} \end{array}\right\} \tag{6.53}$$

All terms on the right-hand side of Equation 6.53 are known. After performing the matrix multiplication, the right-hand side can be replaced by a column matrix $\{b\}$. Equation 6.53 can then be written in the standard form for linear matrix equations as

$$[A]\{x\}=\{b\} \tag{6.54}$$

where [A] is the square coefficient matrix on the left-hand side in Equation 6.53 and contains contributions from both $[F]$ and $[G]$ matrices. The matrix $\{x\}$ contains all the unknown nodal displacements and tractions on the boundary for the system, and $\{b\}$ is a known matrix. The matrix $[A]$ is fully populated and unsymmetric, as are the matrices $[F]$ and $[G]$ in Equation 6.48. Equation 6.54 may be solved using a direct or an iterative solver for fully populated, unsymmetric matrices.

It is clear from Equation 6.48 that the terms of the $[F]$ and $[G]$ matrices are not of comparable magnitude. The introduction of terms from both $[F]$ and $[G]$ matrices in matrix $[A]$ may lead to ill-conditioning of matrix $[A]$. The solution of Equation 6.54 may then lead to numerical difficulties as well as erroneous results. The contributions from the $[G]$ matrix to the matrix $[A]$ are commonly scaled as

$$\left[\begin{array}{c|c} [F_{uu}] & \alpha[-G_{uu}] \\ \hline [F_{ku}] & \alpha[-G_{ku}] \end{array}\right]\left\{\begin{array}{c} \{u_u\} \\ \hline \{t_u/\alpha\} \end{array}\right\}=\{b\} \tag{6.55}$$

where α is a scaling parameter. For elasticity problems, α is taken to be the shear modulus.

6.11 DISPLACEMENT AT POINTS IN THE INTERIOR

The solution of Equation 6.54, or more precisely Equation 6.55, yields the unknown displacements $\{u_u\}$ and the unknown tractions $\{t_u\}$. Together with the known $\{u_k\}$ and $\{t_k\}$ matrices, the displacements $\{u\}$ and $\{t\}$ on the entire boundary Γ are known. It is, however, often desirable to know the displacements at one or several points within the interior of the body. Recall that Equation 6.47, with an appropriate value of ε, is valid both inside the body and on its boundary. For a point P inside the boundary, $\varepsilon=\frac{1}{2}$. Equation 6.47 can then be used directly to obtain the displacement at any interior location P. Since the load point P is inside the boundary and the field point Q is on the boundary, the quantity r in the fundamental solutions U_{ij}^* and T_{ij}^* in Equations 6.15 and 6.16 is never zero. Thus, no singular integrals are involved in calculating displacements at the interior points. The matrices $[U^*]$ and $[T^*]$ in Equation 6.47 must be computed anew for each interior location P, for which displacements are required. Thus, a significant computational effort is expended if the displacements at a large number of points within the interior are required. This is unlike the finite element method, where the solution of the finite element matrix equations at once yields the displacements of all interior as well as boundary points of the object.

6.12 BODY FORCES

The body forces acting on the elastic object under consideration have been neglected so far to enable us to obtain a boundary-only system of integral equations. The contribution due to body forces (gravity and thermally induced forces, etc.) results in an additional volume integral term given as (from Equation 6.31)

$$\int_{\Omega} F_i(Q)U_{ij}^*(P,Q)\, d\Omega \tag{6.56}$$

To evaluate this volume integral, the domain of the body must be discretized into integration cells. Equation 6.56 does not contain the unknown displacement and traction components of vector $\{x\}$ of Equation 6.54. The contribution from this term augments the right-hand side matrix $\{b\}$ in Equation 6.54. Thus, while the domain of the body is discretized into integrations cells (much like the finite elements), unlike the finite element method no additional unknowns corresponding to the interior of the body are introduced. Regardless, the appearance of the volume integral and its subsequent evaluation by discretizing the domain reduces the attractiveness of the boundary element method. Some of the limitations of the finite element method associated with meshing arbitrary regions are then believed by most analysts to apply also to the boundary element method. It is noted, however, that ill-shaped integration cells do not seriously affect the quality of results obtained using the boundary elements. On the other hand, finite elements of undesirable shape may lead to serious errors in analysis. Numerous efforts have been made in the literature to eliminate the domain integral due to body forces and replace it by equivalent surface integrals. Three of the prominent techniques in this regard are (1) the particular integral approach, (2) the dual reciprocity approach, and (3) the multiple reciprocity approach. These approaches are quite similar to each other. The particular integral approach appears to be quite straightforward and is explained next in further detail.

6.13 PARTICULAR INTEGRAL APPROACH

As stated in Section 6.12, the inclusion of the effect of body forces through a boundary-only formulation presents significant advantages for the boundary element method. The particular integral approach (Pape and Banerjee, 1987) provides a straightforward yet highly general approach to achieve this objective. This approach merely involves nodal evaluations and does not even require the computation of boundary integrals. The equilibrium statement in Equation 6.8 can be written in terms of displacements through the substitution of stress and strain relations given in Equations 6.10 and 6.11, respectively. The resulting equation, called *Navier's equation of elasticity*, is given as (Boresi et al., 2011)

$$(\lambda+\mu)\mu_{j,ij} + \mu u_{i,jj} + F_i = 0 \tag{6.57a}$$

where λ and μ are the Lamé elastic coefficients. In the operator form, Equation 6.57a may be written as

$$\Lambda(u) + F_i = 0 \tag{6.57b}$$

where Λ is a linear differential operator. From standard calculus (Boyce and DiPrima, 1977), the solution u to a nonhomogeneous differential equation may be expressed as the sum of a complementary solution, u^c, and a particular integral solution, u^p, satisfying

$$\Lambda(u^c) = 0 \tag{6.58}$$

and

$$\Lambda(u^p) = -F_i \tag{6.59}$$

with

$$u_i = u_i^c + u_i^p \tag{6.60}$$

The superscripts c and p denote complementary and particular solutions, respectively. Equation 6.58 is, indeed, the equilibrium equation with zero body force for which we earlier developed the boundary integral solution. The complementary solution u^c therefore satisfies the matrix relation in Equation 6.48 and is written as

$$[F]\{u^c\} = [G]\{t^c\} \tag{6.61}$$

Equation 6.59 may now be solved directly to obtain the particular solution u^p. The particular solution may be obtained easily if the body forces are given in terms of polynomials. For nonuniform distributions of body forces, these may be approximated by polynomials using least squares (Saigal et al., 1990b), Fourier series (Zeng and Saigal, 1991), or other simple techniques. For several cases, the particular solution may even be obtained by inspection. It is noted here that the particular integral for u^p satisfying Equation 6.59 is not unique. Any polynomial or a linear combination of polynomials satisfying Equation 6.59 is a valid particular integral solution. Assuming that the particular integral solution is available, the complementary solution, u^c, is expressed by rearranging Equation 6.60 and writing it in the matrix form as

$$\{u^c\} = \{u\} - \{u^p\} \tag{6.62}$$

Substituting Equation 6.62 into 6.61 and rearranging terms lead to

$$[F]\{u\} = [G]\{t\} + \{f\} \tag{6.63}$$

where

$$\{f\} = [F]\{u^p\} - [G]\{t^p\} \quad \text{and} \quad \{t^p\} = \{t\} - \{t^c\} \tag{6.64}$$

Thus, the body forces using the particular integral approach lead merely to the addition of a force term, $\{f\}$, on the right-hand side of boundary element equations. The matrices $[F]$ and $[G]$ required for computing $\{f\}$ are already available. The particular solutions $\{u^p\}$ and $\{t^p\}$ required to compute $\{f\}$ are direct nodal evaluations and do not require the computations of any volume or boundary integrals.

To illustrate the computation of particular integral solution, consider the common case of gravity loading due to self-weight. This loading is expressed mathematically for two-dimensional analysis as

$$F_1 = 0 \qquad F_2 = -\gamma \tag{6.65}$$

where γ is the weight density. Substituting Equation 6.65 into Navier's equation of elasticity (Equation 6.57), the solution, u^p, using the Galerkin vector (Boresi et al., 2011, Chapter 8) can be written as

$$u_1^p = \frac{-\lambda}{4\mu(\lambda+\mu)}\gamma x_1 x_2 \tag{6.66}$$

and

$$u_2^p = \frac{1}{8\mu(\lambda+\mu)}\gamma\left[(\lambda+2\mu)^2 x_2^2 + \lambda x_1^2\right] \tag{6.67}$$

Substituting the displacement relations in Equations 6.66 and 6.67 into the strain–displacement relations of Equation 6.10 and thence into the stress–strain relations of Equation 6.11, the stresses due to self-weight are obtained as

$$\sigma_{11}^p = \sigma_{12}^p = 0 \qquad \sigma_{22}^p = \gamma x_2 \tag{6.68}$$

Using the relation $t_i = \sigma_{ij} n_j$, the corresponding tractions are obtained as

$$t_1^p = 0 \qquad t_2^p = y x_2 n_2 \tag{6.69}$$

where x_1 and x_2 are the coordinates of the point where these traction quantities are being evaluated. Thus, for self-weight body forces, the particular solutions $\{u^p\}$ and $\{t^p\}$ are obtained using Equations 6.66, 6.67, and 6.69. It is noteworthy that for this case, the matrices $\{u^p\}$ and $\{t^p\}$ are direct nodal evaluations at boundary nodes. It is clear that the particular integral approach provides a powerful tool to account for body forces in boundary elements while retaining the boundary-only character of the method. The particular integral approach has been extended to solve a wide variety of problems that result in volume integrals. These include free vibration analysis (Wang and Banerjee, 1990; Wilson et al., 1990), thermoelastic analysis (Saigal et al., 1990b), and elastoplasticity, among others.

6.14 EVALUATION OF STRESSES AND STRAINS

The stresses and strains within a body are often the engineering quantities of interest in an analysis. A procedure will now be formulated for the boundary element method that yields these quantities given the displacements and tractions on the boundary of the object. A direct procedure may be to substitute the relation for displacement $u_i(P)$ given in Equation 6.47 into the strain-displacement relation of Equation 6.10 to obtain the strains at point P. Substituting these strains into the stress-strain relations in Equation 6.11 can next yield the stresses at point P. It is noted, however, that for determining boundary stresses, the integral relation in Equation 6.47 contains U^*_{ij}, which for two-dimensional analysis leads to a strongly singular integral due to the presence of $1/r$ terms in the integrand. A differentiation of Equation 6.47 as required by the strain–displacement relation in Equation 6.10 will yield $1/r^2$ terms that lead to a hypersingular integral. Several techniques have been developed recently for the treatment of hypersingular integrals, and much research in this direction is currently ongoing.

A procedure is presented here that avoids the evaluation of hypersingular integrals and allows the computation of boundary stresses directly from boundary displacements and tractions. The displacements and tractions at nodal locations on the boundary are known through the solution of Equation 6.54. These quantities at any other location on the boundary can easily be obtained using shape functions as given in Equations 6.37a and 6.37b. A local coordinate system is now constructed at the point P of interest. The coordinate axes for this local coordinate system lie along the tangents and the normal to the surface at that point, as shown in Figure 6.10. The displacements and tractions at point P with respect to the local coordinate system are now obtained as

$$\{u'\} = [R][N]\{u\} \tag{6.70}$$

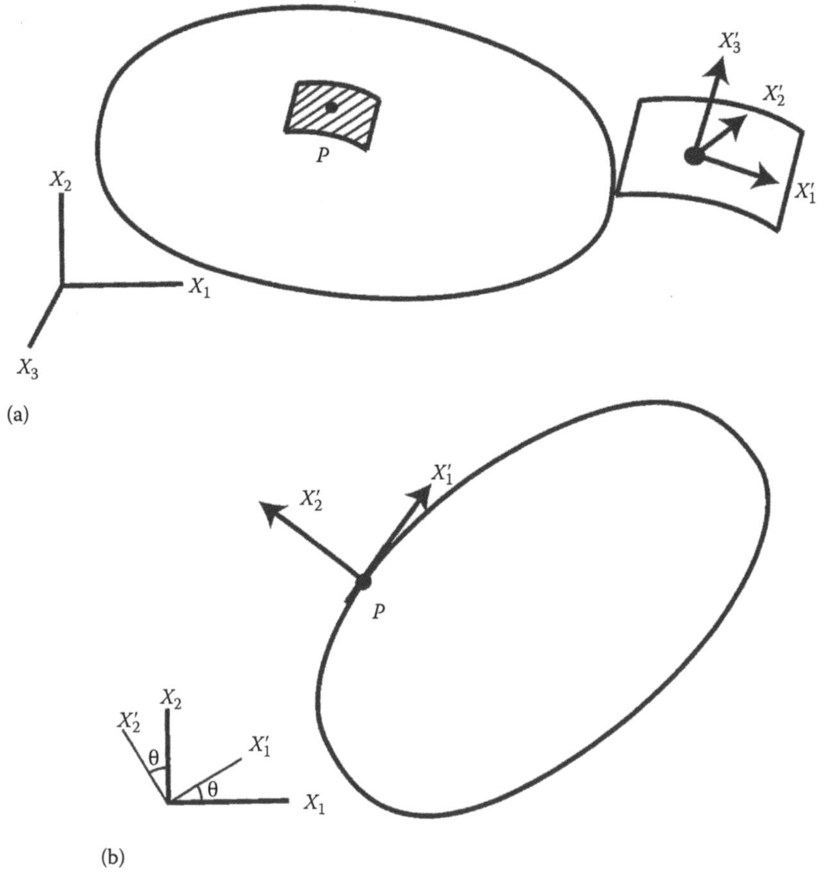

Figure 6.10 Local coordinate system at a point on the boundary. (a) Three-dimensional object. (b) Two-dimensional object.

and

$$\{t'\} = [R][N]\{t\} \tag{6.71}$$

where $\{u\}$ and $\{t\}$ are the nodal displacements vectors, $[N]$ are the shape functions matrices to obtain displacements and tractions at the point P of interest, and $[R]$ is the coordinate transformation matrix. For two-dimensional analysis, the matrix $[R]$ is given as

$$[R] = \begin{bmatrix} \cos\theta & \sin\theta \\ -\sin\theta & \cos\theta \end{bmatrix} \tag{6.72}$$

where θ is the angle between the coordinate axes x_1 and x_1' (Figure 6.10b).

It is noted that several stress components in the local coordinate systems are already available as

$$\begin{aligned} \sigma'_{13} &= \sigma'_{31} = t'_1 \\ \sigma'_{23} &= \sigma'_{32} = t'_2 \\ \sigma'_{33} &= t'_3 \end{aligned} \tag{6.73}$$

where the primed tractions were calculated in Equation 6.71. The strains in the local coordinate system are computed as

$$\varepsilon'_{ij} = \frac{1}{2}\left(u'_{j,i} + u'_{i,j}\right) \tag{6.74}$$

A typical derivative, $u'_{i,j}$, in Equation 6.74 needs to be computed to obtain the strain $\varepsilon'_{i,j}$. Using the chain rule of differentiation gives us

$$\frac{\partial u'_i}{\partial x'_j} = \frac{\partial u'_i}{\partial \xi}\frac{\partial \xi}{\partial x'_j} = \frac{\partial u'_i}{\partial \xi} J^{-1} \tag{6.75}$$

where $J = \partial x'_j/\partial \xi$ is the Jacobian at point P. The term $\partial u'_i/\partial \xi$ on the right-hand side of Equation 6.75 is computed using shape functions $[N]$. The displacements at point P in the local coordinate system are expressed as

$$u'_i = \sum_{k=1}^{\mathrm{NNODE}} N_k u_i'^{(k)} \tag{6.76}$$

where NNODE represents the number of nodes of the element within which the point P lies and $u_i'^{(k)}$ represents the nodal displacements of that element in the local coordinate system. The derivative $\partial u'_i/\partial \xi$ is then obtained as

$$\frac{\partial u'_i}{\partial \xi} = \sum_{k=1}^{\mathrm{NNODE}} \frac{\partial N_k}{\partial \xi} u_i'^{(k)} \tag{6.77}$$

The Jacobian, J, is similarly written as

$$\frac{\partial x_i'}{\partial \xi} = \sum_{k=1}^{\text{NNODE}} \frac{\partial N_k}{\partial \xi} x_i'^{(k)} \tag{6.78}$$

where $x_j'^{(k)}$ are the nodal coordinates of node k expressed in the local coordinate system.

The derivatives $\partial N_k/\partial \xi$ are obtained through differentiation of relations in Equation 6.36. Using Equations 6.77 and 6.78, the strain components ε_{ij}' can now be obtained easily. In the final step, these strains are substituted in the stress–strain relations given in Equation 6.11 to obtain the remaining stress components in the local coordinate system as

$$\begin{aligned} \sigma_{12}' &= \sigma_{21}' = 2\mu\varepsilon_{12}' \\ \sigma_{11}' &= \frac{1}{1-\upsilon}[\upsilon\sigma_{33}' + 2\mu(\varepsilon_{11}' + \varepsilon_{22}')] \end{aligned} \tag{6.79}$$

where ν is Poisson's ratio. The stress components σ_{ij}' in local coordinate system are all available from Equations 6.73 and 6.79. The stress components σ_{ij} may now be obtained from σ_{ij}' using a tensor coordinate transformation (Chung, 1988).

6.15 CORNER PROBLEM IN THE BOUNDARY ELEMENT METHOD

The presence of sharp corners in the boundary of an object causes difficulties in the boundary element analysis approach. This is primarily because tractions at such sharp corners are not continuous. In two dimensions, for example, two additional traction components exist at a node located at a sharp corner compared to when it is located at a smooth corner, as shown in Figure 6.11. The problem may first be realized for the case when displacements $\{u\}$ are prescribed everywhere on the boundary. For this case, from Equation 6.48, it is seen that we will need to solve the matrix equation

$$[G]\{t\} = \{f_u\} \tag{6.80}$$

where $\{f_u\}$ = $[F]\{u\}$ is the force term on the right-hand side due to the prescribed displacements $\{u\}$. The matrix $[G]$, however, is rectangular, as explained earlier, and a solution to Equation 6.80 may or may not be obtained. In general, for two-dimensional analysis, only two unknowns may exist at any node location. These could be displacements, tractions, or a combination of displacement and traction. If there are, indeed, only two unknowns at a node, no problems arise and the boundary element solution is easily obtained. Although this is always true at a smooth boundary, more than two unknowns may exist at a sharp corner. At such corners, six variables—two displacement and four traction components—exist. If four of these are prescribed through boundary conditions, leaving two of them unknown, no problems exist. If, on the other hand, less than four of these variables are prescribed, the boundary element method fails to yield a solution. Consider, for example, the sharp corner shown in Figure 6.12. The displacements on both faces meeting at the corner node are prescribed. Thus, the corner node in this situation will have four unknown variables, two traction components from each of the faces meeting at the sharp corner. This will result in more unknowns than the number of equations available from the overall boundary element system. A number of

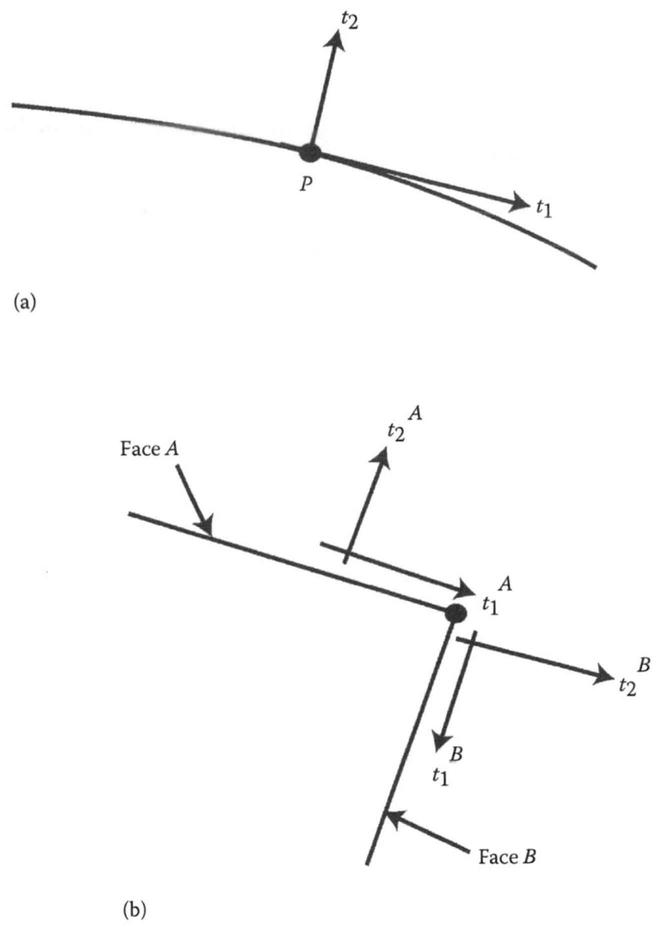

Figure 6.11 Traction components at a node on a smooth boundary and at a sharp corner. (a) Smooth boundary. (b) Sharp corner.

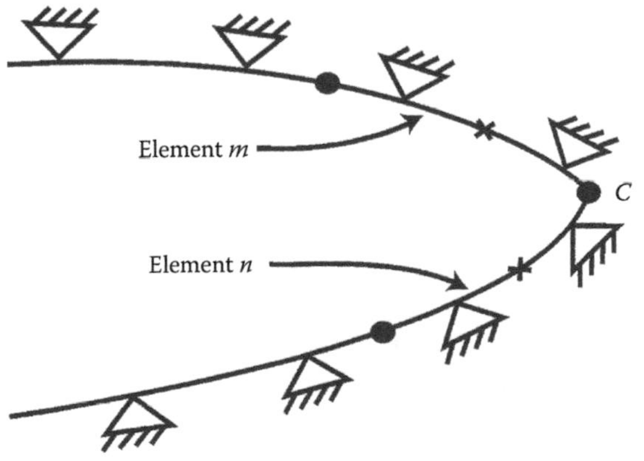

Figure 6.12 Sharp corner with "corner problem" in boundary element approach.

remedies to the corner problem have been proposed. Some of them are ad hoc in nature, while others are based on advanced concepts in elasticity. A few of these remedies are described here.

Corner smoothing: The sharp corner is eliminated by smoothing the corner. The resulting geometry is smooth; thus, no problem exists. This procedure may be unreliable for cases where the presence of corner is important to the overall response of the body.

Double-node representation: Two nodes, instead of one, are provided at the sharp corner. Thus, for the sharp corner shown in Figure 6.12, element m will have a node C_1 at the corner C, while the element n will have a node C_2 at the same corner. The two nodes C_1 and C_2 are carried in the assembled boundary element equations (Equation 6.48) as two separate nodes; thus, each can have its own set of tractions. It is noted that each row in the matrices $[F]$ and $[G]$ of Equation 6.48 corresponds to evaluation of the boundary integrals in Equation 6.47 with respect to a node on the boundary being a load point. Thus, the two rows corresponding to nodes C_1 and C_2 lying at the same location will be identical, leading to a singular $[F]$ matrix. To avoid this situation, the two nodes are typically separated by a small distance ε. This distance must be carefully prescribed. A small value may lead to a singular $[F]$ matrix, while a large value may alter the physical geometry of the object.

Discontinuous boundary elements: Special boundary elements, known as *discontinuous boundary elements*, may be formulated for use in the vicinity of the corner point. These special elements have one or both of their end nodes not lie at the ends of the elements but at a small distance inside. This requires the use of modified shape functions to account for the new location of the end node. Since the geometry of the boundary is also interpolated using the same end nodes and the same shape functions, the use of discontinuous boundary elements may lead to geometries with gaps and overlaps at the end. It is also possible to formulate a boundary element using a different set of shape functions and end nodes for interpolating the geometry and the solution variables (tractions and displacements). Discontinuous boundary elements were popular in the earlier implementations of boundary elements but have largely been abandoned, due to their poor performance.

Finite element at comer: The corner problem in boundary elements may also be resolved by providing a finite element at the corner as shown in Figure 6.13. This approach is not very popular and may lead to ambiguity in the choice of normal required in formulation of the finite element.

Several other approaches have been proposed in the boundary element literature. Typically, these approaches provide two additional equations based on advanced elasticity concepts (see, e.g., Chaudonneret, 1978). These equations augment the boundary element equations of Equation 6.48 to account for the special behavior at the corner. The inclusion of these

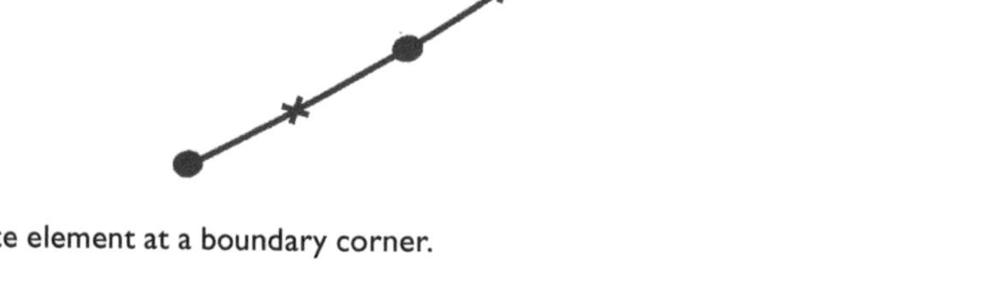

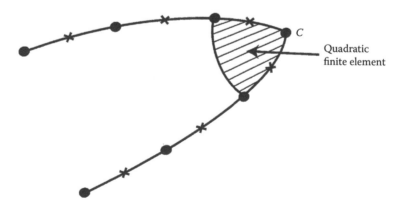

Figure 6.13 Finite element at a boundary corner.

additional equations complicates the assembly procedure and is not very desirable. Most commercial implementations use the double-node representation of the sharp corner to treat the corner problem.

6.16 CLOSING REMARKS

The boundary element method has emerged as a powerful numerical analysis tool in the last three decades. Many engineers erroneously believe the method to be applicable only to linear problems, perhaps due to the early difficulties with nonlinear boundary element formulations. The method has been applied for the solution of a wide variety of linear and nonlinear problems in materials (Mukherjee, 1982) and displacements (Banerjee and Cathie, 1980; Banerjee and Raveendra, 1986; Banerjee et al., 1989). Initial formulations of nonlinear problems required the discretization of the domain of the body into integration cells. Techniques such as the particular integral approach and the multiple reciprocity method have been extended to derive boundary-only integral relations for nonlinear problems. The boundary element method has had special success in far-field problems of geomechanics (Brebbia, 1988b), acoustics (Banerjee et al., 1988; Bernhard and Keltie, 1989), and electromagnetics (Brebbia, 1988a) due to its ability to account exactly for the far-field effects (Kaljevic et al., 1992). The method has also been applied to the analysis of inclusion problems (Banerjee and Henry, 1991, 1992), contact problems (Simunovic and Saigal, 1995), shape sensitivity analysis (Kane and Saigal, 1988; Saigal, 1989; Saigal et al., 1990a, Saigal and Chandra, 1991), thermoelasticity (Rizzo and Shippy, 1977; Saigal et al., 1990b), stochastic analysis (Ettouney et al., 1989; Kaljevic and Saigal, 1993, 1995b), fluid flow (Brebbia, 1989a), and the highly nonlinear problems of metalworking (Chandra and Mukherjee, 1984a, 1984b), including sheet metal forming, extrusion (Chandra and Saigal, 1991), and so on. While the boundary element method certainly does not enjoy the popularity of the finite element method with analysts, it has been shown to be a powerful method in its own right. Application of the boundary element method is largely restricted to bulky objects. Its application to thin geometries, especially in parts where thin geometries exist next to bulky parts, may lead to ill-conditioned matrices. This can be understood by looking at any individual row of the matrix $[F]$ in Equation 6.48 to correspond to a load point P on the boundary. For thin geometries, two adjacent nodes may lie approximately at the same location (especially in comparison with nodes for a neighboring bulky object). These two nodes will then yield nearly identical rows leading to a singular $[F]$ matrix. The problem may be resolved by providing very fine mesh in the bulky region, comparable in density to the mesh of the thin geometry. Another approach to avoid this problem is through the use of multiple zones (Kane et al., 1990), as shown in Figure 6.14. Separate boundary element relations are written for zones ABF, $BCEF$, and CDE, respectively. The overall boundary element matrices are assembled by considering continuity of displacements and tractions at interface surfaces between two zones. In such a procedure, the equations from a zone representing a thin geometry may be scaled appropriately to avoid numerical complications. An additional advantage of the zoning procedure lies in the fact that it imparts a sparse, banded structure to the overall boundary element matrices.

The absence of symmetry in the boundary element matrices has also been a cause for the boundary element method not attaining the popularity that was initially anticipated. This should cease to be a limitation, as the numerical analysis community continues to develop ever more robust algorithms for the solution of nonsymmetric systems of equations. Also, a number of symmetric boundary element formulations have been developed lately. Although

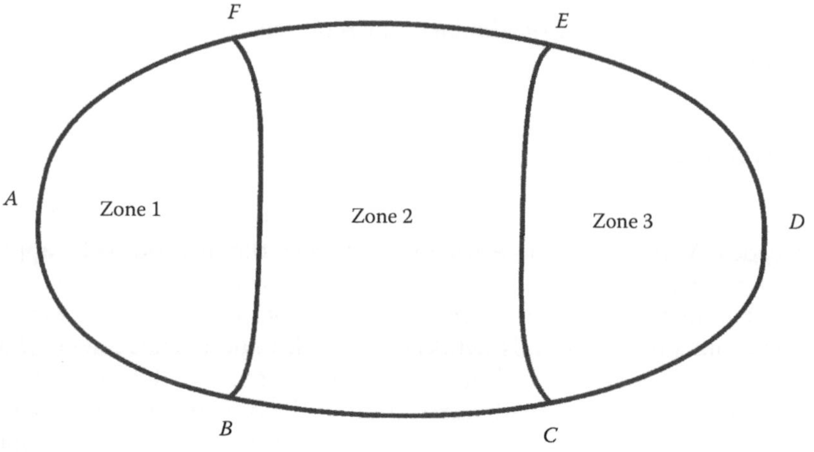

Figure 6.14 Multiple zones for boundary element analysis of a body.

these new formulations are not widely used, they have the potential to further enhance the attractiveness of the boundary element method.

REFERENCES

Ahmad, S., and Banerjee, P. K. 1986. Free-vibration analysis by BEM using particular integrals, *ASCE J. Eng. Mech.*, **112**, 682–695.

Balakrishna, C., Gray, L. J., and Kane, J. H. 1994. Efficient analytical integration of symmetric Galerkin boundary integrals over curved elements: Elasticity formulation, *Comput. Methods Appl. Mech. Eng.*, **117**, 157–179.

Banerjee, P. K. 1994. *Boundary Element Methods in Engineering*, McGraw-Hill, London.

Banerjee, P. K., Ahmad, S., and Wang, H. C. 1988. A new BEM formulation for the acoustic Eigen frequency analysis, *Int. J. Numer. Methods Eng.*, **26**, 1299–1309.

Banerjee, P. K., and Cathie, D. N. 1980. A direct formulation and numerical implementation of the boundary element method for two-dimensional problems of elastoplasticity, *Int. J. Mech. Sci.*, **22**, 233–245.

Banerjee, P. K., and Henry, D. 1991. Analysis of three-dimensional solids with holes by BEM, *Int. J. Numer. Methods Eng.*, **31**, 369–384.

Banerjee, P. K., and Henry, D. 1992. Analysis of 3-dimensional solids with inclusions by BEM, *Int. J. Solids Struct.*, **29**, 2423–2440.

Banerjee, P. K., Henry, D., and Raveendra, S. T. 1989. Advanced BEM for inelastic analysis of solids, *Int. J. Mech. Sci.*, **31**, 309–322.

Banerjee, P. K., and Raveendra, S. T. 1986. Advanced boundary element method for two- and three-dimensional problems of elastoplasticity, *Int. J. Numer. Methods in Eng.*, **23**, 985–1002.

Banerjee, P. K., and Wilson, R. B. (eds.) 1989. *Industrial Applications of Boundary Element Methods*, Elsevier Applied Science, New York.

Bernhard, R. J., and Keltie, R. F. (eds.) 1989. *Numerical Techniques in Acoustic Radiation*, ASME Press, New York.

Boresi, A. P., Chong, K. P., and Lee, J. D. 2011. *Elasticity in Engineering Mechanics*, 3rd ed., Wiley, New York.

Boyce, W. E., and DiPrima, R. C. 1977. *Elementary Differential Equations and Boundary Value Problems*, 3rd ed., Wiley, New York.

Brebbia, C. A. (ed.) 1988a. *Boundary Elements X, Vol. 2, Heat Transfer, Fluid Flow and Electrical Applications*, Computational Mechanics Publications, Ashurst, Southampton, England.

Brebbia, C. A. (ed.) 1988b. *Boundary Elements X, Vol. 4., Geomechanics, Wave Propagation and Vibrations*, Computational Mechanics Publications, Ashurst, Southampton, England.

Chandra, A., and Mukherjee, S. 1984a. A finite element analysis of metal forming problems with an elastic-viscoplastic material model, *Int. J. Numer. Methods Eng.*, **20**, 1613–1628.

Chandra, A., and Mukherjee, S. 1984b. A finite element analysis of metal forming processes with thermomechanical coupling, *Int. J. Mech. Sci.*, **26**, 661–676.

Chandra, A., and Saigal, S. 1991. A boundary element analysis of the axisymmetric extrusion processes, *Int. J. Solids Struct*, **26**, 1–13.

Chaudonneret, M. 1978. On the discontinuity of the stress vector in the boundary integral equation method for elastic analysis, in *Recent Advances in Boundary Element Methods*, Brebbia, C. A. (ed.), TA347.B69.R42, Pentech Press, London, pp. 185–194.

Chung, T. J. 1988. *Continuum Mechanics*, Prentice Hall, Upper Saddle River, NJ.

Churchill, R. V., and Brown, J. W. 1987. *Fourier Series and Boundary Value Problems*, McGraw-Hill, New York.

Ettouney, M., Benaroya, H., and Wright, H. 1989. Boundary element methods in probabilisitic structural analysis, *Appl. Math. Model.*, **13**, 432–441.

Gray, L. J., and Paulino, G. H. 1997. Symmetric Galerkin boundary integral formulation for interface and multi-zone problems, *Int. J. Numer. Methods Eng.*, **16**, 3085–3101.

Hartmann, F., Katz, C., and Protopsaltis, B. 1985. Boundary elements and symmetry, *Ing. Arch.*, **55**, 440–449.

Henry, D., and Banerjee, P. K. 1988a. A new boundary element formulation for two and three-dimensional problems of thermoelasticity using particular integrals, *Int. J. Numer. Methods Eng.*, **26**, 2061–2077.

Henry, D., and Banerjee, P. K. 1988b. A new boundary element formulation for two and three-dimensional problems of elastoplasticity using particular integrals, *Int. J. Numer. Methods Eng.*, **26**, 2079–2096.

Henry, D. P., Jr., Pape, D. A., and Banerjee, P. K. 1987. A new axisymmetric BEM formulation for body forces using particular integrals, *ASCE J. Eng. Mech.*, **113**, 671–688.

Kaljevic, I., and Saigal, S. 1993. Stochastic boundary elements in elastostatics, *Comput. Methods Appl. Mech. Eng.*, **109**, 259–280.

Kaljevic, I., and Saigal, S. 1995a. Analysis of symmetric domains in advanced applications with the boundary element method, *Int. J. Numer. Methods Eng.*, **38**, 2373–2388.

Kaljevic, I., and Saigal, S. 1995b. Stochastic boundary elements for two-dimensional potential flow in homogeneous domains, *Int. J. Solids Struct.*, **32**, 1873–1892.

Kaljevic, I., Saigal, S., and Ali, A. 1992. An infinite boundary element formulation for three-dimensional potential problems, *Int. J. Numer. Methods Eng.*, **35**, 2079–2100.

Kane, J. H., Kumar, B. L. K., and Saigal, S. 1990. An arbitrary condensing, noncondensing solution strategy for large scale, multi-zone boundary element analysis, *Comput. Methods Appl. Mech. Eng.*, **79**(2), 219–244.

Kane, J. H., and Saigal, S. 1988. Design sensitivity analysis of solids using BEM, *ASCE J. Eng. Mech.*, **114**, 1703–1722.

Layton, J. B., Ganguly, S., Balakrishna, C., and Kane, J. H. 1997. A symmetric Galerkin multi-zone boundary element formulation, *Int. J. Numer. Methods Eng.*, **16**, 2913–2932.

Love, A. E. H. 2013. *A Treatise on the Mathematical Theory of Elasticity*, Cambridge University Press, New York.

Maier, G., Novati, G., and Sirtori, S. 1990. On symmetrization in boundary element elastic and elastoplastic analysis, in *Discretization Methods in Structural Mechanics*, Kuhn, G., and Mang, H. (eds.), Springer-Verlag, Berlin, Germany, pp. 191–200.

Mukherjee, S. 1982. *Boundary Element Methods in Creep and Fracture*, Elsevier Applied Science, Barking, Essex, England.

Neves, A. C., and Brebbia, C. A. 1991. Multiple reciprocity boundary element method in elasticity: A new approach for transforming domain integrals to the boundary, *Int. J. Numer. Methods Eng.*, **31**(4), 709–727.

Niku, S. M., and Brebbia, C. A. 1988. Dual reciprocity boundary element formulation for potential problems with arbitrarily distributed sources, *Eng. Anal.*, **5**, 46–48.

Norris, C. H., Wilbur, J. B., and Utku, S. 1976. *Elementary Structural Analysis*, 3rd ed., McGraw-Hill, New York.

Nowak, A. J., and Neves, A. C. (eds.) 1994. *The Multiple Reciprocity Boundary Element Method*, Wessex Institute of Technology, Ashurst, Southampton, England.

Pape, D. A., and Banerjee, P. K. 1987. Treatment of body forces in 2-D elastostatic BEM using particular integrals, *ASME J. Appl. Mech.*, **54**, 866–871.

Partridge, P. W., and Brebbia, C. A. 1990. Computer implementation of the BEM dual reciprocity method for the solution of general field equations, *Commun. Appl. Numer. Methods*, **6**, 83–92.

Partridge, P. W., Brebbia, C. A., and Wrobel, L. C. (eds.) 1991. *Dual Reciprocity Boundary Element Method*, Springer Science & Business Media, Springer, Netherlands.

Rizzo, F. J., and Shippy, D. J. 1977. An advanced boundary integral equation method for three-dimensional thermoelasticity, *Int. J. Numer. Methods Eng.*, **11**, 1753–1768.

Saigal, S. 1989. Treatment of body forces in axisymmetric boundary element design sensitivity formulation, *Int. J. Solids Struct.*, **25**, 947–959.

Saigal, S., Aithal, R., and Dyka, C. T. 1990a. Boundary element design sensitivity analysis of symmetric bodies, *AIAA J.*, **28**(1), 180–183.

Saigal, S., and Chandra, A. 1991. Shape sensitivities and optimal configurations for heat diffusion problems: A BEM approach, *ASME J. Heat Transfer*, **113**, 287–295.

Saigal, S., Gupta, A., and Cheng, J. 1990b. Stepwise linear regression particular integrals for uncoupled thermoelasticity with boundary elements, *Int. J. Solids Struct.*, **26**, 471–482.

Saigal, S., and Kane, J. H. 1990. Boundary element shape optimization system for aircraft structural components, *AIAA J.*, **28**, 1203–1204.

Schey, H. M. 2005. *Div, Grad, Curl, and All That: An Informal Text on Vector Calculus*, W.W. Norton, New York.

Simunovic, S., and Saigal, S. 1995. A linear programming formulation for incremental contact analysis, *Int. J. Numer. Methods Eng.*, **38**, 2703–2725.

Sladek, V., and Sladek, J. 1996. Multiple reciprocity method in BEM formulations for solution of plate bending problems, *Eng. Anal. Boundary Elements*, **17**, 161–173.

Sokolnikoff, I. 1956. *Mathematical Theory of Elasticity*, McGraw-Hill, New York.

Somigliana, C. 1886. Sópra l'equilìbriodi un còrpo elàstico isotropo, *Nuovo Cimento*, pp. 17–19.

Stewart, J. 2010. *Calculus: Early Transcendentals*, Brooks/Cole Cengage Learning, Belmont, CA.

Wang, H. C., and Banerjee, P. K. 1990. Generalized axisymmetric free-vibration analysis by BEM, *Int. J. Numer. Methods Eng.*, **29**, 985–1001.

Wilson, R. B., Miller, N. M., and Banerjee, P. K. 1990. Calculations of natural frequencies and mode shapes for three-dimensional structures by BEM, *Int. J. Numer. Methods Eng.*, **29**, 1737–1757.

Wrobel, L. C., and Brebbia, C. A. 1987. Dual reciprocity boundary element formulation for nonlinear diffusion problems, *Comput. Methods Appl. Mech. Eng.*, **65**(2), 147–164.

Zeng, X., and Saigal, S. 1991. A Fourier expansion based particular integral approach for non-boundary loadings in boundary element method, *Commun. Appl. Numen. Methods*, 7, 453–464.

BIBLIOGRAPHY

Beer, G., and Meek, J. L. 1981. The coupling of boundary and finite element methods for infinite domain problems in elasto-plasticity, in *Boundary Element Methods*, Brebbia, C. A. (ed.), Springer Berlin, Heidelberg, Germany, pp. 575–591.

Brebbia, C. A., Telles, J. C. F., and Wrobel, L. 2012. *Boundary Element Techniques: Theory and Applications in Engineering*, Springer Verlag Berlin, Heidelberg, Germany.

Crouch, S. L., Starfield, A. M., and Rizzo, F. J. 1983. Boundary element methods in solid mechanics, *J. Appl. Mech.*, 50, 704.

Manolis, G. D., and Beskos, D. E. 1988. *Boundary Element Methods in Elastodynamics*, Unwin Hyman, London.

Partridge, P. W., and Brebbia, C. A., and Wrobel, L. C. (eds.) 2012. *Dual Reciprocity Boundary Element Method*, Springer Science & Business Media.

Sauter, S. A., and Schwab, C. 2010. Boundary element methods, in *Boundary Element Methods*, Springer Berlin, Heidelberg, Germany, pp. 183–287.

Chapter 7

Meshless methods of analysis

7.1 INTRODUCTION

A series of meshless methods for analysis have been introduced in the literature, mostly in the late 1990s. The development of these methods may have been motivated by the shortcomings arising in the difficulty of use of the finite element methods by analysts. Some of the shortcomings were outlined in Section 6.1. Methods in the category of meshless methods include smooth particle hydrodynamics (SPH) (Swegle et al., 1994), the element-free Galerkin (EFG) method (Belytschko et al., 1994a; Lu et al., 1994), reproducing kernel particle methods (RKPMs) (Liu, 1995; Liu and Chen, 1995; Liu et al., 1996), partition of unity methods (Melenk and Babuska, 1996), h-p clouds (Duarte and Oden, 1996), the meshless local Petrov–Galerkin (MLPG) method (Atluri, 2004; Atluri and Zhu, 1998, 2000a), the natural element method (Sukumar et al., 1998), the natural neighbor Galerkin methods (Sukumar et al., 2001), and the method of finite spheres (De and Bathe, 2000, 2001). Except for the SPH methods, these methods are all based, in principle, on the finite element method. Several additional meshless methods based on the boundary element method have also been proposed. These include boundary contour methods (Mukherjee et al., 1997; Nagarajan et al., 1994, 1996), the boundary node method (Mukherjee and Mukherjee, 1997a; Chati and Mukherjee, 2000; Chati et al., 1999), and the local boundary integral equation method (Atluri et al., 2000). Most of these methods are of recent origin and are still undergoing development. Some applications of meshless methods in modeling life-cycle engineering appeared in a book by Chong et al. (2002). A popular meshless method is the EFG method. It has been developed largely in the context of elastostatics. Chapters 4 and 6 have both been presented for use in elastostatics analysis. Continuing in that vein, the EFG method is also presented in this chapter for linear elastic analysis of solid bodies.

The step of meshing is a cumbersome, time-consuming step, especially for arbitrarily shaped three-dimensional objects. Meshless methods have been developed with the intent to eliminate the need to define a mesh for the object being analyzed. None of the methods developed to date, however, are truly meshless, as they still have the need to define a background grid. A significant shortcoming of these methods is that they involve a heavy computational effort compared to the finite element method. It is to be understood that these methods are all of recent origin and have not enjoyed the growth that the finite element method has seen over the last six decades. Despite their known shortcomings, these methods hold significant promise, as developments for their refinement and better understanding continue. The EFG method is covered first in detail in this chapter, followed by an introduction to some of the other meshless analysis methods.

7.2 EQUATIONS OF ELASTICITY

The analysis of linear elastic behavior of solid bodies (Section 6.3) is considered to illustrate the EFG formulation. The stress components σ_{ij} must satisfy the differential equations of equilibrium in a domain, Ω, of the body under consideration

$$\sigma_{ij,j} + F_i = 0 \tag{7.1}$$

where F_i is the component of the body force acting on the object. The previous field equations are to be solved given certain boundary conditions as

$$\begin{aligned} u_i &= \bar{u}_i && \text{on } \Gamma_u \\ t_i &= \bar{t}_i && \text{on } \Gamma_t \end{aligned} \tag{7.2}$$

where the tractions on a surface with normal n_j are given as $t_i = \sigma_{ij}n_j$, and $\bar{u}_i$ and $\bar{t}_i$ are the prescribed values of displacements and tractions on boundaries Γ_u and Γ_t, respectively. $\Gamma_u \cup \Gamma_t = \Gamma$ is the boundary of the subject domain Ω. The stress components σ_{ij} in Equation 7.1 are symmetric (i.e., $\sigma_{ij} = \sigma_{ji}$). The kinematic equations, relating the strain components ε_{ij} to the displacements u_i, are

$$\varepsilon_{ij} = \frac{1}{2}(u_{i,j} + u_{j,i}) \tag{7.3}$$

where a comma (,) denotes differentiation; thus, $u_{i,j} = \partial u_i/\partial x_j$. The constitutive relations for a linear elastic material are given as

$$\sigma_{ij} = \lambda\varepsilon_{kk}\delta_{ij} + 2\mu\varepsilon_{ij} = E_{ijkl}\varepsilon_{kl} \tag{7.4}$$

where λ and μ are Lamé's constants and E_{ijkl} is the constitutive tensor. Equations 7.1 to 7.4 summarize the equations of elastostatics for a linear, isotropic, homogeneous elastic object.

7.3 WEAK FORMS OF THE GOVERNING EQUATIONS

The first law of thermodynamics states that for a body in equilibrium and subjected to arbitrary virtual displacement δv_i, the variation in work of the external forces δW is equal to the variation of internal energy δU. These quantities are written as

$$\delta W - \delta U = \delta W_S + \delta W_B - \delta U = 0 \tag{7.5}$$

$$\delta W_S = \int_{\Gamma_t} \bar{t}_i \, \delta v_i \, d\Gamma \tag{7.6}$$

$$\delta W_b = \int_{\Omega} F_i \, \delta v_i \, d\Omega \tag{7.7}$$

and

$$\delta U = \int_{\Omega} \sigma_{ij}\, \delta\varepsilon_{ij}\, d\Omega \tag{7.8}$$

where $\delta\varepsilon_{ij} = \frac{1}{2}(\delta v_{i,j} + \delta v_{j,i})$ is the strain due to the virtual displacement field δv_i. Since $\sigma_{ij} = \sigma_{ji}$, the product $\sigma_{ij}\, \delta\varepsilon_{ij}$ may be expressed using Equation 7.3 as $\sigma_{ij}\, \delta v_{ij}$.

In the context of computational mechanics, the displacement field u_i that leads to the stress field σ_{ij} is referred to as the *trial function* and the virtual displacement field δv_i is referred to as the *test function*.

Both sides of the equilibrium relation in Equation 7.1 are multiplied by the test function and integrated over the domain Ω to give

$$\int_{\Omega} (\sigma_{ij,j} + F_i)\, \delta v_i\, d\Omega = 0 \tag{7.9}$$

Using the result $\sigma_{ij,j}\, \delta v_i = (\sigma_{ij}\, \delta v_i)_{,j} - \sigma_{ij}\, \delta v_{i,j}$ from differentiation by parts, we get

$$\int_{\Omega} (\sigma_{ij}\, \delta v_i)_{,j}\, d\Omega - \int_{\Omega} \sigma_{ij}\, \delta v_{i,j}\, d\Omega + \int_{\Omega} F_i\, \delta v_i\, d\Omega = 0 \tag{7.10}$$

The divergence theorem of calculus (Schey, 2005) for a vector field with components R_j is written as

$$\int_{\Omega} R_{j,k}\, d\Omega = \int_{\Gamma} R_j n_k\, d\Gamma \tag{7.11}$$

where Γ is the bounding surface for the volume Ω and n_k is the outward normal to the surface dΓ. Applying divergence theorem to the first term in Equation 7.10 and noting that $\sigma_{ij} n_j = t_i$, we obtain

$$\int_{\Gamma} t_i\, \delta v_i\, d\Gamma - \int_{\Omega} \sigma_{ij}\, \delta v_{i,j}\, d\Omega + \int_{\Omega} F_i\, \delta v_i\, d\Omega = 0 \tag{7.12}$$

The test functions are specified such that they satisfy the essential boundary conditions. Thus, $\delta v_i = 0$ everywhere on Γ except on the portion Γ_t where $t_i = \bar{t}_i$ from Equation 7.2, and Equation 7.12 may be written to obtain a weak form of the governing equations as

$$W_1: \quad \int_{\Gamma_t} t_i\, \delta v_i\, d\Gamma + \int_{\Omega} F_i\, \delta v_i\, d\Omega - \int_{\Omega} \sigma_{ij}\, \delta v_{i,j}\, d\Omega = 0 \tag{7.13}$$

It is noted that Equation 7.13 is the same relation as the virtual work relation in Equations 7.5 to 7.8. Equation 7.13 is adequate when the test function u_i satisfies the essential boundary conditions of Equation 7.2; i.e., $u_i = \bar{u}_i$ on Γ_u. Alternate forms are resorted to when these conditions are not met. A Lagrange multiplier form (see Belytschko et al., 1994a) may be written by appropriately augmenting Equation 7.13 to give the weak form

$$\begin{aligned} W_2: \quad & \int_{\Omega} \sigma_{ij}\, \delta v_{i,j}\, d\Omega - \int_{\Omega} F_i\, \delta v_i\, d\Omega - \int_{\Gamma_t} \bar{t}_i\, \delta v_i\, d\Gamma \\ & - \int_{\Gamma_u} (u_i - \bar{u}_i)\, \delta\lambda_i\, d\Gamma - \int_{\Gamma_u} \lambda_i\, \delta v_i\, d\Gamma = 0 \end{aligned} \tag{7.14}$$

where λ_i and $\delta\lambda_i$ are the Lagrange multiplier and its variation, respectively. From heuristic arguments, Lu et al. (1994) concluded that the physical meaning of the Lagrange multipliers λ_i is the traction t_i along the boundary Γ_u. Thus, replacing λ_i by t_i in Equation 7.14, an additional weak form is obtained as

$$W_3: \quad \int_{\Omega} \sigma_{ij}\, \delta v_{i,j}\, d\Omega - \int_{\Omega} F_i\, \delta v_i\, d\Omega - \int_{\Gamma_t} \bar{t}_i\, \delta v_i\, d\Gamma - \int_{\Gamma_u} (u_i - \bar{u}_i)\, \delta t_i\, d\Gamma - \int_{\Gamma_u} t_i\, \delta v_i\, d\Gamma = 0 \tag{7.15}$$

The essential boundary conditions can also be accounted for by means of a penalty formulation (Belytschko et al., 1994b; Hegen, 1996; Zhu and Atluri, 1998) to give

$$W_4: \quad \int_{\Omega} \sigma_{ij}\, \delta v_{i,j}\, d\Omega - \int_{\Omega} F_i\, \delta v_i\, d\Omega - \int_{\Gamma_t} \bar{t}_i\, \delta v_i\, d\Gamma + \alpha \int_{\Gamma_u} (u_i - \bar{u}_i)\, \delta v_i\, d\Gamma = 0 \tag{7.16}$$

where $\alpha \gg 1$ is a penalty parameter to enforce $u_i = \bar{u}_i$ on Γ_u. The choice of the weak form employed in a formulation depends on the strategy adopted therein to satisfy the essential boundary conditions.

7.4 MOVING LEAST SQUARES APPROXIMATIONS

In the EFG method, as well as several other meshless methods, the test and trial functions are represented over the domain of the body using moving least squares (MLS) approximants. The details for such approximations were presented in Lancaster and Salkauskas (1980). To define an MLS approximation over domain Ω, a set of nodal points $\underset{\sim}{x}_i$, $i = 1, 2, \ldots, N$ are used. The MLS approximant u^h of the function u at a location $\underset{\sim}{x}$ is given by

$$u^h(\underset{\sim}{x}) = \sum_{j=1}^{m} p_j(\underset{\sim}{x}) a_j(\underset{\sim}{x}) = \lfloor p(\underset{\sim}{x}) \rfloor \{a(\underset{\sim}{x})\} \tag{7.17}$$

where

$$\lfloor p(\underset{\sim}{x}) \rfloor = \lfloor p_1(\underset{\sim}{x}), p_2(\underset{\sim}{x}), \ldots, p_m(\underset{\sim}{x}) \rfloor \tag{7.18}$$

and

$$\{a(x)\}^{\mathrm{T}} = \lfloor a_1(\underset{\sim}{x}), a_2(\underset{\sim}{x}), \ldots, a_m(\underset{\sim}{x}) \rfloor \tag{7.19}$$

$\{a(\underset{\sim}{x})\}$ is a vector of coefficients that depend on the location $\underset{\sim}{x}$ and $\lfloor p(\underset{\sim}{x}) \rfloor$ is a vector of monomials called *basis functions*. The monomials are chosen so that they provide a complete basis of order m. For a two-dimensional domain, complete linear and quadratic bases are provided by the following:

$$\text{Linear:} \quad \lfloor p(\underset{\sim}{x}) \rfloor = \lfloor 1, x, y \rfloor \qquad m = 3 \tag{7.20}$$

and

$$\text{Quadratic:} \quad \lfloor p(\underset{\sim}{x}) \rfloor = \lfloor 1, x, y, x^2, xy, y^2 \rfloor \qquad m = 6 \tag{7.21}$$

The approximation error over the entire domain may be characterized by a weighted discrete L_2 norm defined at the N nodal locations as (Lancaster and Salkauskas, 1980)

$$\begin{aligned} J(\{a(\underset{\sim}{x})\}) &= \sum_{i=1}^{N} [u^h(\underset{\sim}{x}_i) - u(\underset{\sim}{x}_i)]^2 w_i(\underset{\sim}{x}) \\ &= \sum_{i=1}^{N} \left[\sum_{s=1}^{m} p_s(\underset{\sim}{x}_i) a_s(\underset{\sim}{x}) - u(\underset{\sim}{x}_i) \right]^2 w_i(\underset{\sim}{x}) \end{aligned} \tag{7.22}$$

where $u(\underset{\sim}{x}_i)$ are the generalized displacements at nodal locations $\underset{\sim}{x}_i$ and $w_i(\underset{\sim}{x})$ are weight functions. Obviously, for $u^h(\underset{\sim}{x})$ to be a good approximation for $u(\underset{\sim}{x})$, the error $J(\{a(\underset{\sim}{x})\})$ must be a minimum. The coefficients $\{a(\underset{\sim}{x})\}$ in Equation 7.22 are still undetermined. These coefficients need to be selected in such a way that the error $J\{a(\underset{\sim}{x})\}$ is minimized. This leads to the following m conditions corresponding to the m unknown coefficients:

$$\frac{\partial J\{a(\underset{\sim}{x})\}}{\partial a_k(\underset{\sim}{x})} = 0 \qquad k = 1, 2, \ldots, m \tag{7.23}$$

Substituting for J from Equation 7.22, we get

$$\frac{\partial J}{\partial a_k} = \sum_{i=1}^{N} 2[u^h(\underset{\sim}{x}_i) - u(\underset{\sim}{x}_i)] \frac{\partial u^h(\underset{\sim}{x}_i)}{\partial a_k} w_i(\underset{\sim}{x}) \tag{7.24}$$

where the derivative

$$\begin{aligned}\frac{\partial u^h(\underset{\sim}{x}_i)}{\partial a_k} &= \frac{\partial}{\partial a_k}\left[\sum_{s=1}^{m} p_s(\underset{\sim}{x}_i)a_s(\underset{\sim}{x})\right] \\ &= \sum_{s=1}^{m} p_s(\underset{\sim}{x}_i)\frac{\partial a_s(\underset{\sim}{x})}{\partial a_k(\underset{\sim}{x})}\end{aligned} \tag{7.25}$$

Since $\partial a_s(\underset{\sim}{x})/\partial a_k(\underset{\sim}{x}) = 1$ when $s = k$ and 0 otherwise, the summation above reduces to $p_k(\underset{\sim}{x}_i)$. Thus,

$$\frac{\partial u^h(\underset{\sim}{x}_i)}{\partial a_k} = p_k(\underset{\sim}{x}_i) \tag{7.26}$$

Substituting Equation 7.26 into Equation 7.24 and plugging the result in Equation 7.25, upon cancellation of the factor of 2 gives, with Equation 7.22,

$$\sum_{i=1}^{N}[u^h(\underset{\sim}{x}_i) - u(\underset{\sim}{x}_i)]p_k(\underset{\sim}{x}_i)w_i(\underset{\sim}{x}) = \sum_{i=1}^{N}\left[\sum_{s=1}^{m} p_s(\underset{\sim}{x}_i)a_s(\underset{\sim}{x}) - u(\underset{\sim}{x}_i)\right]p_k(\underset{\sim}{x}_i)w_i(\underset{\sim}{x}) = 0 \tag{7.27}$$

Upon rearranging terms, we obtain

$$\sum_{i=1}^{N}\left[\sum_{s=1}^{m} p_s(\underset{\sim}{x}_i)a_s(\underset{\sim}{x})\right]p_k(\underset{\sim}{x}_i)w_i(\underset{\sim}{x}) = \sum_{i=1}^{N} u(\underset{\sim}{x}_i)p_k(\underset{\sim}{x}_i)w_i(\underset{\sim}{x}) \tag{7.28}$$

Equation 7.28 represents m equations corresponding to $k = 1, 2, \ldots, m$. To aid the beginner in writing the matrix form of Equation 7.28, the expanded form of right-hand side of Equation 7.28 is written as, for $k = 1$,

$$u(\underset{\sim}{x}_1)p_1(\underset{\sim}{x}_1)w_1(\underset{\sim}{x}) + u(\underset{\sim}{x}_2)p_1(\underset{\sim}{x}_2)w_2(\underset{\sim}{x}) + \cdots + u(\underset{\sim}{x}_N)p_1(\underset{\sim}{x}_N)w_N(\underset{\sim}{x}) \tag{7.29}$$

and for $k = 2, \ldots, m$,

$$u(\underset{\sim}{x}_1)p_2(\underset{\sim}{x}_1)w_1(\underset{\sim}{x}) + u(\underset{\sim}{x}_2)p_2(\underset{\sim}{x}_2)w_2(\underset{\sim}{x}) + \cdots + u(\underset{\sim}{x}_N)p_2(\underset{\sim}{x}_N)w_N(\underset{\sim}{x}) \tag{7.30}$$

$$\vdots$$

$$u(\underset{\sim}{x}_1)p_m(\underset{\sim}{x}_1)w_1(\underset{\sim}{x}) + u(\underset{\sim}{x}_2)p_m(\underset{\sim}{x}_2)w_2(\underset{\sim}{x}) + \cdots + u(\underset{\sim}{x}_N)p_m(\underset{\sim}{x}_N)w_N(\underset{\sim}{x}) \tag{7.31}$$

The matrix form of expressions in Equations 7.29 to 7.31 can now easily be seen to be

$$\underset{m\times N}{\begin{bmatrix} p_1(\underset{\sim}{x}_1) & p_1(\underset{\sim}{x}_2) & \cdots & p_1(\underset{\sim}{x}_N) \\ p_2(\underset{\sim}{x}_1) & p_2(\underset{\sim}{x}_2) & \cdots & p_2(\underset{\sim}{x}_N) \\ \vdots & & & \\ p_m(\underset{\sim}{x}_1) & p_m(\underset{\sim}{x}_2) & \cdots & p_m(\underset{\sim}{x}_N) \end{bmatrix}} \tag{7.32}$$

$$\underset{N\times N}{\begin{bmatrix} w_1(\underset{\sim}{x}) & & & [0] \\ & w_2(\underset{\sim}{x}) & & \\ [0] & & \ddots & \\ & & & w_N(\underset{\sim}{x}) \end{bmatrix}} \underset{N\times 1}{\begin{Bmatrix} u(\underset{\sim}{x}_1) \\ u(\underset{\sim}{x}_2) \\ \vdots \\ u(\underset{\sim}{x}_N) \end{Bmatrix}}$$

or in compact matrix notation as

$$\text{RHS} = [P]^{\mathrm{T}}[W]\{u\} \tag{7.33}$$

Aided by the expressions in Equations 7.32 and 7.33, Equation 7.28 can be written in matrix form as

$$[P]^{\mathrm{T}}_{m\times N}[W]_{N\times N}[P]_{N\times m}\{a(\underset{\sim}{x})\}_{m\times 1} = [P]^{\mathrm{T}}_{m\times N}[W]_{N\times N}\{u\}_{N\times 1} \tag{7.34}$$

$$[P]^{\mathrm{T}} = \begin{bmatrix} p_1(\underset{\sim}{x}_1) & p_1(\underset{\sim}{x}_2) & \cdots & p_1(\underset{\sim}{x}_N) \\ p_2(\underset{\sim}{x}_1) & p_2(\underset{\sim}{x}_2) & \cdots & p_2(\underset{\sim}{x}_N) \\ \vdots & & & \\ p_m(\underset{\sim}{x}_1) & p_m(\underset{\sim}{x}_2) & & p_m(\underset{\sim}{x}_N) \end{bmatrix} \tag{7.35}$$

$$[W] = \begin{bmatrix} w_1(\underset{\sim}{x}) & \cdots & & [0] \\ \vdots & w_2(\underset{\sim}{x}) & & \vdots \\ & & \ddots & \\ [0] & \cdots & & w_N(\underset{\sim}{x}) \end{bmatrix} \tag{7.36}$$

$$\{a(\underset{\sim}{x})\} = \begin{Bmatrix} a_1(\underset{\sim}{x}) \\ a_2(\underset{\sim}{x}) \\ \vdots \\ a_m(\underset{\sim}{x}) \end{Bmatrix} \tag{7.37}$$

$$\{u\} = \begin{Bmatrix} u(\underset{\sim}{x}_1) \\ u(\underset{\sim}{x}_2) \\ \vdots \\ u(\underset{\sim}{x}_N) \end{Bmatrix} \tag{7.38}$$

Introducing the short-hand notation,

$$[A(\underset{\sim}{x})]_{m \times m} = [P]^{\mathrm{T}}[W][P] \tag{7.39}$$

The unknown coefficients $\{a(\underset{\sim}{x})\}$ are obtained from Equation 7.34 as

$$\{a(\underset{\sim}{x})\} = [A(\underset{\sim}{x})]^{-1}[P]^{\mathrm{T}}[W]\{u\} \tag{7.40}$$

Substituting Equation 7.40 into Equation 7.17, we obtain

$$\{u^h\} = [P]^{\mathrm{T}}[A(\underset{\sim}{x})]^{-1}[P]^{\mathrm{T}}[W]\{u\} \tag{7.41}$$

with

$$\{u^h\} = \begin{Bmatrix} u^h(\underset{\sim}{x}_1) \\ u^h(\underset{\sim}{x}_2) \\ \vdots \\ u^h(\underset{\sim}{x}_N) \end{Bmatrix} \tag{7.42}$$

From Equation 7.40, we note that

$$\begin{aligned} a_j(x) &= [[A(\underset{\sim}{x})]^{-1}[P]^T[W]]_{j^{\text{th}}\text{row}}\{u\} \\ &= \sum_{i=1}^{N}[[A(\underset{\sim}{x})]^{-1}[P]^T[W]]_{ji}u_i \end{aligned} \tag{7.43}$$

where $u_i = u(\underset{\sim}{x}_i)$. Substituting in Equation 7.17, we obtain

$$u^h(\underset{\sim}{x}) = \sum_{j=1}^{m} p_j(\underset{\sim}{x}) \sum_{i=1}^{N}[[A(\underset{\sim}{x})]^{-1}[P]^{\mathrm{T}}[W]]_{ji}u_i \tag{7.44}$$

The order of the two summations in Equation 7.44 can be changed to obtain

$$\begin{aligned} u^h(\underset{\sim}{x}) &= \sum_{i=1}^{m} u_i \sum_{j=1}^{m} p_j(\underset{\sim}{x})[[A(\underset{\sim}{x})]^{-1}[P]^{\mathrm{T}}[W]]_{ji} \\ &= \sum_{i=1}^{N} u_i \phi_i(\underset{\sim}{x}) \end{aligned} \tag{7.45}$$

with

$$\phi_i(\underset{\sim}{x}) = \sum_{j=1}^{m} p_j(\underset{\sim}{x})[[A(\underset{\sim}{x})]^{-1}[P]^{\mathrm{T}}[W]]_{ji} \tag{7.46}$$

where $\phi_i(\underset{\sim}{x})$ are the shape functions obtained from MLS approximation. This approximation is possible only when a unique solution exists for the system of Equation 7.40. The matrix $[A(\underset{\sim}{x})]$ of Equation 7.39 must be invertible. This is the case if and only if the rank of matrix $[P]$ equals m, the number of basis functions employed.

7.5 CHARACTERISTICS OF MLS APPROXIMATION

The expression in Equation 7.45 provides an approximation for the displacement field using the shape functions $\phi_i(\underset{\sim}{x})$ defined in Equation 7.46. Even when the basis functions $p_j(\underset{\sim}{x})$ in Equation 7.46, which also appear in the matrix $[P]$ in that equation, are polynomials, the MLS approximation in Equation 7.45 is not a polynomial. It can be shown (see Nayroles et al., 1992) that if $u(\underset{\sim}{x})$ is a polynomial, it is reproduced exactly by the approximation of $u^h(\underset{\sim}{x})$ in Equation 7.45. Thus, for any polynomial function $g(\underset{\sim}{x})$,

$$\sum_{i=1}^{N} \phi_i(\underset{\sim}{x}) g(\underset{\sim}{x}_i) = g(\underset{\sim}{x}) \tag{7.47}$$

For $g(\underset{\sim}{x}) = 1$, we obtain

$$\sum_{i=1}^{N} \phi_i(\underset{\sim}{x}) = 1 \tag{7.48}$$

Equation 7.48, called the *partition of unity* (Melenk and Babuska, 1996), reproduces a rigid-body field exactly. The necessary condition for convergence in a Galerkin formulation is the completeness of the interpolations for trial and test functions. The fundamental requirement for completeness is the ability of the functions to represent a constant-valued field as is satisfied by the MLS shape functions in Equation 7.48.

Unlike the shape functions employed in the finite element method, the EFG shape functions do not satisfy the *Kronecker delta criterion*, also termed the *selectivity property*, so that $\phi_i(\underset{\sim}{x}_j) \neq \delta_{ij}$. This property of the EFG shape functions causes difficulty in the direct imposition of the essential boundary conditions (i.e., $u_i = \bar{u}_i$ on Γ_u; see Section 7.9).

The smoothness of the EFG shape functions ϕ_i is determined by the smoothness of the basis functions p_j and the weight functions w_i. Let $C^k(\Omega)$ denote a set of functions that have continuous derivatives up to order k on Ω. If $p_j(\underset{\sim}{x}) \in C^s(\Omega)$ and $w_j(\underset{\sim}{x}) \in C^r(\Omega)$, then $\phi_i(\underset{\sim}{x}) \in C^t(\Omega)$, where t = mm(s, r).

As seen from the expression for $\phi_i(\underset{\sim}{x})$ in Equation 7.46, the inversion of matrix $[A]$ must be performed at each nodal location to obtain the shape function for that location. This is a computationally burdensome step in the EFG formulation. An alternative procedure that avoids the inversion step is based on the orthogonalization of the basis functions with respect to the values of weight functions in $\underset{\sim}{x}$ (Lu et al., 1994). Although the orthogonalization procedure is preferred from the point of view of accuracy, it involves the same order of costs as the matrix inversion.

To obtain the strain-displacement matrix $[B]$ as shown in Section 7.7, partial derivatives of the shape function $\phi_i(\underset{\sim}{x})$ are required. Differentiating Equation 7.46 with respect to the spatial coordinate $\underset{\sim}{x}_k$, we get

$$\phi_{i,k} = \sum_{j=1}^{m} p_{j,k}[[A]^{-1}[C]]_{ji} + p_j[[A]_{,k}^{-1}[C]+[A]^{-1}[C]_{,k}]_{ji} \tag{7.49}$$

where $[C] = [P]^T[W]$ and the index following a comma denotes a spatial derivative. To obtain the derivative $[A]_{,k}^{-1}$, the identity $[A][A]^{-1} = [I]$ is differentiated, where $[I]$ is the identity matrix. Thus,

$$[A][A]_{,k}^{-1} + [A]_{,k}[A]^{-1} = [0] \tag{7.50}$$

leading to

$$[A]_{,k}^{-1} = -[A]^{-1}[A]_{,k}[A]^{-1} \tag{7.51}$$

As seen from Equation 7.51, no further matrix inversions are required to obtain $[A]_{,k}^{-1}$ as well as the derivative $\phi_{i,k}$ of the shape function.

7.6 MLS WEIGHT FUNCTIONS

The MLS weight functions are introduced into the EFG formulation via the L_2 norm defined in Equation 7.22, which is minimized to obtain the unknown coefficients $\{a(\underset{\sim}{x})\}$. These weight functions constitute the matrix $[W]$ and are used in defining the EFG shape functions n Equation 7.46. As stated in Section 7.5, these functions, together with the basis functions $p_j(\underset{\sim}{x})$, determine the differentiability of the EFG shape functions $\phi_i(\underset{\sim}{x})$. The weight function $w_i(\underset{\sim}{x})$, corresponding to a node i, is defined such that it is a monotonically decreasing function of $\| \underset{\sim}{x} - \underset{\sim}{x}_i \|$. The domain in which the value of the weight function is nonzero is called the *support* of the weight function. From Equation 7.46 it can be seen that $\phi_i(\underset{\sim}{x}) = 0$ when $w_i(\underset{\sim}{x}) = 0$. The fact that $\phi_i(\underset{\sim}{x})$ vanishes when $\underset{\sim}{x}$ does not lie in the support of nodal point $\underset{\sim}{x}_i$ imparts a local character to the MLS approximation.

Several different weight functions have been used in the EFG formulations. Some of the more prominent ones are presented here. In the expression for weight functions, $d_i = \| \underset{\sim}{x} - \underset{\sim}{x}_i \|$ is the distance from node $\underset{\sim}{x}_i$ to $\underset{\sim}{x}$, and r_i is the size of the support for the weight function w_i and determines the support at node $\underset{\sim}{x}_i$.

1. Exponential weight function:

$$w_i(\underset{\sim}{x}) = w(r_i) = \begin{cases} \dfrac{\exp[-(d_i/c_i)^{2ki}] - \exp[-(r_i/c_i)^{2ki}]}{1-\exp[-(r_i/c_i)^{2ki}]} & d_i \le r_i \\ 0 & d_i > r_i \end{cases} \tag{7.52}$$

where c_i is a constant controlling the shape of the weight function (Belytschko et al., 1994a; Hegen, 1996) and k_i is a constant exponent generally chosen to be unity.

2. Spline weight function:

$$w_i(\underset{\sim}{x}) = w(r_i) = \begin{cases} 1 - 6\left(\dfrac{d_i}{r_i}\right)^2 + 8\left(\dfrac{d_i}{r_i}\right)^3 - 3\left(\dfrac{d_i}{r_i}\right)^4 & 0 \le d_i \le r \\ 0 & d_i > r_i \end{cases} \tag{7.53}$$

3. B-spline weight function:

$$w_i(\underset{\sim}{x}) = w(r_i) = \begin{cases} \dfrac{2}{3} - 4\xi^2 + 4\xi^3 & \xi \le \dfrac{1}{2} \\ \dfrac{4}{3} - 4\xi + 4\xi^2 - \dfrac{4}{3}\xi^3 & \dfrac{1}{2} \le \xi \le 1 \\ 0 & \xi > 1 \end{cases} \tag{7.54}$$

where $\xi = d_i/r_i$.

4. Conical weight function:

$$w_i(\underset{\sim}{x}) = w(r_i) = \begin{cases} 1 - (d_i/r_i)^{2ki} & d_i \le r_i \\ 0 & d_i > r_i \end{cases} \tag{7.55}$$

where k_i is a real constant generally chosen to be unity.

5. Singular weight function:

$$w_i(\underset{\sim}{x}) = w(r_i) = s_i(\underset{\sim}{x}) \| x - \underset{\sim}{x}_i \|^{-2\alpha i} \tag{7.56}$$

where α_i is a real constant and $s_i(\underset{\sim}{x})$ is one of the weight functions defined previously. Generally, $\alpha_i = 1$.

Other weight functions are possible and have been employed in meshless analyses. In defining the weight functions, the size of its support, r_i, must be carefully specified. A necessary condition for an MLS approximation is that at least m (= number of basis functions employed) weight functions are nonzero to ensure the regularity of matrix [A]. The support size, r_i, then must be large enough to have a sufficient number of nodes covered in the domain of each nodal point with which the weight function $w_i(\underset{\sim}{x})$ is associated (i.e., $N \ge m$). A large value of r_i will cover a large number of nodes and lead to an increase in the cost of computations. A small size of the support of a weight function is economical but may lead to a nonsingular definition of the matrix [A]. A smaller support domain also produces a relatively complex shape function, due to the almost singular shape of the weight function.

The support of a weight function at a node is also termed the *domain of influence* of that node in the meshless analysis literature. In the descriptions of weight functions earlier, a circular domain of influence with radius r_i is assumed. A rectangular domain of influence can also be employed. For rectangular domains, product weights may be employed as $w_i(\underset{\sim}{x}) = w(r_x)w(r_y)$, where $r_x = \|x - x_i\|$ and $r_y = \|y - y_i\|$.

7.7 DISCRETE EFG FORMULATION

The discrete equations describing the behavior of the solid object are obtained using the weak form of the governing equations. Since the shape functions $\phi_i(\underset{\sim}{x})$ do not possess the selectivity property, the trial function u does not satisfy the essential boundary conditions. Hence, one of the weak forms W2, W3, or W4 appropriate for such conditions as described in Section 7.3 must be used. The weak form W2 given in Equation 7.14 is employed here to demonstrate the procedure.

Following Equation 7.45, the trial and test functions are approximated using the MLS shape functions as

$$u = \sum_{a=1}^{N} \phi_a(\underset{\sim}{x}) u_a \tag{7.57}$$

and

$$\delta v = \sum_{b=1}^{N} \phi_b(\underset{\sim}{x}) \delta \upsilon_b \tag{7.58}$$

Both the x and y components of test and trial functions are approximated similarly. The Lagrange multiplier λ and its variation $\delta\lambda$ are expressed as

$$\lambda(\underset{\sim}{x}) = \sum_{a=1}^{k} \psi_a(\xi(\underset{\sim}{x}))\lambda_a \tag{7.59}$$

and

$$\delta\lambda(\underset{\sim}{x}) = \sum_{b=1}^{k} \psi_b(\xi(\underset{\sim}{x}))\delta\lambda_b \tag{7.60}$$

where $\xi(\underset{\sim}{x})$ is the normalized are length along the boundary and ψ_a are approximation functions for λ on Γ_u. For example, Lagrange interpolants may be used to represent ψ_a.

A quadratic variation of λ along the boundary may be represented by Lagrange shape functions ψ_i, $i = 1, 2, 3$ as

$$\psi_1(\xi) = -\frac{1}{2}\xi(1-\xi) \tag{7.61}$$

$$\psi_2(\xi) = \frac{1}{2}\xi(1+\xi) \tag{7.62}$$

and

$$\psi_3(\xi) = (1+\xi)(1-\xi) \tag{7.63}$$

The displacement field, $\underset{\sim}{u} = \lfloor u_x \; u_y \rfloor$, consisting of the x- and y-displacement components for a two-dimensional analysis, is written in terms of nodal displacements using Equation 7.45 as

$$\begin{Bmatrix} u_x \\ u_y \end{Bmatrix} = \begin{bmatrix} \phi_1 & 0 & \phi_2 & 0 & \cdots & \phi_N & 0 \\ 0 & \phi_1 & 0 & \phi_2 & \cdots & 0 & \phi_N \end{bmatrix} \begin{Bmatrix} u_x^1 \\ u_y^1 \\ u_x^2 \\ u_y^2 \\ \vdots \\ u_x^N \\ u_y^N \end{Bmatrix} = [\Phi]\{u\}.s \tag{7.64}$$

The two-dimensional strain field is next written for small displacement analysis using Equation 7.3 as

$$\begin{Bmatrix} \varepsilon_x \\ \varepsilon_y \\ \gamma_{xy} \end{Bmatrix} = \begin{Bmatrix} \dfrac{\partial u_x}{\partial x} \\ \dfrac{\partial u_y}{\partial y} \\ \dfrac{\partial u_x}{\partial y} + \dfrac{\partial u_y}{\partial x} \end{Bmatrix} \tag{7.65}$$

where $\gamma_{xy} = 2\varepsilon_{xy}$ is referred to as *engineering shear strain*. Substituting in Equation 7.65 from Equation 7.64, the strain field is written as

$$\begin{Bmatrix} \varepsilon_x \\ \varepsilon_y \\ \gamma_{xy} \end{Bmatrix} = \begin{bmatrix} \phi_{1,x} & 0 & \phi_{2,x} & 0 & \cdots & \phi_{N,x} & 0 \\ 0 & \phi_{1,y} & 0 & \phi_{2,y} & \cdots & 0 & \phi_{N,y} \\ \phi_{1,y} & \phi_{1,x} & \phi_{2,y} & \phi_{2,x} & \cdots & \phi_{N,y} & \phi_{N,x} \end{bmatrix} \begin{Bmatrix} u_x^1 \\ u_y^1 \\ u_x^2 \\ u_y^2 \\ \vdots \\ u_x^N \\ u_y^N \end{Bmatrix} \tag{7.66}$$

which is expressed in shorthand matrix notation as

$$\{\varepsilon\} = [B]\{u\} \tag{7.67}$$

$$\{\varepsilon\} = \begin{Bmatrix} \varepsilon_x \\ \varepsilon_y \\ \gamma_{xy} \end{Bmatrix} \qquad \{u\} = \begin{Bmatrix} u_x^1 \\ u_y^1 \\ u_x^2 \\ u_y^2 \\ \vdots \\ u_x^N \\ u_y^N \end{Bmatrix} \tag{7.68}$$

$$[B] = [[B_1][B_2]\cdots[B_N]] \tag{7.69}$$

and

$$[B_a] = \begin{bmatrix} \phi_{a,x} & 0 \\ 0 & \phi_{a,y} \\ \phi_{a,y} & \phi_{a,x} \end{bmatrix} \tag{7.70}$$

The Lagrange multiplier field is expressed in the matrix notation using Equation 7.59 as

$$\begin{Bmatrix} \lambda_x \\ \lambda_y \end{Bmatrix} = \begin{bmatrix} \psi_1 & 0 & \psi_2 & 0 & \cdots & \psi_k & 0 \\ 0 & \psi_1 & 0 & \psi_2 & \cdots & 0 & \psi_k \end{bmatrix} \begin{Bmatrix} \lambda_x^1 \\ \lambda_y^1 \\ \lambda_x^2 \\ \lambda_y^2 \\ \vdots \\ \lambda_x^k \\ \lambda_y^k \end{Bmatrix} = [N]\{\lambda\} \tag{7.71}$$

where

$$[N] = [[N_1][N_2]\cdots[N_k]] \tag{7.72}$$

$$[N_b] = \begin{bmatrix} \psi_b & 0 \\ 0 & \psi_b \end{bmatrix} \tag{7.73}$$

and

$$\{\lambda\} = \begin{Bmatrix} \lambda_x^1 \\ \lambda_y^1 \\ \lambda_x^2 \\ \lambda_y^2 \\ \vdots \\ \lambda_x^k \\ \lambda_y^k \end{Bmatrix} \tag{7.74}$$

In Equation 7.74, k is the number of boundary nodes on which the essential boundary conditions are prescribed. Similarly, the variational quantities are expressed using the shape-function-based descriptions as

$$\{\delta\varepsilon\} = [B]\{\delta v\} \tag{7.75}$$

and

$$\begin{Bmatrix} \delta\lambda_x \\ \delta\lambda_y \end{Bmatrix} = [N]\{\delta\lambda\} \tag{7.76}$$

with

$$\{\delta v\} = \begin{Bmatrix} \delta v_x^1 \\ \delta v_y^1 \\ \delta v_x^2 \\ \delta v_y^2 \\ \vdots \\ \delta v_x^N \\ \delta v_y^N \end{Bmatrix} \quad \{\delta\lambda\} = \begin{Bmatrix} \delta\lambda_x^1 \\ \delta\lambda_y^1 \\ \delta\lambda_x^2 \\ \delta\lambda_y^2 \\ \vdots \\ \delta v_x^k \\ \delta v_y^k \end{Bmatrix} \tag{7.77}$$

The stress–strain relation of Equation 7.4 can be written in the matrix form as

$$\{\sigma\} = [D]\{\varepsilon\} \tag{7.78}$$

where for two-dimensional plane stress analysis,

$$[D] = \frac{E}{1-v^2}\begin{bmatrix} 1 & v & 0 \\ v & 1 & 0 \\ 0 & 0 & (1-v)/2 \end{bmatrix} \tag{7.79}$$

in which E and ν are the Young's modulus and Poisson's ratio, respectively.

For two dimensions, the weak form W_2 in Equation 7.14 is now expressed in matrix notation as

$$\int_\Omega \begin{Bmatrix} \delta\varepsilon_x \\ \delta\varepsilon_y \\ \delta\gamma_{xy} \end{Bmatrix}^T \begin{Bmatrix} \sigma_x \\ \sigma_y \\ \sigma_{xy} \end{Bmatrix} d\Omega - \int_\Omega \begin{Bmatrix} \delta v_x \\ \delta v_y \end{Bmatrix}^T \begin{Bmatrix} F_x \\ F_y \end{Bmatrix} d\Omega - \int_{\Gamma_t} \begin{Bmatrix} \delta v_x \\ \delta v_y \end{Bmatrix}^T \begin{Bmatrix} \bar{t}_x \\ \bar{t}_y \end{Bmatrix} d\Gamma$$
$$-\int_{\Gamma_u} \begin{Bmatrix} \delta\lambda_x \\ \delta\lambda_y \end{Bmatrix}^T \begin{Bmatrix} u_x \\ u_y \end{Bmatrix} d\Gamma - \int_{\Gamma_u} \begin{Bmatrix} \delta v_x \\ \delta v_y \end{Bmatrix}^T \begin{Bmatrix} \lambda_x \\ \lambda_y \end{Bmatrix} d\Gamma + \int_{\Gamma_u} \begin{Bmatrix} \delta\lambda_x \\ \delta\lambda_y \end{Bmatrix}^T \begin{Bmatrix} \bar{u}_x \\ \bar{u}_y \end{Bmatrix} d\Gamma = 0 \tag{7.80}$$

A few key terms in Equation 7.80 are expanded explicitly to show the process followed to obtain the system of EFG matrix equations. The first term in Equation 7.80 is written as

$$\int_\Omega \begin{Bmatrix} \delta\varepsilon_x \\ \delta\varepsilon_y \\ \delta\gamma_{xy} \end{Bmatrix}^T \begin{Bmatrix} \sigma_x \\ \sigma_y \\ \sigma_{xy} \end{Bmatrix} d\Omega = \int_\Omega \{\delta v\}^T[B]^T[D][B]\{u\}\, d\Omega \tag{7.81}$$

where substitutions have been made from relations in Equations 7.75, 7.67, and 7.78 in obtaining the right-hand side of Equation 7.81. The right-hand side of Equation 7.81 is next written as

$$\int_\Omega \{\delta v\}^T[B]^T[D][B]\{u\}\, d\Omega = \{\delta v\}^T[K]\{u\} \tag{7.82}$$

where the stiffness matrix $[K]$ is

$$[K] = \int_\Omega [B]^T[D][B]\, d\Omega \tag{7.83}$$

The integrand in Equation 7.83 is expanded using the relation in Equation 7.69 as

$$[B]^T[D][B] = \begin{bmatrix} [B_1]^T \\ [B_2]^T \\ \vdots \\ [B_N]^T \end{bmatrix}_{2N\times 3} [D]_{3\times 3}\,[[B_1][B_2]\cdots[B_N]]_{3\times 2N}$$

$$= \begin{bmatrix} [B1]^T[D][B_1] & [B_1]^T[D][B_2] & \cdots & [B_1]^T[D][B_N] \\ [B2]^T[D][B_1] & [B_2]^T[D][B_2] & \cdots & [B_2]^T[D][B_N] \\ \vdots & & & \\ [B_N]^T[D][B_1] & [B_N]^T[D][B_2] & \cdots & [B_N]^T[D][B_N] \end{bmatrix} \tag{7.84}$$

The stiffness matrix $[K]$ is seen in Equation 7.84 to be composed of submatrices $[K_{ab}]$, where

$$\begin{gathered} [K_{ab}] = \int_{\Omega} [B_a]^T[D][B_b]\, d\Omega \\ a = 1,2,\ldots N; \qquad b = 1,2,\ldots,N \end{gathered} \tag{7.85}$$

The submatrix $[K_{ab}]$ in Equation 7.85 premultiplies the displacement vector $\{u_b\}$ in Equation 7.82. The second term in Equation 7.80 is now expanded as

$$\int_{\Omega} \begin{Bmatrix} \delta v_x \\ \delta v_y \end{Bmatrix}^T \begin{Bmatrix} F_x \\ F_y \end{Bmatrix} d\Omega = \int_{\Omega} \{\delta v\}^T[\Phi]^T \begin{Bmatrix} F_x \\ F_y \end{Bmatrix} d\Omega \tag{7.86}$$

where the variational displacement field, $\delta v = \lfloor \delta v_x \delta v_y \rfloor$, is expressed similar to the expression for displacement field in Equation 7.64. The right-hand side of Equation 7.86 may be written as

$$\int_{\Omega} \{\delta v\}^T[\Phi]^T \begin{Bmatrix} F_x \\ F_y \end{Bmatrix} d\Omega = \{\delta v\}^T\{F\} \tag{7.87}$$

where the force vector $\{F\}$ is written as

$$\{F\} = \int_{\Omega} [\Phi]^T \begin{Bmatrix} F_x \\ F_y \end{Bmatrix} d\Omega \tag{7.88}$$

The integrand in Equation 7.88 is further expanded by substituting for $[\Phi]$ from Equation 7.64 to obtain

$$[\Phi]^T \begin{Bmatrix} F_x \\ F_y \end{Bmatrix} = \begin{Bmatrix} \phi_1 & 0 \\ 0 & \phi_1 \\ \phi_2 & 0 \\ 0 & \phi_2 \\ \vdots & \vdots \\ \phi_N & 0 \\ 0 & \phi_N \end{Bmatrix} \begin{Bmatrix} F_x \\ F_y \end{Bmatrix} = \begin{Bmatrix} \phi_1 \begin{Bmatrix} F_x \\ F_y \end{Bmatrix} \\ \phi_2 \begin{Bmatrix} F_x \\ F_y \end{Bmatrix} \\ \vdots \\ \phi_N \begin{Bmatrix} F_x \\ F_y \end{Bmatrix} \end{Bmatrix} \tag{7.89}$$

The force vector $\{F\}$ is seen from Equation 7.89 to be composed of subvectors $\{f_b^{\Omega}\}$, where

$$\{f_b^{\Omega}\} = \int_{\Omega} \phi_b \begin{Bmatrix} F_x \\ F_y \end{Bmatrix} d\Omega \qquad b = 1,2,\ldots,N \tag{7.90}$$

The remaining terms in Equation 7.80 may be expanded similarly to obtain

$$\{\delta v\}^{\mathrm{T}}[[K_{ab}]\{u_b\} - \{f_b^{\Omega}\} - \{f_b^{\Gamma}\} + [G_{ab}]\{\lambda_b\}] + \{\delta\lambda\}^{\mathrm{T}}[[G_{ab}]^{\mathrm{T}}\{u^b\} - \{q_b\}] = 0 \tag{7.91}$$

Since $\{\delta v\}$ and $\{\delta\lambda\}$ are arbitrary variations, we obtain two matrix equations by equating the terms that postmultiply each of them separately to $\{0\}$ as

$$[K_{ab}]\{u_b\} - \{f_b^{\Omega}\} - \{f_b^{\Gamma}\} + [G_{ab}]\{\lambda_b\} = \{0\} \tag{7.92}$$

and

$$[G_{ab}]^{\mathrm{T}}\{u_b\} - \{q_b\} = \{0\} \tag{7.93}$$

Writing these equations together in matrix form, we have

$$\begin{bmatrix} [K_{ab}] & [G_{ab}] \\ [K_{ab}]^{\mathrm{T}} & [0] \end{bmatrix} \begin{Bmatrix} \{u_b\} \\ \{\lambda_b\} \end{Bmatrix} = \begin{Bmatrix} \{f_b\} \\ \{q_b\} \end{Bmatrix} \tag{7.94}$$

with

$$\{f_b\} = \{f_b^{\Omega}\} + \{f_b^{\Gamma}\} \tag{7.95}$$

$$[G_{ab}] = -\int_{\Gamma} \phi_a [N_b] d\Gamma \tag{7.96}$$

$$\{f_b^{\Gamma}\} = \int_{\Gamma_t,} \phi_b \begin{Bmatrix} \bar{t}_x \\ \bar{t}_y \end{Bmatrix} d\Gamma \tag{7.97}$$

and

$$\{q_b\} = -\int_{\Gamma_u} \psi_b \begin{Bmatrix} \bar{u}_x \\ \bar{u}_y \end{Bmatrix} d\Gamma \tag{7.98}$$

Finally, the contributions of the form of Equation 7.94 for each nodal point are assembled to obtain a global system of equations as

$$\begin{bmatrix} [K] & [G] \\ [G]^T & [0] \end{bmatrix} \begin{Bmatrix} \{u\} \\ \{\lambda\} \end{Bmatrix} = \begin{Bmatrix} \{f\} \\ \{q\} \end{Bmatrix} \tag{7.99}$$

In Equation 7.99, $\{u\}$ and $\{\lambda\}$ are the vectors of unknown nodal displacements and Lagrange multipliers, respectively. The system of Equation 7.99 may be solved using standard linear system solvers to obtain these unknown quantities.

7.8 NUMERICAL IMPLEMENTATION

The computation of the matrices $[K]$, $[G]$, $\{f\}$, and $\{q\}$ in Equation 7.99 requires evaluation of various domain and boundary integrals, such as in Equations 7.85, 7.90, and 7.95 to 7.98. To evaluate the domain integrals, the domain Ω is discretized into volume cells. Similarly, to evaluate the boundary integrals on portions Γ_u and Γ_t of the boundary, these portions are discretized into boundary cells. The volume and boundary cells for an arbitrary two-dimensional body are shown in Figure 7.1. Numerical integration (e.g., Gauss quadrature) is then performed over each of these cells to obtain the cell contributions $[K_c]$, $[G_c]$, $\{f_c\}$, and $\{q_c\}$.

The procedure to obtain the cell contributions $[K_c]$, $[G_c]$, $\{f_c\}$, and $\{q_c\}$ is outlined in the following. A typical volume cell with a 4 × 4 distribution of Gauss integration points is shown in Figure 7.2. The computation steps performed at a typical integration (quadrature) point x_q in a cell are as follows:

Step 1. Determine all nodes $\underset{\sim}{x}_i$ such that the integration point $\underset{\sim}{x}_q$ lies within the support of the weight function at $\underset{\sim}{x}_i$[i.e., $w_i(\underset{\sim}{x}_q) > 0$].
Step 2. At each of the node $\underset{\sim}{x}_i$ found in step 1, compute the shape functions $\phi_i(\underset{\sim}{x}_q)$ and their derivatives $\phi_{i,j}(\underset{\sim}{x}_q)$.
Step 3. Compute the contributions of $\underset{\sim}{x}_q$ to the integrals in Equations 7.85 and 7.90.

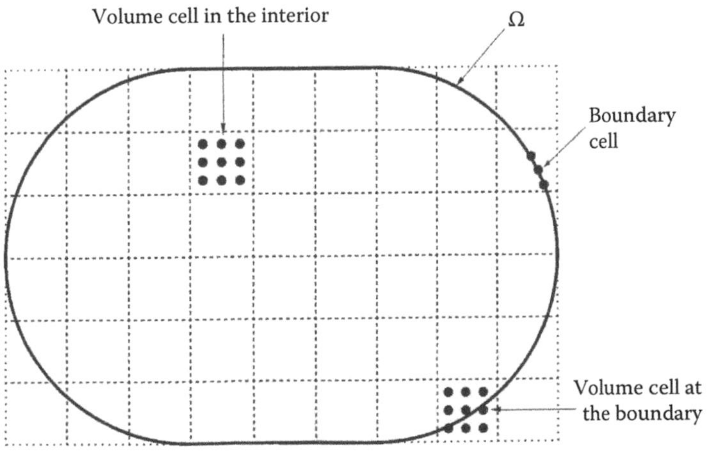

Figure 7.1 Background cells for numerical integration in the EFG method.

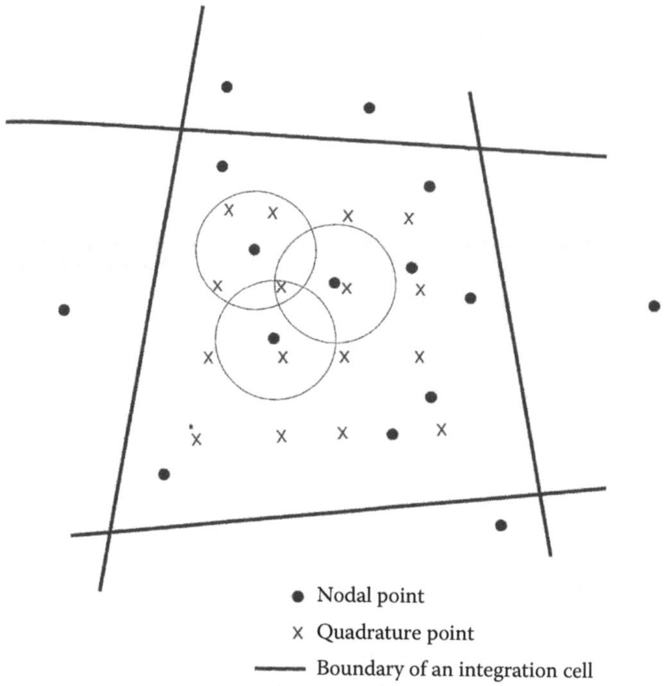

Figure 7.2 Typical integration cell, nodal points, and quadrature points in the EFG method.

Step 4. Assemble these contributions from step 3 into cell contributions $[K_c]$ and $\{f_c\}$, respectively, by suitably multiplying them with Gaussian (or other) weight functions and adding together.

Step 5. Assemble the cell contributions in overall matrices $[K]$ and $\{f\}$, respectively.

Cell integrations for boundary cells are performed using similar steps to obtain contributions $[G_c]$, $\{f_c\}$, and $\{q_c\}$, which are assembled into the overall matrices $[G]$, $\{f\}$, and $\{q\}$.

As stated in Section 7.4, the MLS shape functions are not piecewise polynomials. This requires a relatively dense distribution of quadrature points in each cell for accurate evaluation of the respective integrals. Various suggestions for the number of integration points needed for accurate evaluation of integrals have been made in the literature. For example (Hegen, 1996), a total of n_{int} integration points may be used in the domain with

$$n_{\text{int}} = \mu N \tag{7.100}$$

where N is the number of nodal points in the domain and $5 \leq \mu \leq 8$. For objects with domains that have curved boundaries, a background cell structure will produce cells on the boundary for which the integration points may not lie within the domain of the boundary. One such cell is shown in Figure 7.1, where for a 3 × 3 Gauss integration distribution of quadrature points, four such points fall outside the domain Ω. In several implementations (Belytschko et al., 1994a; Lu et al., 1994), an integration point is simply skipped when it falls outside the domain of the body. In other implementations (Hegen, 1996), the background cells are constructed using isoparametric mapping such that the cells on the boundary conform to the curved shape of the boundary.

The desire to remove the need for a mesh in finite element analyses has been a major motivating factor for the development of meshless methods. The use of background integration cells as described previously thus constitutes a major criticism of the EFG and other meshless methods. The background cells are viewed as a type of mesh similar to the mesh for finite element analysis. Several attempts have been reported in the literature (Beissel and Belytschko, 1996; Chen et al., 2001) to develop a node-based integration that does not require the background cells. Nodal integration usually introduces numerical instabilities. Stabilization terms may be added to improve the performance of nodal integration. For problems that do not contain unstable modes in their original solution, the addition of stabilization actually decreases the accuracy. The stabilized conforming nodal integration scheme (Chen et al., 2001) has shown a great deal of promise, but the scheme requires the Vornoi tessellation (O'Rourke, 2000) of a set of nodes. The tessellation does appear like a background mesh, even though it does not perform the same function as the finite element mesh.

7.9 TREATMENT OF BOUNDARY CONDITIONS

As shown in Section 7.4, the MLS shape functions do not satisfy the Kronecker delta condition [i.e., $\phi_i(\underset{\sim}{x}_j) \neq \delta_{ij}$], so that

$$\begin{aligned} u^h(\underset{\sim}{x}_p) &= \sum_{i=1}^{N} \phi_i(\underset{\sim}{x}_p) u_i \\ &\neq u_p \end{aligned} \tag{7.101}$$

If u_p were the prescribed value at a node $\underset{\sim}{x}_p$ on Γ_u, the MLS approximant does not reproduce this value at $\underset{\sim}{x}_p$. This causes a serious problem in terms of direct enforcement of essential boundary conditions. A number of techniques are available to circumvent this difficulty. However, each has its limitations and no method to date has received universal acceptance. Some of the techniques available to account for the essential boundary conditions in an EFG formulation are outlined briefly here.

Lagrange multiplier approach: This approach employs Lagrange multipliers for the enforcement of essential boundary conditions. The approach was described in detail earlier and results in the system of equations shown in Equation 7.99. This approach has several limitations: (1) It leads to larger system matrices since additional unknowns corresponding to the Lagrange multipliers are introduced; (2) the bandedness of the system is sacrificed due to the appearance of matrices $[G]$ and $[G]^T$; and (3) the matrix, which has to be inverted, possesses zeros on its main diagonals. Solvers with variable bandwidth that do not take advantage of positive definiteness of the matrix need to be employed (Lu et al., 1994).

Penalty parameter approach: The essential boundary conditions are incorporated into weak form W4 of the governing equations using a penalty parameter α as shown in Equation 7.16. An analysis on W4 is performed similar to that performed on weak form W2 in Section 7.6. This results in a set of equations given as

$$[[K] + \alpha[K^u]]\{u\} = \{f\} + \alpha\{f^u\} \tag{7.102}$$

Additional terms appearing with the penalty parameter are

$$\left[K_{ab}^{u}\right] = \int_{\Gamma_u} \phi_a [S] \phi_j \, d\Gamma \tag{7.103}$$

and

$$\left\{f_b^u\right\} = \int_{\Gamma_u} \phi_b [S] \begin{Bmatrix} \overline{u}_x \\ u_y \end{Bmatrix} d\Gamma \tag{7.104}$$

where

$$[S] = \begin{bmatrix} S_x & 0 \\ 0 & S_y \end{bmatrix} \tag{7.105a}$$

$$S_i = \begin{cases} 1 \ \text{if } u_i \text{ is prescribed on } \Gamma_u \\ 0 \ \text{if } u_i \text{ is not prescribed on } \Gamma_u \end{cases} \tag{7.105b}$$

A large value of the penalty parameter in the range $\alpha = 10^3$ to 10^7 is employed. The solution of the system of equations may be sensitive to the value of the penalty parameter α.

Transformation methods: As explained in Section 7.4, the vector $\{u\}$ in Equation 7.99 contains generalized nodal displacements, not the actual nodal displacements, as a result of which the essential boundary conditions cannot be imposed directly.

A transformation is sought that converts the generalized displacements $\{u\}$ to the actual displacements $\{\hat{u}\}$ at the nodal locations. Requiring that the approximation in Equation 7.45 yields the actual displacement $\hat{u}_{ij}$ at location $\underset{\sim}{x}_J$, we obtain

$$\begin{aligned} \hat{u}_{ij} &= \sum_{I=1}^{N} u_{iI} \phi_I(\underset{\sim}{x}_J) \qquad J = 1, 2, \ldots, N \\ &= \sum_{I=1}^{N} \Lambda_{IJ} u_{iI} \end{aligned} \tag{7.106}$$

where i is the coordinate direction and

$$\Lambda_{IJ} = \phi_I(\underset{\sim}{x}_J) \tag{7.107}$$

In matrix form, we can write

$$\{\hat{u}_i\} = [\Lambda]^T \{u_i\} \tag{7.108a}$$

or

$$(u_i) = [\Lambda]^{-T} \{\hat{u}_i\} \tag{7.108b}$$

In component form,

$$u_{iI} = \sum_{K=1}^{N} \Lambda_{IK}^{-T} \hat{u}_{iK} \tag{7.109}$$

Substituting this relation back into the approximation in Equation 7.45 yields

$$\begin{aligned} u_i^h(\underset{\sim}{x}) &= \sum_{I=1}^{N} \phi_I(\underset{\sim}{x}) u_{iI} \\ &= \sum_{I=1}^{N} \phi_I(\underset{\sim}{x}) \sum_{K=1}^{N} \Lambda_{IK}^{-T} \hat{u}_{iK} \\ &= \sum_{K=1}^{N} \Psi_K(\underset{\sim}{x}) \hat{u}_{iK} \end{aligned} \tag{7.110}$$

where

$$\Psi_K(\underset{\sim}{x}) = \sum_{I=1}^{N} \Lambda_{KI}^{-1} \phi_I(\underset{\sim}{x}) \tag{7.111}$$

Thus, a new interpolation is obtained in Equation 7.110, expressed in terms of the actual displacements {û} and modified shape functions $\Psi_K(\underset{\sim}{x})$. This new interpolation does satisfy the Kronecker delta property as follows. From Equation 7.111, $\Psi_K(\underset{\sim}{x}_j)$ is obtained by substituting $\underset{\sim}{x} = \underset{\sim}{x}_j$ as

$$\Psi_K(\underset{\sim}{x}_j) = \sum_{I=1}^{N} \Lambda_{KI}^{-1} \phi_I(\underset{\sim}{x}_j) \tag{7.112}$$

But from Equation 7.107, $\phi_I(\underset{\sim}{x}_j) = \Lambda_{IJ}$; thus,

$$\Psi_K(\underset{\sim}{x}_j) = \sum_{I=1}^{N} \Lambda_{KI}^{-1} \Lambda_{IJ} = \delta_{KJ} \tag{7.113}$$

which is the selectivity property of the interpolant. The matrices [K] and {f} in Equation 7.99 are now computed using the interpolation functions $\Psi_K(\underset{\sim}{x})$ of Equation 7.110. Since the essential boundary conditions can now be enforced directly, the matrices [G] and {q} in Equation 7.99 are no longer required.

The transformation method described previously requires the inversion of a $N \times N$ matrix [Λ]. A mixed transformation method may be devised (Chen et al., 2000a) in which only generalized displacements corresponding to the nodes at the essential boundary are transformed to the corresponding actual displacements. The generalized displacements at the remaining nodes are not transformed. This mixed transformation still allows direct enforcement of the essential boundary conditions. Only a $N_u \times N_u$ matrix now needs to be inverted, where N_u is the number of nodes on the essential boundary.

Coupled finite element–EFG method approach: To take advantage of the direct enforcement of essential boundary conditions available in the finite element method, several formulations coupling the finite element method with the EFG method have been developed (Belytschko et al., 1995c; Chen et al., 1995; Hegen, 1996; Liu et al., 1996). A layer of finite elements is provided along the entire boundary or at least along that portion of the boundary on which essential boundary conditions need to be enforced. The rest of the body is discretized by a distribution of EFG nodal points as shown in Figure 7.3. The essential boundary conditions are then imposed as they are in standard finite element analysis.

The MLS formulation becomes a finite element interpolation when the domain of the influence coincides with the element. Hegen (1996) took advantage of this characteristic to formulate the finite element–EFG coupling. Another formulation (Belytschko et al., 1995c) utilizes blending functions, R, to combine the two analysis methods as

$$u = R(\underset{\sim}{x})u^{\mathrm{EFG}}(\underset{\sim}{x}) + [1 - R(\underset{\sim}{x})]u^{\mathrm{FE}}(\underset{\sim}{x}) \tag{7.114}$$

with the blending function given as

$$R(\underset{\sim}{x}) = \sum_J N_J(\underset{\sim}{x}) \qquad \underset{\sim}{x}_J \in \Gamma_E \tag{7.115}$$

where Γ_E is the boundary of the portion of the domain modeled using EFG. The introduction of finite elements into a meshless formulation constitutes a major criticism of this approach. It is felt that the meshless character of analysis is compromised by requiring even a portion of the domain to be discretized using finite elements.

Boundary singular kernel method: The Kronecker delta property of $\phi_i(\underset{\sim}{x}_j)$ may be achieved by employing singular kernel functions in obtaining the MLS interpolant (Lancaster and Salkauskas, 1980). This idea was employed by Kaljevic and Saigal (1997) to obtain an EFG formulation that allows direct imposition of essential boundary conditions. The technique results in satisfying the essential boundary conditions at discrete nodal points on the

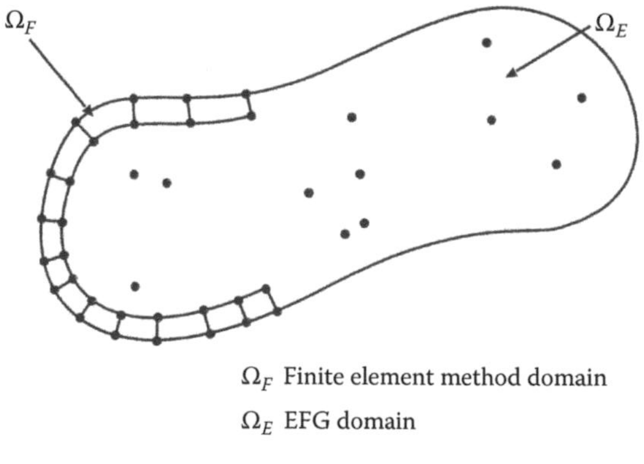

Figure 7.3 Object with coupled finite elements and EFG domains.

boundary. The portions of the boundary lying between these nodes, however, do not satisfy the requisite conditions.

Several other methods for enforcing essential boundary conditions in meshless analyses have also been proposed (see Mukherjee and Mukherjee, 1997b, and the review by Belytschko et al., 1996).

7.10 OTHER METHODS FOR MESHLESS ANALYSIS

The essence of the meshless EFG method described in Section 7.9 lies in the approximation of the response $u(\underset{\sim}{x})$ using MLS-based nodal shape functions as in Equation 7.46. Other meshless methods may be devised by arriving at similar node-based shape functions or node-based representations of displacement u. Consider the exact representation of a function obtained from using the sifting property of the Dirac delta function. A function $f(x)$ may be represented as

$$f(x) = \int_{-\infty}^{\infty} \delta(x-s) f(s)\, ds \tag{7.116}$$

where $\delta(x - s)$ is the Dirac delta function that satisfies the property

$$\int_{-\infty}^{\infty} \delta(x-s)\, ds = 1 \tag{7.117}$$

A smoothed representation of $f(x)$ may be obtained by replacing the Dirac delta function in Equation 7.116 by a smooth function, say $\phi(x - s)$, as

$$f_s(x) = \int_{-\infty}^{\infty} \phi(x-s) f(s)\, ds \tag{7.118}$$

where $f_s(x)$ is a smoothed representation of $f(x)$ but is not equal to it. The function $\phi(x - s)$ must satisfy the property satisfied by $\delta(x - s)$ in Equation 7.117: that

$$\int_{-\infty}^{\infty} \phi(x-s)\, ds = 1 \tag{7.119}$$

The SPH method is one of the earliest particle or mesh-free methods. It was developed initially for the study of problems in astrophysics (Gingold and Monaghan, 1977; Lucy, 1977). The method was later extended to apply to problems in solid mechanics (Libersky and Petschek, 1991; Libersky et al., 1993). In the SPH nomenclature, the function $\phi(x - s)$ is termed the *kernel function*. Thus, the response $u(x)$ of a solid object is given by using the smoothed representation of Equation 7.118 as

$$u(x) = \int_{-\infty}^{\infty} \phi(x-s)\, u(s)\, ds \tag{7.120}$$

A positive function such as the Gaussian function or the B-spline function is usually employed as the kernel function. In one dimension, these functions are given as follows:

1. Gaussian function:

$$\phi(x,s) = g(x,h) = \frac{1}{(\pi h^2)^{n/2}} \exp\left(-\frac{x^2}{h^2}\right) \qquad 1 \le n \le 3 \tag{7.121}$$

where h is a parameter called the *smoothing length.*
2. B-spline function:

$$\phi(x,s) = b(x,h) = \frac{2}{3h} \begin{cases} 1 - \frac{3}{2}q^2 + \frac{3}{4}q^3 & q \le 1 \\ \frac{1}{4}(2-q)^3 & 1 \le q \le 2 \\ 0 & q \ge 2 \end{cases} \tag{7.122}$$

where $q = \|x - s\|/h$, h is the smoothing length and the radius of the support of this kernel is $2h$.

A major difficulty in SPH methods arises from the fact that even when the kernel function satisfies Equation 7.48, its discrete counterpart does not, so that

$$\sum_I \phi(x - s_I)\Delta s_I \ne 1 \tag{7.123}$$

In the discrete form, therefore, the kernel function does not satisfy the partition of unity condition. Referring to the discussion following Equation 7.48, the kernel function violates the completeness condition and is not able to represent constant-valued fields. Completeness can, however, be restored in a kernel approximation through a correction transformation. A correction for the constant reproducing condition was given by Shepard (1968). Completeness corrections may also be applied to the derivatives of the approximating kernel functions. Several corrections have been proposed in the literature to provide for derivative reproducing conditions. Prominent among these corrections are those of Monaghan (1988, 1992), Johnson and Beissel (1996), Randles and Libersky (1996), and Krongauz and Belytschko (1997, 1998).

The SPH method suffers from several drawbacks. The motion of the particles becomes unstable under certain tensile stress states. Several strategies to circumvent this limitation have been proposed (e.g., Dyka and Ingel, 1995; Dyka et al., 1995; Morris, 1996; Randles and Libersky, 1996). The SPH formulation leads to undesirable zero-energy modes that pollute the solution. A stress point approach (Vignjevic et al., 2000) has been proposed for the elimination of such modes. Finally, the SPH method does not allow direct imposition of essential boundary conditions. Several remedies to this difficulty have also been proposed (Morris et al., 1997; Randles and Libersky, 1996; Takeda et al., 1994).

Another meshless method is the RKPM. Liu and Chen (1995) noted that the smoothed representation of Equation 7.120 employed by the SPH method leads to several difficulties. When applied to discretized domains, it causes amplitude and phase errors in addition to the deterioration of solution at domain boundaries. They proposed the addition of a correction factor, C, to improve the smoothed representation as

$$u(x) = \int_{-\infty}^{\infty} C(x, x-s)\phi(x-s)u(s)\,ds \tag{7.124}$$

The correction factor C may be given as

$$C(x, x-s) = \sum_{j=0}^{m} a_j(x)p_j(x,s) \tag{7.125}$$

where $a_j(x)$ are the unknown coefficients and $p_j(x, s)$ are the basis functions. If the basis functions are chosen to be polynomials, the RKPM reduces to the EFG method described earlier. The RKPM method has been applied to a number of linear and nonlinear analysis problems which are reviewed extensively in Li and Liu (2001).

A recent meshless method is the natural neighbor Galerkin method (Sukumar et al., 1998, 2001). The method requires a Vornoi tessellation (O'Rourke, 2000) of the domain of the solid object. As with other meshless methods, the interpolation scheme is expressed as

$$u^h(\underset{\sim}{x}) = \sum_{i=1}^{N} \phi_i(\underset{\sim}{x})u_i \tag{7.126}$$

where $\phi_i(\underset{\sim}{x})$ are the nodal shape functions. These shape functions are computed based on the concept of natural neighbors and the corresponding interpolation introduced by Sibson (1980). An improvement in this interpolation employs the non-Sibsonian interpolation due to Belikov et al. (1997). The advantages of these natural neighbors–based interpolations include satisfaction of the partition of unity condition, allowing the constant reproducing conditions, and satisfaction of the selectivity property, allowing direct imposition of essential boundary conditions. The major apparent disadvantage lies in the fact that a Delaunay triangulation of the object is required to obtain the Vornoi tessellation on which the interpolation functions are based. This triangulation is viewed as being similar to the finite element mesh, thus diminishing the meshless character of the method.

A meshless method quite similar to the EFG method but with an improved local (node-based) character is the MLPG method (Atluri and Zhu, 1998, 2000b). Recall that in the weak formulations, the same interpolation of Equation 7.48 is employed for both the trial and test functions. In the MLPG formulation, the same trial functions are employed, while the weight functions of the MLS approximation with compact local support are used as the test functions. This imparts a strong local character to the formulation, requiring integration at each nodal point over only the support of that nodal point.

7.11 GENERALIZATIONS

The meshless method and finite element method are almost identical in most aspects. In Section 4.6, we demonstrate that the knowledge provided in Section 4.1 to Section 4.5 can be generalized from two-dimensional space to three-dimensional space, from static to dynamic, from elastic material to viscoelastic material or elastic-plastic material, from small strain theory to finite strain theory, and from local theory to nonlocal theory. Similarly, one may make the same statements in meshless method. However, it is noticed that the way meshless method treat the essential boundary conditions is very different from finite element method. Therefore, it is not a straightforward matter to extend the meshless method to dynamic case.

7.11.1 Dynamic problems

To extend the meshless method from static case to dynamics case, we first recall and rewrite the global system of equations, Equation 7.99, for static elasticity as

$$\begin{bmatrix} \mathbf{K} & \mathbf{G} \\ \mathbf{G}^T & 0 \end{bmatrix} \begin{Bmatrix} \mathbf{U} \\ \mathbf{\Lambda} \end{Bmatrix} = \begin{Bmatrix} \mathbf{F} \\ \mathbf{q} \end{Bmatrix} \tag{7.127}$$

where **U** and Λ are the vectors of nodal displacements and the Lagrangian multipliers, respectively; **F** is the forcing term due to body forces and applied surface tractions; **q** is the forcing term due to displacement-specified boundary conditions (cf. Equations 7.94–7.98). In fact, Equation 7.127 reflects one of a very few major differences between the finite element method and the meshless method.

For the sake of discussion, the dynamic governing equations of linear viscoelastic materials can be expressed (cf. Equation 4.102) as

$$\mathbf{M}\ddot{\mathbf{U}} + \mathbf{C}\dot{\mathbf{U}} + \mathbf{K}\mathbf{U} + \mathbf{G}\mathbf{\Lambda} = \mathbf{F} \tag{7.128}$$

$$\mathbf{G}^T\mathbf{U} = \mathbf{q} \tag{7.129}$$

To solve Equations 7.128 and 7.129, one needs to solve the following system of linear equations for **A** and **B** first (Chen et al., 2002, 2006):

$$\begin{bmatrix} \mathbf{K}^T & \mathbf{G} \\ \mathbf{G}^T & 0 \end{bmatrix} \begin{Bmatrix} \mathbf{A} \\ \mathbf{B} \end{Bmatrix} = \begin{Bmatrix} \mathbf{I} \\ 0 \end{Bmatrix} \tag{7.130}$$

where matrices **A** and **B** have the same rank as $\mathbf{K}^T$ and $\mathbf{G}^T$, respectively. Notice that Equation 7.130 implies

$$\mathbf{K}^T\mathbf{A} + \mathbf{G}\mathbf{B} = \mathbf{I} \;\Rightarrow\; \mathbf{A}^T\mathbf{K} + \mathbf{B}^T\mathbf{G}^T = \mathbf{I} \tag{7.131}$$

$$\mathbf{G}^T\mathbf{A} = 0 \;\Rightarrow\; \mathbf{A}^T\mathbf{G} = 0 \tag{7.132}$$

Multiply Equation 7.128 by $\mathbf{A}^T$ and further derive as

$$\begin{aligned} &\mathbf{A}^T\mathbf{M}\ddot{\mathbf{U}} + \mathbf{A}^T\mathbf{C}\dot{\mathbf{U}} + \mathbf{A}^T\mathbf{K}\mathbf{U} + \mathbf{A}^T\mathbf{G}\mathbf{\Lambda} - \mathbf{A}^T\mathbf{F} \\ &= \mathbf{A}^T\mathbf{M}\ddot{\mathbf{U}} + \mathbf{A}^T\mathbf{C}\dot{\mathbf{U}} + (\mathbf{I} - \mathbf{B}^T\mathbf{G}^T)\mathbf{U} - \mathbf{A}^T\mathbf{F} \\ &= \mathbf{A}^T\mathbf{M}\ddot{\mathbf{U}} + \mathbf{A}^T\mathbf{C}\dot{\mathbf{U}} + \mathbf{U} - \mathbf{A}^T\mathbf{F} - \mathbf{B}^T\mathbf{q} = 0 \end{aligned} \tag{7.133}$$

In other words, the dynamic equation for viscoelastic materials based on meshless method can be rewritten as

$$\mathbf{A}^T\mathbf{M}\ddot{\mathbf{U}}+\mathbf{A}^T\mathbf{C}\dot{\mathbf{U}}+\mathbf{U}=\mathbf{A}^T\mathbf{F}+\mathbf{B}^T\mathbf{q} \tag{7.134}$$

It is seen that Equation 7.134 has the same form as that from finite element method. Also, there is no extra equation for boundary conditions.

In the following, we are going to prove that if Equation 7.134 is satisfied, then the boundary condition, $\mathbf{G}^T\mathbf{U} = \mathbf{q}$, is automatically satisfied.

$$\begin{aligned}
0 &= \mathbf{A}^T\mathbf{M}\ddot{\mathbf{U}}+\mathbf{A}^T\mathbf{C}\dot{\mathbf{U}}+\mathbf{U}-\mathbf{A}^T\mathbf{F}-\mathbf{B}^T\mathbf{q}\\
&= \mathbf{A}^T\{\mathbf{M}\ddot{\mathbf{U}}+\mathbf{C}\dot{\mathbf{U}}-\mathbf{F}\}+\mathbf{U}-\mathbf{B}^T\mathbf{q}\\
&= \mathbf{A}^T\{-\mathbf{K}\mathbf{U}-\mathbf{G}\mathbf{\Lambda}\}+\mathbf{U}-\mathbf{B}^T\mathbf{q}\\
&= (-\mathbf{I}+\mathbf{B}^T\mathbf{G}^T)\mathbf{U}+\mathbf{U}-\mathbf{B}^T\mathbf{q}\\
&= \mathbf{B}^T\mathbf{G}^T\mathbf{U}-\mathbf{B}^T\mathbf{q}\\
&= \mathbf{B}^T\{\mathbf{G}^T\mathbf{U}-\mathbf{q}\}
\end{aligned} \tag{7.135}$$

which implies

$$\mathbf{G}^T\mathbf{U}=\mathbf{q} \tag{7.136}$$

It means that in the process of solving Equation 7.134, there is no concern about the displacement-specified boundary conditions.

7.11.2 Heun's method

One may rewrite Equation 7.134 as

$$\begin{aligned}
\ddot{\mathbf{U}} &= -\mathbf{M}^{-1}\mathbf{C}\dot{\mathbf{U}}-\mathbf{M}^{-1}(\mathbf{A}^T)^{-1}\mathbf{U}+\mathbf{M}^{-1}\mathbf{F}+\mathbf{M}^{-1}(\mathbf{A}^T)^{-1}\mathbf{B}^T\mathbf{q}\\
&\equiv \mathbf{Y}(\mathbf{U},\mathbf{V},t)
\end{aligned} \tag{7.137}$$

where $\mathbf{V}\equiv\dot{\mathbf{U}}$. Now define

$$\mathbf{X}\equiv\begin{Bmatrix}\mathbf{U}\\ \mathbf{V}\end{Bmatrix} \tag{7.138}$$

Then the governing equation of the system can be expressed as

$$\begin{aligned}
\dot{\mathbf{X}} &= \begin{Bmatrix}\dot{\mathbf{U}}\\ \dot{\mathbf{V}}\end{Bmatrix}=\begin{Bmatrix}\mathbf{V}\\ \mathbf{Y}(\mathbf{U},\mathbf{V},t)\end{Bmatrix}\\
&\equiv \mathbf{Z}(\mathbf{X},t)
\end{aligned} \tag{7.139}$$

Heun's method in solving Equation 7.139 may be described as follows.

Assuming at $t^n = n\ \Delta t$, we know $\mathbf{X}^n$ and $\dot{\mathbf{X}}^n = \mathbf{Z}(\mathbf{X}^n, t^n)$. Then detailed procedures are as follows:

Step 1:

$$\tilde{\mathbf{X}}^{n+1} = \mathbf{X}^n + \Delta t\ \dot{\mathbf{X}}^n \tag{7.140}$$

Step 2:

$$\dot{\tilde{\mathbf{X}}}^{n+1} = \mathbf{Z}(\tilde{\mathbf{X}}^{n+1}, t^{n+1}) \tag{7.141}$$

Step 3:

$$\bar{\dot{\mathbf{X}}} = \frac{1}{2}(\dot{\tilde{\mathbf{X}}}^{n+1} + \dot{\mathbf{X}}^n) \tag{7.142}$$

Step 4:

$$\mathbf{X}^{n+1} = \mathbf{X}^n + \Delta t\ \bar{\dot{\mathbf{X}}} \tag{7.143}$$

Step 5:

$$\dot{\mathbf{X}}^{n+1} = \mathbf{Z}(\mathbf{X}^{n+1}, t^{n+1}) \tag{7.144}$$

These five steps indicate that information at $t^n = n\ \Delta t$ enables us to obtain information at $t^{n+1} = (n + 1)\ \Delta t$.

7.12 CLOSING REMARKS

Starting in the mid-1990s, there has been lot of activity toward the development of meshless methods of analysis. Several significant advances have been made and have been reviewed by Belytschko et al. (1996) and Li and Liu (2001). Meshless methods offer the possibility of considerable simplification of the analysis process by eliminating the need for a structured mesh. Among several others, two application problems that have been envisioned to see significant simplification due to the success of meshless methods are (1) the arbitrary propagation of cracks in solid bodies and (2) solid bodies undergoing severe deformations, such as in the case of metal forming. Indeed, several of the earlier developments concentrated on demonstrating the effectiveness of mesh-free methods in crack propagation (Belytschko et al., 1995a, 1995b). Their success in modeling large deformation and other nonlinear problems has also been amply demonstrated (Chen et al., 1996, 1997, 1998). Several factors, however, continue to hinder the widespread acceptance of these methods for analysis. These include (1) the lack of a truly meshless method—the use of background cells or Delaunay triangulation to discretize the domain reduces the attractiveness of these methods, and (2) the difficulties encountered in the imposition of essential boundary conditions.

Although it was believed initially that meshless methods do not suffer from locking problems, some recent studies have indeed discovered such problems with these methods (Kanok-Nukulchai et al., 2001). Several recent developments, such as the natural neighbors method (Moran and Yoo, 2001), offer the hope of improved performance of these methods. A large number of impressive applications, including sensitivity analysis and shape optimization (Kim et al., 2001), stochastic analysis (Rahman and Rao, 2001), analysis of geologic materials (Belytschko et al., 2000), dynamic fracture (Belytschko and Tabarra, 1996), and multiple-length-scale problems (Chen et al., 2000b), have been reported in the literature. The entire field of meshless analysis, however, is in its infancy. It will require several years of continued development to bring these methods to the maturity of the finite element method, to verify the claims of efficiency and ease made on behalf of these methods, and to realize the promised potential of these methods.

Meshless analysis methods are extremely computationally burdensome, requiring 3 to 10 times the computational resources of the finite element method. No guidelines exist at present to suggest a desirable distribution of nodal points in the domain. Several analysts have, in fact, created background finite element meshes to provide nodal distributions for their meshless analysis. Crack propagation problems, although demonstrated successfully, have required an unusually dense distribution of nodes to allow crack propagation. While this situation could be improved significantly by adaptively providing more nodes near the current location of a crack tip, systematic formulations for such adaptive analyses are not yet available. From the point of view of elimination of the tedium due to the complexity of meshing, meshless methods are required especially for three-dimensional analyses. However, the success of these methods has been demonstrated most often for two-dimensional cases, and only a few three-dimensional studies have been reported (Barry and Saigal, 1999). The analysis of a complex three-dimensional industrial component has not yet been attempted. Meshless methods will gain popularity as their various limitations are addressed and as they begin to be used for practical industrial problems.

REFERENCES

Atluri, S. N., 2004. *The Meshless (MLPG) Method*, Tech Science Press, Encino, CA.

Atluri, S. N., Sladek, J., Sladek, V., and Zhu, T. 2000. The local boundary integral equation (LBIE) and its meshless implementation for linear elasticity, *Comput. Mech.*, **25**, 180–198.

Atluri, S. N., and Zhu, T. 1998. A new meshless local Petrov–Galerkin (MLPG) approach in computational mechanics, *Comput. Mech*, **22**, 117–127.

Atluri, S. N., and Zhu, T. 2000a. New concepts in meshless methods, *Int. J. Numer. Methods Eng.*, **47**, 537–556.

Atluri, S. N., and Zhu, T.-L. 2000b. The meshless local Petrov–Galerkin (MLPG) approach for solving problems in elastostatics, *Comput. Mech.*, **25**, 169–179.

Barry, W., and Saigal, S. 1999. A three-dimensional element-free Galerkin elastic and elastoplastic formulation, *Int. J. Numer. Methods Eng.*, **46**, 671–693.

Beissel, S., and Belytschko, T. 1996. Nodal integration of the element-free Galerkin method, *Comput. Methods Appl. Mech. Eng.*, **139**, 49–74.

Belikov, V., Ivanov, V., Kontorovich, V., Korytnok, S., and Semenov, A. 1997. The non-Gibsonian interpolation: A new method of interpolation of the values of a function on an arbitrary set of points, *Comput. Math. Math. Phys.*, **37**(1), 9–15.

Belytschko, T., Lu, Y. Y., and Gu, L. 1994a. Element free Galerkin methods, *Int. J. Numer. Methods Eng.*, **37**, 229–256.

Belytschko, T., Lu, Y. Y., and Gu, L. 1994b. Fracture and crack growth by element-free Galerkin methods, *Model. Simul. Sci. Comput. Eng.*, **2**, 519–534.

Belytschko, T., Lu, Y. Y, and Gu, L. 1995a. Crack propagation by element-free Galerkin methods, *Eng. Fracture Mech.*, **51**, 295–315.

Belytschko, T., Lu, Y. Y, and Gu, L. 1995b. Element-free Galerkin methods for static and dynamic fracture, *Int. J. Solids Struct.*, **32**, 2547–2570.

Belytschko, T., Organ, D., and Krongauz, Y. 1995c. A coupled finite element–element-free Galerkin method, *Comput. Mech.*, **17**, 186–195.

Belytschko, T., Kronagauz, Y, Organ, D., Fleming, M., and Krysl, P. 1996. Meshless methods: An overview and recent developments, *Comput. Methods Appl. Mech. Eng.*, **129**, 3–47.

Belytschko, T., Organ, D., and Gerlach, C. 2000. Element-free Galerkin methods for dynamic fracture in concrete, *Comput. Methods Appl. Mech. Eng.*, **187**, 385–399.

Belytschko, T., and Tabbara, M. 1996. Dynamic fracture using element-free Galerkin methods. *J. Comput. Appl. Math.*, **39**, 923–938.

Chati, M. K., and Mukherjee, S. 2000. The boundary node method for three-dimensional problems in potential theory, *Int. J. Numer. Methods Eng.*, **47**, 1523–1547.

Chati, M. K., Mukherjee, S., and Mukherjee, Y. X. 1999. The boundary node method for three-dimensional linear elasticity, *Int. J. Numer. Methods Eng.*, **46**, 1163–1184.

Chen, Y., Uras, R. A., and Liu, W.K. 1995. Enrichment of the finite element method with the reproducing kernel particle method, in *Current Topics in Computational Mechanics*, **305**, Cory, J. J. F., and Gordon, J. L. (eds.), ASME-PVP, Honolulu, HI, 253–258.

Chen, Y., Lee, J. D., and Eskandarian, A. 2002. Dynamic meshless method applied to nonlocal crack problems, *Theor. Appl. Fract. Mech*, **38**, 293–300.

Chen, Y., Lee, J. D., and Eskandarian, A. 2006. *Meshless Methods in Solid Mechanics*, Springer, New York.

Chen, J. S., Pan, C., and Wu, C. T. 1997. Large deformation analysis of rubber based on a reproducing kernel particle method, *Comput. Mech.*, **19**, 153–168.

Chen, J. S., Pan, C., Wu, C. T., and Roque, C. 1996. Reproducing kernel particle methods for large deformation analysis of nonlinear structures, *Comput. Methods Appl. Mech. Eng.*, **139**, 195–227.

Chen, J. S., Roque, C., Pan, C., and Button, S. T. 1998. Analysis of metal forming process based on meshless method, *J. Mater. Process. Technol.*, **80–81**, 642–646.

Chen, J. S., Wang, H. P., Yoon, S., and You, Y. 2000a. Some recent improvement in meshfree methods for incompressible finite elasticity boundary value problems with contact, *Comput. Mech.*, **25**, 137–156.

Chen, J. S., Wu, C. T., and Belytschko, T. 2000b. Regularization of material instabilities by meshfree approximations with intrinsic length scales, *Int. J. Numer. Methods Eng.*, **47**, 1303–1322.

Chen, J. S., Wu, C. T., Yoon, S., and You, Y. 2001. A stabilized conforming nodal integration for Galerkin meshfree methods, *Int. J. Numer. Methods Eng.*, **50**, 435–466.

Chong, K. P., Saigal, S., Thynell, S., and Morgan, H. S. (eds.) 2002. *Modeling and Simulation-Based Life Cycle Engineering*, Spon Press, New York.

De, S., and Bathe, K. J. 2000. The method of finite spheres. *Comput. Mech.*, **25**, 329–345.

De, S., and Bathe, K. J. 2001. Displacement/pressure mixed interpolation in the method of finite spheres, *Int. J. Numer. Methods Eng.*, **51**, 275–292.

Duarte, C. A., and Oden, J. T. 1996. H-P clouds: An H-P meshless method, *Numer. Methods Partial Differential Equations*, **12**, 673–705.

Dyka, C. T., and Ingel, R. P. 1995. An approach for tension instability in smoothed particle hydrodynamics, *Comput. Struct.*, **57**, 573–580.

Dyka, C. T., Randles, P. W., and Ingel, R. P. 1995. Stress points for tensor instability in SPH, *Int. J. Numer. Methods Eng.*, **40**, 2325–2341.

Gingold, R. A., and Monaghan, J. J. 1977. Smoothed particle hydrodynamics: Theory and application to non-spherical stars, *Mon. Not. R. Astron. Soc.*, **181**, 375–389.

Hegen, D. 1996. Element-free Galerkin methods in combination with finite element approaches, *Comput. Methods Appl. Mech. Eng.*, **135**, 143–166.

Johnson, G. R., and Beissel, S. R. 1996. Normalized smoothing functions for SPH impact computations, *Int. J. Numer. Methods Eng.*, **39**, 2725–2741.

Kaljevic, I., and Saigal, S. 1997. An improved element free Galerkin formulation, *Int. J. Numer. Methods Eng.*, **40**, 2953–2974.

Kanok-Nukulchai, W., Barry, W., Saran-Yasoontorn, K., and Bouillard, P. H. 2001. On elimination of shear locking in the element-free Galerkin method, *Int. J. Numer. Methods Eng.*, **52**, 705–725.

Kim, N. H., Choi, K. K., and Chen, J. S. 2001. Die shape design optimization of sheet metal stamping process using meshfree method, *Int. J. Numer. Methods Eng.*, **51**(12), 1385–1405.

Krongauz, Y., and Belytschko, T. 1997. Consistent pseudo-derivatives in meshless methods, *Comput. Methods Appl. Mech. Eng.*, **146**, 371–386.

Krongauz, Y., and Belytschko, T. 1998. EFG approximation with discontinuous derivatives, *Int. J. Numer. Methods Eng.*, **41**, 1215–1233.

Lancaster, P., and Salkauskas, K. 1980. Surface generated by moving least square methods, *Math. Comput*, **37**, 141–158.

Li, S., and Liu, W. K. 2001. Meshfree and particle methods and their applications, http://www.ce.berkeley.edu/~shaofan/review.pdf.

Libersky, L. D. and Petschek, A. G. 1991. Smooth particle hydrodynamics with strength of materials, in *Advances in the Free-Lagrange Method*, Vol. 359, Harold E. Trease, Martin F. Fritts, W. Patrick Crowley, Lecture Notes in Physics, Springer-Verlag, New York, pp. 248–257.

Libersky, L. D., Petschek, A. G., Carney, T. C., Hipp, J. R., and Allahdadi, F. A. 1993. High strain Lagrangian in hydrodynamics: A three-dimensional SPH code for dynamic material response, *J. Comput. Phys.*, **109**, 67–75.

Liu, W. K. 1995. An introduction to wavelet reproducing kernel particle methods, *USACM Bull.*, 8, 3–16.

Liu, W. K., and Chen, Y. 1995. Wavelet and multiple scale reproducing kernel method, *Int. J. Numer. Methods Fluids*, **21**, 901–933.

Liu, W. K., Chen, Y., Chang, C. T., and Belytschko, T. 1996. Advances in multiple scale kernel particle methods, *Comput. Mech.*, **18**, 73–111.

Liu, W. K., Chen, Y., and Uras, R. A. 1995. Enrichment of the finite element method with the reproducing kernel particle method, in *Current Topics in Computational Mechanics*, Vol. 305, Cory, J. J. F., and Gordon, J. L. (eds.), ASME-PVP, pp. 253–258.

Lu, Y. Y., Belytschko, T., and Gu, L. 1994. A new implementation of the element free Galerkin method, *Comput. Methods Appl. Mech. Eng.*, **113**, 397–414.

Lucy, L. B. 1977. A numerical approach to the testing of the fission hypothesis, *Astrophys. J.*, **82**, 1013.

Melenk, J. M., and Babuska, I. 1996. The partition of unity finite element method: Basic theory and application, *Comput. Methods Appl. Mech. Eng.*, **139**, 289–314.

Monaghan, J. J. 1988. An introduction to SPH, *Comput. Phys. Commun.*, **48**, 89–96.

Monaghan, J. J. 1992. Smoothed particle hydrodynamics, *Annu. Rev. Astron. Astrophys.*, **30**, 543–574.

Moran, B., and Yoo, J. 2001. Meshless methods for life-cycle engineering simulation: Natural neighbor methods, in *Modeling and Simulation-Based Life-Cycle Engineering*, Chong, K. P., Morgan, H. S., and Saigal, S. (eds.), E&FN Spon, London, 91–105.

Morris, J. P. 1996. Stability properties of SPH, *Publ. Astron. Soc. Aust.*, **13**, 97.

Morris, J. P., Fox, P. J., and Zhu, Y. 1997. Modeling low Reynolds number incompressible flows using SPH, *J. Comput. Phys.*, **136**, 214–226.

Mukherjee, Y. X., and Mukherjee, S. 1997a. The boundary node method in potential problems, *Int. J. Numer. Methods Eng.*, 40, 797–815.

Mukherjee, Y. X., and Mukherjee, S. 1997b. On boundary conditions in the element-free Galerkin method, *Comput. Mech.*, **19**, 264–270.

Mukherjee, Y. X., Mukherjee, S., Shi, X., and Nagarajan, A. 1997. The boundary contour method for three-dimensional linear elasticity with a new quadratic boundary element, *Eng. Anal. Boundary Elements*, **20**, 35–44.

Nagarajan, A., Lutz, E., and Mukherjee, S. 1994. A novel boundary element method for linear elasticity with no numerical integration for two-dimensional and line integrals for three-dimensional problems, *ASME J. Appl. Mech.*, **61**, 264–269.

Nagarajan, A., Mukherjee, S., and Lutz, E. 1996. The boundary contour method for three-dimensional linear elasticity, *ASME J. Appl. Mech.*, **63**, 278–286.

Nayroles, B., Touzot, G., and Villon, P. 1992. Generalization of the finite element method: Diffuse approximation and diffuse elements, *Comput. Mech.*, **10**, 307–318.

O'Rourke, J. 2000. *Computational Geometry in C*, Cambridge University Press, Cambridge.
Rahman, S., and Rao, B. N. 2001. A perturbation method for stochastic meshless analysis in elastostatics, *Int. J. Numer. Methods Eng.*, **50**, 1969–1992.
Randles, P. W., and Libersky, L. D. 1996. Smoothed particle hydrodynamics: Some recent improvements and applications, *Comput. Methods Appl. Mech. Eng.*, **139**, 375–408.
Schey, H. M. 2005. *Div, Curl, and All That*, W. W. Norton, New York.
Shepard, D. 1968. A two-dimensional interpolation function for irregularly spaced points, in *Proc. ACM National Conference*, pp. 517–524.
Sibson, R. 1980. A vector identity for the Dirichlet tessellation, *Math. Proc. Cambridge Philos. Soc*, **87**, 151–155.
Sukumar, N., Moran, B., and Belytschko, T. B. 1998. The natural element method in solid mechanics, *Int. J. Numer. Methods Eng.*, **43**, 839–887.
Sukumar, N., Moran, B., Semenov, Y., and Belikov, B. 2001. Natural neighbour Galerkin methods, *Int. J. Numer. Methods Eng.*, **50**, 1–27.
Swegle, J. W., Attaway, S. W., Heinstein, M. W., Mello, F. J., and Hicks, L. 1994. An analysis of smoothed particle hydrodynamics, *Sandia Report SAND-93-2513*, Sandia National Laboratories, Albuquerque, NM.
Takeda, H., Miyama, S. M., and Sekiya, M. 1994. Numerical simulation of viscous flow by smoothed particle hydrodynamics, *Prog. Theor. Phys.*, **116**, 123–134.
Vignjevic, R., Campbell, J., and Libersky, L. 2000. A treatment of zero-energy modes in the smoothed particle hydrodynamics method, *Comput. Methods Appl. Mech. Eng.*, **184**, 67–85.
Zhu, T., and Atluri, S. N. 1998. A modified collocation method and a penalty formulation for enforcing the essential boundary conditions in the element free Galerkin method, *Comput. Mech.*, **21**, 211–222.

BIBLIOGRAPHY

Belinha, J. 2014. Meshless methods in biomechanics-bone tissue remodelling analysis, in *Lecture Notes in Computational Vision and Biomechanics*, Vol. 16, Tavares, J. M. R. S., and Natal Jorge, R. M. (eds.), Springer, the Netherlands, p. 320.
Chau, K. T. 2018. *Theory of Differential Equations in Engineering and Mechanics*, CRC Press, Boca Raton, FL.
Li, S., and Liu, W. K. 2002. Meshfree and particle methods and their applications, *Appl. Mech. Rev.*, **55**(1), 1–34.
Liu, G. R., and Chen, X. L. 2001. A mesh-free method for static and free vibration analyses of thin plates of complicated shape, *J. Sound Vibr.*, **241**, 839–855.
Zhang, L. T., Liu, W. K., Li, S. F., Qian, D., and Hao, S. 2003. Survey of multi-scale meshfree particle methods, in *Meshfree Methods for Partial Differential Equations*, Griebel, M., and Schweitzer, M. A. (eds.), Springer Berlin Heidelberg, pp. 441–457.

Chapter 8

Multiphysics in molecular dynamics simulation

James D. Lee[1], *Jiaoyan Li*[2], *Zhen Zhang*[1], *and Kerlin P. Robert*[1]

[1]Department of Mechanical and Aerospace Engineering,
The George Washington University, Washington, DC
[2]School of Engineering, Brown University, Providence, Rhode Island

8.1 INTRODUCTION

Molecular dynamics (MD) simulation is computer simulation method that allows one to predict the dynamical evolution of a material system that consists of interacting particles, such as atoms and molecules. In an MD system, particles interact via analytical potential functions or molecular mechanics force fields. By integrating Newton's equations of motion for all particles simultaneously, the trajectories of atoms and molecules are determined. Since the 1980s, the development of physically sound interatomic potentials that go beyond simple pair–additive interactions and describe chemical reactions correctly has brought in impressive modeling capabilities and critical breakthroughs in materials research (Brenner et al., 1998; Erkoc, 1997; Garrison and Srivastava, 1995; Sinnott and Brenner, 2012; Srolovitz and Vitek, 2012). Interatomic potentials express the chemical bonding (e.g., bond stretching) and other interactions (e.g., van der Waals force) among particles. Besides interatomic potentials, the motions of particles are also affected by thermostat algorithm and external force field.

In MD, temperature is not an independent variable as in continuum mechanics (CM), but related to the microscopic random motions of atoms. Currently, popular techniques to control temperature include velocity rescaling, the Berendsen thermostat (Berendsen et al., 1984), the Andersen thermostat (Andersen, 1980), the Nosé–Hoover thermostat (Hoover, 1985; Nosé, 1984a, 1984b), Nosé–Hoover chains (Martyna et al., 1992), and Langevin dynamics (Lemons and Gythiel, 1997). Velocity rescaling is straightforward to implement. However, this thermostat does not allow the proper temperature fluctuations, cannot remove localized correlation, and is not time reversible. As a result, it is good for use in initialization state of a system, or say to warm up a system. The Berendsen thermostat is actually a specialized velocity rescaling thermostat, such that it has same advantages and disadvantages as velocity rescaling. Andersen thermostat maintains constant temperature condition for a material system by reassigning velocities of atoms or molecules that have collisions based on Maxwell-Boltzmann statistics. Even though the algorithm allows sampling from the canonical ensemble, the dynamics in fact is not physical. Langevin dynamics allows controlling the temperature of the canonical ensemble by the use of stochastic differential equations where a friction force term and a random force term are introduced. Compared to Langevin dynamics, Nosé–Hoover thermostat is deterministic, time reversible, and easy to implement. It has been widely used to calculate material properties. However, Nosé–Hoover thermostat is not suitable for a nanomaterial system whose temperature varies spatially and temporally during the simulation with the imposition of a temperature gradient. Li and Lee (2014a) reformulated the Nosé–Hoover thermostat to locally regulate temperatures at many distinct regions without introducing the physical linear and angular momenta and finally extend the feedback temperature force to a more general level.

Electromagnetic (EM) field is a typical external force field that would affect the particles' trajectories and physical quantities (e.g., energy) of a material system. One good example is microwave heating. Nowadays, everyone uses a microwave to heat or cook food every day. Besides household uses, microwave heating is also an important industrial process (Metaxas and Meredith, 1983). There has been an increasing focus on the application of microwave radiation for a variety of physicochemical purposes, such as synthesis of catalysts, separation processes, and the enhancement of reaction rates (Roussy and Pearce, 1995). Properties of variant materials in EM fields have been studied by MD simulations. English and his colleagues have conducted a series of MD studies for materials in EM fields, including water (English and MacElroy, 2003), TiO_2 (English et al., 2006), hen egg white lysozyme (English and Mooney, 2007), and ionic liquid (English et al., 2011). Wood et al. (2014) studied high-energy-density material α-HMX excited thermally and via electric fields at various frequencies using MD. Yang et al. (2009) studied the dielectric properties of aqueous solutions of a polar solvent (*N*,*N*-dimethylformamide), which is widely used in the production of acrylic fibers and plastics.

It is worthwhile to mention that new materials, new interatomic potentials, new physical phenomena, and new methodologies are emerging continuously (Bi et al., 2016a, 2016b; Cabriolu and Li, 2015; Filla et al., 2017; Qin et al., 2017).

The appearance of thermostat and EM force field enables one to study multiple simultaneous physical phenomena and also investigate material properties beyond mechanical properties, such as thermal properties and EM properties. In this chapter, the focus is placed on multiphysics in MD simulations, including mechanical, thermal, and EM applications. We will start from the basic theory, including governing equations, several types of interatomic potentials, and proof of objectivity. Following, MD simulation of multiphysics is formulated. It includes reformulation of Nosé–Hoover thermostat, proof of the objectivity of Nosé–Hoover thermal velocity and virial stress tensor, and the introduction of Maxwell's equations and Lorentz force at nanoscale. At last, Heun's method is introduced to numerically solve the time-dependent equations in MD.

8.2 GOVERNING EQUATIONS

When MD was originally conceived (Alder and Wainwright, 1959; Rahman, 1964), the fundamental idea was to determine the trajectories of atoms or molecules by numerically solving the Newton's equations of motion for a system of interacting particles. To begin with, let Newton's law be expressed as

$$m^i \dot{\mathbf{v}}^i = \mathbf{f}^i + \boldsymbol{\varphi}^i, \qquad i = [1,2,3,....,N] \tag{8.1}$$

where N is the total number of atoms in the system; m^i is the mass of *atom i*; $\mathbf{r}^i$, $\mathbf{v}^i, \equiv \dot{\mathbf{r}}^i$, and $\dot{\mathbf{v}}^i$ are the position, velocity, and acceleration of *atom i*, respectively; in this chapter, $\mathbf{f}^i$ and $\boldsymbol{\varphi}^i$ are reserved for the interatomic force acting on *atom i* and for forces other than interatomic force acting on *atom i*, respectively. In other words, $\boldsymbol{\varphi}^i$, known as body force, could be any combination of (1) applied force, (2) Lorentz force and fictitious force due to the translation and rotation of the coordinate system, and (3) thermal force due to the presence of thermostat. Also, it is noticed that the mass of *atom i* is denoted as m^i. This means each atom is identified by its own mass—this opens the door for a material system consisting of different kinds of atoms.

The interatomic force $\mathbf{f}^i$ acting on *atom i* can be expressed as

$$\mathbf{f}^i = -\frac{\partial U}{\partial \mathbf{r}^i} \tag{8.2}$$

It means that there exists a scalar-valued function U known as the interatomic potential, which is a function of position vectors of all atoms in the system, i.e.,

$$U = U(\mathbf{r}^1, \mathbf{r}^2, \mathbf{r}^3, \ldots, \mathbf{r}^N) \triangleq U(\mathbf{r}) \tag{8.3}$$

where $\mathbf{r}$ represents the totality of $\{\mathbf{r}^1, \mathbf{r}^2, \mathbf{r}^3, \ldots, \mathbf{r}^N\}$. The interatomic force of *atom i* is then calculated as the negative derivative of U with respect to the position vector of *atom i*.

The total energy of the system T^{total} is equal to the sum of kinetic energy K and potential energy U, i.e.,

$$\begin{aligned} T^{total} &= K + U \\ &= \sum_{i=1}^{N} \frac{1}{2} m^i \mathbf{v}^i \cdot \mathbf{v}^i + U(\mathbf{r}) \end{aligned} \tag{8.4}$$

The work done by the body force $\varphi \triangleq \{\varphi^1, \varphi^2, \varphi^3, \ldots, \varphi^N\}$ can be calculated as an integral

$$W(t) = \sum_{i=1}^{N} \int_0^t \boldsymbol{\varphi}^i(\tau) \cdot \mathbf{v}^i(\tau) d\tau \tag{8.5}$$

It is seen that the integrand is the inner product of the body force on *atom i* and velocity of *atom i*, which is equal to the rate of work done on *atom i*. One may readily show that

$$\frac{dT^{total}}{dt} = \frac{d(K+U)}{dt} = \sum_{i=1}^{N} \left\{ m^i \dot{\mathbf{v}}^i \cdot \mathbf{v}^i + \frac{\partial U}{\partial \mathbf{r}^i} \cdot \mathbf{v}^i \right\} = \sum_{i=1}^{N} \left\{ \left(m^i \dot{\mathbf{v}}^i - \mathbf{f}^i \right) \cdot \mathbf{v}^i \right\} \tag{8.6}$$

and

$$\frac{dW}{dt} = \sum_{i=1}^{N} \frac{d}{dt} \int_0^t \boldsymbol{\varphi}^i(\tau) \cdot \mathbf{v}^i(\tau) d\tau = \sum_{i=1}^{N} \boldsymbol{\varphi}^i(t) \cdot \mathbf{v}^i(t) \tag{8.7}$$

The governing equation, Equation 8.1, dictates that

$$\frac{dT^{total}}{dt} = \frac{dW}{dt} \tag{8.8}$$

If the body force is zero, i.e., the system is isolated from its environment, then Equation 8.8 says

$$\frac{dT^{total}}{dt} = \frac{d(K+U)}{dt} = 0 \quad \Rightarrow \quad K+U = \text{constant} \tag{8.9}$$

This is actually the *law of conservation of energy* in MD. One recalls that in CM, in addition to Newton's law or, say, *law of balance of linear momentum*, there is an independent *law of conservation of energy*. It is because, in CM, temperature is an independent field variable; but in MD, temperature is a dependent variable—it depends on the velocities of atoms in the system; therefore, there is no need for an independent energy equation. We will discuss this point in later sections. Now one may consider that Equations 8.1 and 8.2 are the balance law of linear momentum and constitutive equation, respectively, in MD. In a way, Equations 8.1 and 8.2 define MD.

8.3 INTERATOMIC POTENTIALS FOR IONIC CRYSTALS

The general, interatomic potential in MD can be written as

$$U = \frac{1}{2!}\sum_{i,j=1}^{N} U^{ij}(\mathbf{r}^i,\mathbf{r}^j) + \frac{1}{3!}\sum_{i,j,k=1}^{N} U^{ijk}(\mathbf{r}^i,\mathbf{r}^j,\mathbf{r}^k) + \frac{1}{4!}\sum_{i,j,k,l=1}^{N} U^{ijkl}(\mathbf{r}^i,\mathbf{r}^j,\mathbf{r}^k,\mathbf{r}^l) + \ldots. \tag{8.10}$$

which is just a more detailed description of Equation 8.3. The terms on the right-hand side may be referred to as two-body potential, three-body potential, four-body potential, and so on. It is understood that, in the summation, all indices must be distinct. For example, $j \neq i$ for the two-body potential; $j \neq i$, $k \neq i$ and $k \neq i$ for the three-body potential. In other words, $U^{ijkl....} = 0$ if any two indices are equal. Therefore, Equation 8.10 may also be expressed as

$$U = \sum_{j>i}^{N} U^{ij}(\mathbf{r}^i,\mathbf{r}^j) + \sum_{k>j>i}^{N} U^{ijk}(\mathbf{r}^i,\mathbf{r}^j,\mathbf{r}^k) + \sum_{l>k>j>i}^{N} U^{ijkl}(\mathbf{r}^i,\mathbf{r}^j,\mathbf{r}^k,\mathbf{r}^l) + \ldots. \tag{8.10*}$$

Before we introduce a few potential energies popularly used in MD simulation, we define the relative position vector $\mathbf{r}^{ij}$ and separation distance r^{ij}, between atom i and atom j, as follows

$$\begin{aligned} \mathbf{r}^{ij} &\equiv \mathbf{r}^i - \mathbf{r}^j \\ r^{ij} &\equiv \left\|\mathbf{r}^i - \mathbf{r}^j\right\| = \sqrt{\left(r_x^i - r_x^j\right)^2 + \left(r_y^i - r_y^j\right)^2 + \left(r_z^i - r_z^j\right)^2} \end{aligned} \tag{8.11}$$

* Equations 8.10 *simply means Equations 8.10 but in a different form. This rule applies to all equations with * in Chapters 8 and 9.

It is worthwhile to show that

$$\frac{\partial r^{ij}}{\partial r_x^i} = \frac{\partial \left\{ \left(r_x^i - r_x^j\right)^2 + \left(r_y^i - r_y^j\right)^2 + \left(r_z^i - r_z^j\right)^2 \right\}^{1/2}}{\partial r_x^i}$$
$$= \left\{ \left(r_x^i - r_x^j\right)^2 + \left(r_y^i - r_y^j\right)^2 + \left(r_z^i - r_z^j\right)^2 \right\}^{-1/2} \left(r_x^i - r_x^j\right) \tag{8.12}$$
$$= \frac{r_x^i - r_x^j}{r^{ij}}$$

Now one may readily verify that

$$\frac{\partial r^{ij}}{\partial \mathbf{r}^i} = \frac{\mathbf{r}^{ij}}{r^{ij}}, \quad \frac{\partial r^{ij}}{\partial \mathbf{r}^j} = -\frac{\mathbf{r}^{ij}}{r^{ij}} \tag{8.13}$$

Two frequently used two-body potentials, also known as pair potentials, are introduced as follows.

8.3.1 Lennard-Jones potential

Lennard-Jones potential (Kittel, 2005; Lennard-Jones, 1924) can be expressed as

$$U^{ij}(r^{ij}) = 4\varepsilon\left[\left(\frac{\sigma}{r^{ij}}\right)^{12} - \left(\frac{\sigma}{r^{ij}}\right)^{6}\right] \tag{8.14}$$

The interatomic force can be calculated as

$$\mathbf{f}^i = -\frac{\partial U^{ij}}{\partial \mathbf{r}^i} = -\mathbf{f}^j = \frac{\partial U^{ij}}{\partial \mathbf{r}^j} = 24\varepsilon\left\{2\frac{\sigma^{12}}{r_{ij}^{13}} - \frac{\sigma^6}{r_{ij}^7}\right\}\frac{\mathbf{r}^{ij}}{r_{ij}} \tag{8.15}$$

It is seen that (1) when the separation distance is very small, the first term dominates and the force is repulsive because $\mathbf{f}^i$ is in the direction of $\mathbf{r}^{ij} = \mathbf{r}^i - \mathbf{r}^j$; (2) when the separation distance is large, the second term dominates and the force is attractive; (3) the equilibrium position is at where $\mathbf{f}^i = 0$ and it is obtained as $r^o = 2^{1/6}\sigma$. However, one should not be misled by the last statement, which was obtained for the case that only two atoms are considered. Actually, one should consider all the atoms in the system in principle, at least all the nearest-neighbors for reasonably accurate solution. Notice that there are 12 nearest-neighbors in the face centered cubic crystal structure and Kittel (2005) obtained $r^o/\sigma = 1.09$.

8.3.2 Coulomb–Buckingham potential

The Coulomb–Buckingham potential is a typical pair potential (Chen and Lee, 2011; Kittel, 2005) and can be expressed as

$$U^{ij}(r^{ij}) = \frac{q^i q^j}{r^{ij}} + A^{ij}\exp(-r^{ij}/B^{ij}) - \frac{C^{ij}}{(r^{ij})^6} + \frac{D^{ij}}{(r^{ij})^{12}} \tag{8.16}$$

where q^i and q^j are the electric charge of *atom i* and *atom j*, respectively; A^{ij}, B^{ij}, C^{ij}, and D^{ij} are the material constants associated with *i*th kind atom and *j*th kind of atom. The first term on the right-hand side of Equation 8.16 is the famous Coulomb potential between two charges, and the next three terms specify the Buckingham potential. The third term may be called the remedy term for the Buckingham potential, which is usually neglected. The Coulomb–Buckingham potential is suitable to describe the atomic interactions of ionic crystals, such as rock salt crystals, perovskite crystals, and wurtzite crystals. Here, we put indices *i* and *j* even as superscripts on the potential function *U* to emphasize that not only the material constants but also the function form depend on the types of atoms involved.

The interatomic force can be readily calculated as

$$\mathbf{f}^i = -\mathbf{f}^j = \left\{ q^i q^j (r^{ij})^{-2} + \frac{A^{ij}}{B^{ij}} e^{-r^{ij}/B^{ij}} - 6C^{ij}(r^{ij})^{-7} + 12C^{ij}(r^{ij})^{-13} \right\} \frac{\mathbf{r}^{ij}}{r^{ij}} \tag{8.17}$$

In coarse-grained MD, which will be discussed in the next chapter, sometimes it is needed to employ the dynamical matrix. In the following, we show how to calculate *stiffness matrix* in MD (Li and Lee, 2014b).

Here, for simplicity, we only consider the case of pair potential. For *atom i* and *atom j*, define

$$\begin{aligned} &\mathbf{r} \equiv \mathbf{r}^i - \mathbf{r}^j, \qquad r \equiv \left\| \mathbf{r}^i - \mathbf{r}^j \right\|, \qquad U^{ij}(r) = U^{ji}(r) \triangleq U(r) \\ &\phi(r) \equiv -\frac{1}{r}\frac{dU}{dr}, \qquad \Phi(r) \equiv \frac{1}{r}\frac{d\phi}{dr} \end{aligned} \tag{8.18}$$

We now rewrite the governing equations in MD using tensor notations as

$$\begin{aligned} m^i \ddot{u}^i_\alpha &= f^i_\alpha + \phi^i_\alpha \\ &= \phi r^{ij}_\alpha + \phi^i_\alpha \end{aligned} \tag{8.19}$$

where $\mathbf{u}^i \equiv \mathbf{r}^i - \mathbf{R}^i$ is the displacement vector of *atom i*; $\mathbf{R}^i$ is the initial position of *atom i*; notice that $\ddot{\mathbf{u}}^i = \ddot{\mathbf{r}}^i = \dot{\mathbf{v}}^i$. To illustrate a point, we assume that all the atomic displacements involved are small. Now we do the Taylor series expansion of the interatomic force about the initial position

$$\begin{aligned} f^i_\alpha &= \phi r^{ij}_\alpha \Big|_{\substack{\mathbf{r}^i=\mathbf{R}^i \\ \mathbf{r}^j=\mathbf{R}^j}} + \sum_\beta \frac{\partial\left[\phi r^{ij}_\alpha\right]}{\partial r^i_\beta}\Big|_{\substack{\mathbf{r}^i=\mathbf{R}^i \\ \mathbf{r}^j=\mathbf{R}^j}} \left(r^i_\beta - R^i_\beta\right) + \sum_\beta \frac{\partial\left[\phi r^{ij}_\alpha\right]}{\partial r^j_\beta}\Big|_{\substack{\mathbf{r}^i=\mathbf{R}^i \\ \mathbf{r}^j=\mathbf{R}^j}} \left(r^j_\beta - R^j_\beta\right) + \ldots \\ &= f^i_\alpha(0) + \sum_\beta \frac{\partial\left[\phi r^{ij}_\alpha\right]}{\partial r^i_\beta}\Big|_{\substack{\mathbf{r}^i=\mathbf{R}^i \\ \mathbf{r}^j=\mathbf{R}^j}} u^i_\beta + \sum_\beta \frac{\partial\left[\phi r^{ij}_\alpha\right]}{\partial r^j_\beta}\Big|_{\substack{\mathbf{r}^i=\mathbf{R}^i \\ \mathbf{r}^j=\mathbf{R}^j}} u^j_\beta + \ldots \end{aligned} \tag{8.20}$$

and notice that

$$\begin{aligned} \frac{\partial\left[\phi r^{ij}_\alpha\right]}{\partial r^i_\beta} &= \frac{\partial\phi}{\partial r^i_\beta} r^{ij}_\alpha + \phi\frac{\partial r^{ij}_\alpha}{\partial r^i_\beta} = \frac{\partial\phi}{\partial r}\frac{r^{ij}_\beta}{r} r^{ij}_\alpha + \phi\delta_{\alpha\beta} \equiv \Phi r^{ij}_\alpha r^{ij}_\beta + \phi\delta_{\alpha\beta} \\ \frac{\partial\left[\phi r^{ij}_\alpha\right]}{\partial r^j_\beta} &= \frac{\partial\phi}{\partial r^j_\beta} r^{ij}_\alpha + \phi\frac{\partial r^{ij}_\alpha}{\partial r^j_\beta} = -\frac{\partial\phi}{\partial r}\frac{r^{ij}_\beta}{r} r^{ij}_\alpha - \phi\delta_{\alpha\beta} \equiv -\left\{\Phi r^{ij}_\alpha r^{ij}_\beta + \phi\delta_{\alpha\beta}\right\} \end{aligned} \tag{8.21}$$

The stiffness matrix can now be obtained as

$$-K_{\alpha\beta}^{ij} \equiv \left(\Phi r_{\alpha}^{ij} r_{\beta}^{ij} + \phi\delta_{\alpha\beta}\right)\Big|_{\substack{\mathbf{r}^i=\mathbf{R}^i \\ \mathbf{r}^j=\mathbf{R}^j}} \tag{8.22}$$

Then, Equation 8.20 can be rewritten as

$$f_{\alpha}^{i} = f_{\alpha}^{i}(0) - \sum_{\beta} K_{\alpha\beta}^{ij} u_{\beta}^{i} + \sum_{\beta} K_{\alpha\beta}^{ij} u_{\beta}^{j} + \ldots.. \tag{8.23}$$

This means that, for a system of N atoms with pair potential and small displacements, the governing equation can be expressed in the following matrix form

$$\mathbf{M}\ddot{\mathbf{u}} + \mathbf{K}\mathbf{u} = \mathbf{f}^{o} + \boldsymbol{\varphi} \tag{8.24}$$

which is exactly equivalent to the governing equations in classical mechanical vibration.

8.4 INTERATOMIC POTENTIALS FOR 2D MATERIALS

2D materials are single-layered materials. Free-standing 2D crystals were believed to be unstable at nonzero temperatures (Landau, 1937; Lifshitz and Pitaevskii, 1980; Peierls, 1935). This point of view has been disproved since Geim and his colleagues discovered a simple but novel method to isolate single atomic layers of graphene from graphite in 2004 (Novoselov et al., 2004). Discovery of graphene had a spillover effect. So far, 2D materials have attracted increasing attention because of their extraordinary physical and chemical properties and have developed a multidiscipline cutting across physics, chemistry, engineering and biology.

8.4.1 Graphene

Carbon nanostructures are leading materials in the nanotechnology field, and graphene is the first member in the 2D material family. It is the building block of other carbon materials with different dimensionalities, like buckyball, carbon nanotube, and three-dimensional graphite, as shown in Figure 8.1 (Geim and Novoselov, 2007).

Graphene is a single-layered graphite, consisting of carbon atoms packed into a honeycomb (hexagonal) architecture. The crystal structure of graphene can be looked through from its "mother material"—graphite. The primitive cell of graphite is shown in Figure 8.2a. A few layers of graphene sheets are shown in Figure 8.2b. A single layer of graphene with primitive cell marked in red is shown in Figure 8.2c. The honeycomb lattice constant is a = 0.142 nm, and the distance between layers is c = 0.335 nm. Each primitive cell of graphite has four atoms, which has the four position vectors: $\left(0,\ 0,\ \frac{c}{2}\right)$, $\left(0,\ 0,\ \frac{3c}{2}\right)$, $\left(\frac{a}{2},\ \frac{a}{2\sqrt{3}},\ \frac{c}{2}\right)$, and $\left(\frac{a}{2},\ -\frac{a}{2\sqrt{3}},\ \frac{3c}{2}\right)$. The three translation vectors are $\left(\frac{a}{2},\ -\frac{\sqrt{3}a}{2},\ 0\right)$, $\left(\frac{\sqrt{3}a}{2},\ \frac{a}{2},\ 0\right)$, and

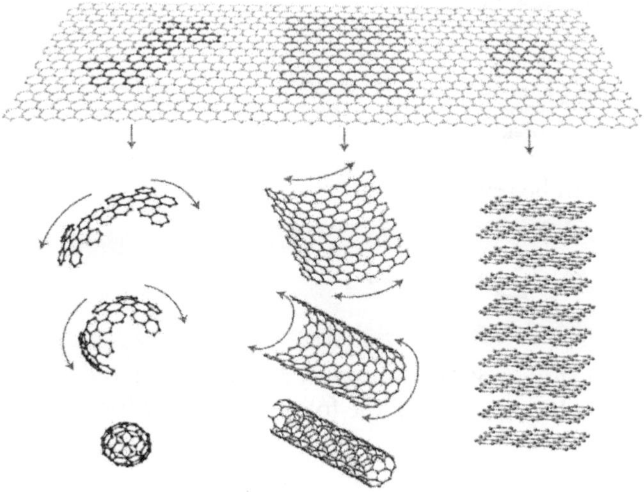

Figure 8.1 **(See color insert.)** Mother of all graphitic forms. Graphene is a 2D building material for carbon materials of all other dimensionalities. It can be wrapped up into a buckyball, rolled into a nanotube or stacked into a 3D graphite.

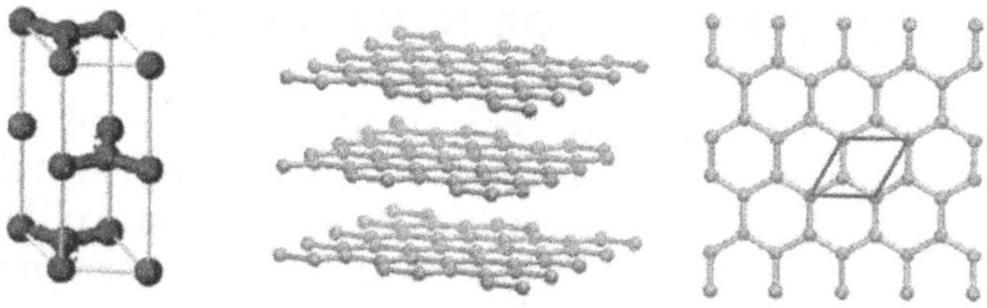

Figure 8.2 **(See color insert.)** (a) Graphite unit cell; (b) few layer graphene sheet; (c) in-plane view of single layer graphene, red line marks the primitive cell.

$(0, 0, 2c)$. As for graphene, only two atoms are in the primitive cell and their coordinates are $(0, 0, 0)$ and $\left(\frac{\sqrt{3}a}{2}, \frac{a}{2}, 0\right)$. The two translation vectors are $\left(\sqrt{3}a, 0, 0\right)$ and $\left(\frac{\sqrt{3}a}{2}, \frac{3a}{2}, 0\right)$.

8.4.2 Tersoff potential

The Tersoff (1988, 1989) potential is usually adopted for the simulation of graphene and it is expressed as

$$U^{ij} = f_C(r^{ij})\left\{f_A(r^{ij}) - b^{ij} f_B(r^{ij})\right\} \tag{8.25}$$

where $f_C(r^{ij})$ is a cutoff function given by

$$f_C(r^{ij}) = \begin{cases} 1 & : \quad r^{ij} < D^{ij} \\ \frac{1}{2}\left\{1 + \cos\left[\pi(r^{ij} - D^{ij})/(S^{ij} - D^{ij})\right]\right\} & : \quad D^{ij} < r^{ij} < S^{ij} \\ 0 & : \quad r^{ij} > S^{ij} \end{cases} \tag{8.26}$$

In Equation 8.26, S^{ij} is the cutoff radius, a material constant for *atom i* and *atom j*. This simply means if the separation distance between *atom i* and *atom j* is greater than the cutoff radius S^{ij}, then the potential energy U^{ij} is vanishing. In other words, Tersoff potential represents a short-range interaction between atoms. The other parameters in Equation 8.25 are explicitly specified as

$$
\begin{aligned}
f_A(r^{ij}) &= A^{ij}\exp(-\lambda^{ij}r^{ij}) \\
f_B(r^{ij}) &= B^{ij}\exp(-\mu^{ij}r^{ij}) \\
b^{ij} &= \left[1+\beta_i^{n_i}(\xi^{ij})^{n_i}\right]^{-1/2n_i} \\
\xi^{ij} &= \sum_{k\neq i,j} f_C(r^{ij})g(\theta^{ijk}) \\
g(\theta^{ijk}) &= 1+c_i^2/d_i^2 - c_i^2/\left[d_i^2+(h_i-\cos\theta^{ijk})^2\right]
\end{aligned}
\tag{8.27}
$$

where θ^{ijk} is the bond angle between the two vectors $\mathbf{r}^{ij}$ and $\mathbf{r}^{jk}$; A^{ij}, B^{ij}, λ^{ij}, μ^{ij}, D^{ij}, β^i, n_i, c_i, d_i, and h_i are material constants. It is noticed that b^{ij} involves the existence of all the other *atom k* within the cutoff radius, and hence, it represents the bond strength due to *many-body effects*.Therefore, rigorously speaking, Tersoff potential should be correctly named as many-body potential. But, when calculating the interatomic forces, three atoms are involved each time. Usually, the material constants related to *atom i* and *atom j* can be simplified as

$$
\begin{aligned}
A^{ij} &= \sqrt{A^iA^j}, \quad & B^{ij} &= \sqrt{B^iB^j} \\
S^{ij} &= \sqrt{S^iS^j}, \quad & D^{ij} &= \sqrt{D^iD^j} \\
\lambda^{ij} &= \sqrt{\lambda^i\lambda^j}, \quad & \mu^{ij} &= \sqrt{\mu^i\mu^j}
\end{aligned}
\tag{8.28}
$$

where A^i, B^i, S^i, D^i, λ^i, and μ^i are the corresponding material constants associated with *atom i*.

To calculate the interatomic forces acting on *atom i*, *atom j*, and *atom k*, to begin with, we find the following derivatives

$$
\begin{aligned}
f_C'(r^{ij}) &\equiv \frac{df_C(r^{ij})}{dr^{ij}} = \begin{cases} 0 & : \quad r^{ij} < D^{ij} \\ -\dfrac{\pi}{2(S^{ij}-D^{ij})}\sin\left[\pi(r^{ij}-D^{ij})/(S^{ij}-D^{ij})\right] & : \quad D^{ij} < r^{ij} < S^{ij} \\ 0 & : \quad r^{ij} > S^{ij} \end{cases} \\
g'(\theta^{ijk}) &\equiv \frac{dg}{d\theta^{ijk}} = 2c_i^2\left[d_i^2+(h_i-\cos\theta^{ijk})^2\right]^{-2}(h_i-\cos\theta^{ijk})\sin\theta^{ijk} \\
\frac{dg}{d\cos\theta^{ijk}} &= -2c_i^2\left[d_i^2+(h_i-\cos\theta^{ijk})^2\right]^{-2}(h_i-\cos\theta^{ijk})
\end{aligned}
\tag{8.29}
$$

Then the atomic forces can be obtained as

$$
\mathbf{f}^i = -\frac{1}{2}\left\{f_C'[f_A - b^{ij}f_B] + f_C[-\lambda^{ij}f_A + \mu^{ij}b^{ij}f_B]\right\}\frac{\mathbf{r}^{ij}}{r^{ij}} + \frac{1}{2}c^{ij}f_C(r^{ij})f_B(r^{ij})\frac{\partial\xi^{ij}}{\partial\mathbf{r}^i}
\tag{8.30}
$$

$$\mathbf{f}^j = \frac{1}{2}\{f_C'[f_A - b^{ij}f_B] + f_C[-\lambda^{ij}f_A + \mu^{ij}b^{ij}f_B]\}\frac{\mathbf{r}^{ij}}{r^{ij}} + \frac{1}{2}c^{ij}f_C(r^{ij})f_B(r^{ij})\frac{\partial \xi^{ij}}{\partial \mathbf{r}^j} \tag{8.31}$$

$$\mathbf{f}^k = \frac{1}{2}c^{ij}f_C(r^{ij})f_B(r^{ij})\frac{\partial \xi^{ij}}{\partial \mathbf{r}^k} \tag{8.32}$$

with

$$\begin{aligned}
\frac{\partial \xi^{ij}}{\partial \mathbf{r}^i} &= \left\{ g f_C'(r^{ik})\frac{\mathbf{r}^{ik}}{r^{ik}} + d^{ij}\frac{\partial \cos\theta^{ijk}}{\partial \mathbf{r}^i} \right\} \\
\frac{\partial \xi^{ij}}{\partial \mathbf{r}^j} &= d^{ij}\frac{\partial \cos\theta^{ijk}}{\partial \mathbf{r}^j} \\
\frac{\partial \xi^{ij}}{\partial \mathbf{r}^k} &= \left\{ -g f_C'(r^{ik})\frac{\mathbf{r}^{ik}}{r^{ik}} + d^{ij}\frac{\partial \cos\theta^{ijk}}{\partial \mathbf{r}^k} \right\}
\end{aligned} \tag{8.33}$$

where

$$\begin{aligned}
c^{ij} &\equiv -0.5\beta_i(\beta_i\xi^{ij})^{n_i-1}\left[1 + (\beta_i\xi^{ij})^{n_i}\right]^{-1/2n_i-1} \\
d^{ij} &\equiv f_C(r^{ik})\frac{-2c_i^2(h_i - \cos\theta^{ijk})}{\left[d_i^2 + (h_i - \cos\theta^{ijk})^2\right]^2}
\end{aligned} \tag{8.34}$$

It is lengthy but straightforward to obtain

$$\begin{aligned}
\frac{\partial \cos\theta^{ijk}}{\partial \mathbf{r}^i} &= \left\{ \frac{1}{r^{ik}} - \frac{\cos\theta^{ijk}}{r^{ij}} \right\}\frac{\mathbf{r}^{ij}}{r^{ij}} + \left\{ \frac{1}{r^{ij}} - \frac{\cos\theta^{ijk}}{r^{ik}} \right\}\frac{\mathbf{r}^{ik}}{r^{ik}} \\
\frac{\partial \cos\theta^{ijk}}{\partial \mathbf{r}^j} &= \frac{\cos\theta^{ijk}}{r^{ij}}\frac{\mathbf{r}^{ij}}{r^{ij}} - \frac{1}{r^{ij}}\frac{\mathbf{r}^{ik}}{r^{ik}} \\
\frac{\partial \cos\theta^{ijk}}{\partial \mathbf{r}^k} &= -\frac{1}{r^{ik}}\frac{\mathbf{r}^{ij}}{r^{ij}} + \frac{\cos\theta^{ijk}}{r^{ik}}\frac{\mathbf{r}^{ik}}{r^{ik}}
\end{aligned} \tag{8.35}$$

It is seen that

$$\frac{\partial \cos\theta^{ijk}}{\partial \mathbf{r}^i} + \frac{\partial \cos\theta^{ijk}}{\partial \mathbf{r}^j} + \frac{\partial \cos\theta^{ijk}}{\partial \mathbf{r}^k} = 0 \tag{8.36}$$

which implies that, among any three atoms, i, j, and k, Tersoff potential yields

$$\mathbf{f}^i + \mathbf{f}^j + \mathbf{f}^k = 0 \tag{8.37}$$

Once more, it says the total interatomic forces are vanishing and it verifies that Newton's third law is automatically satisfied.

8.4.3 Molybdenum disulfide

Molybdenum disulfide (MoS_2) is a graphene-like 2D material. Similar to graphite, originally, it is widely used as lubricants because of the weak van der Waals interlayer interaction. Nowadays, MoS_2 has been attracting plentiful research interests in order to overcome the limitations of graphene and broaden the range of applications of 2D materials.

Single layer MoS_2 has a trilayer structure with one Mo atomic layer sandwiched between two S atomic layers, as shown in Figure 8.3. For a single-layer MoS_2, the top view is similar to graphene, a honeycomb.

The crystal structure of MoS_2 can be described as follows: each primitive cell has three atoms with position vectors: one Mo atom with coordinate (0, 0, 0), and two S atoms with coordinates $\left(0, \frac{c^1}{\sqrt{3}}, c^2\right)$ and $\left(0, \frac{c^1}{\sqrt{3}}, -c^2\right)$. The three base vectors are: $(c^1, 0, 0)$, $\left(-\frac{1}{2}c^1, \frac{\sqrt{3}}{2}c^1, 0\right)$, and $(0, 0, 2c^2 + c^3)$, where $c^1 = 0.316$ nm, $c^2 = [0.242^2 - (c^1)^2/3]^{1/2}$ nm, and $c^3 = 0.350$ nm.

8.4.3.1 Interatomic potentials for Mo–S systems

The Stillinger–Weber (SW) potential (Jiang et al., 2013; Stillinger and Weber, 1985) and the Liang–Phillpot–Sinnott potential (Liang et al., 2009, 2012) may be used to describe the interactions of atoms in the MoS_2 system. For the SW potential, the potential parameters are fitted to an experimentally obtained phonon spectrum, and the resulting empirical potential provides a good description for the energy gap and the crossover in the phonon spectrum. The Liang–Phillpot–Sinnott potential may also be named as AIREBO potential; the fitting scheme used is optimized by appropriate selection of the functions, training databases, initial guesses, and weights on each residual—the four factors that are involved in a weighted nonlinear least-squares fitting. The resulting potential is able to yield good agreement with the structure and energetics of Mo–S systems, including MoS_2 layers. The results are consistent with static potential surface calculations using density-functional theory (Liang et al., 2009).

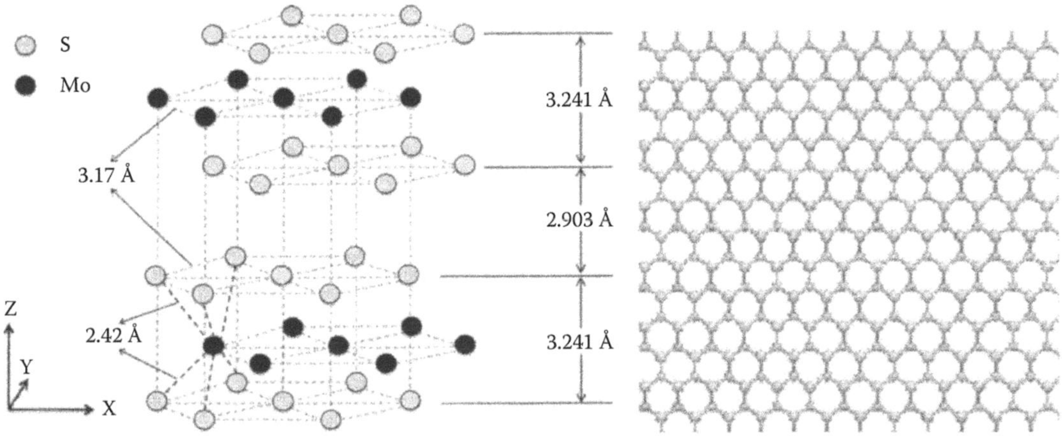

Figure 8.3 Left: Crystal structure of MoS_2. Schematic shows Mo–Mo, S–S, and Mo–S bond distances and S–S planar distances (Stewart and Spearot, 2013). Right: top view of single layer MoS_2.

The potential energy in the AIREBO potential may be expressed as

$$U = \frac{1}{2}\sum_{i \neq j} f_c(r^{ij})\left\{\left(1+\frac{Q}{r^{ij}}\right)Ae^{-\alpha r^{ij}} - b_{ij}Be^{-\beta r^{ij}}\right\} \tag{8.38}$$

with

$$f_c(r^{ij}) = \begin{cases} 1, & r^{ij} \le R_{ij}^{\min} \\ \dfrac{1}{2}\left\{1+\cos\left[\dfrac{\left(r^{ij}-R_{ij}^{\min}\right)\pi}{\left(R_{ij}^{\max}-R_{ij}^{\min}\right)}\right]\right\}, & R_{ij}^{\min} \le r^{ij} \le R_{ij}^{\max} \\ 0, & r^{ij} \ge R_{ij}^{\max} \end{cases} \tag{8.39}$$

$$b_{ij} = \left\{1+P(N_i)+\sum_{k \neq i,j} f_c(r^{ik})G^*\left[\cos(\theta_{ijk})\right]\right\}^{-1/2} \tag{8.40}$$

where U is the binding energy of the system; r^{ij} is the interatomic separation between the *i-atom* and *j-atom*; f_c is the cutoff function; b_{ij} is the bond-order term; P is the coordination term; N_i is the number of neighbors of *i-atom* within the cutoff; G^* is the bond angle term; and θ_{ijk} is the angle between bonds ij and ik. Notice that Q, A, B, α, β, $R^{\min}$, and $R^{\max}$ are pairwise parameters; each type of bond has one set of pairwise parameters. Because of the term $\sum_{k \neq i,j} f_c(r^{ik})G^*[\cos(\theta_{ijk})]$ in b_{ij}, AIREBO potential is known as many-body potential. For more detailed description of the parameters, interested readers may refer to Liang et al. (2009, 2012).

The interatomic forces should always be found as $\mathbf{f}^m = -\dfrac{\partial U}{\partial \mathbf{r}^m}$. The detailed expressions for interatomic forces are given as

$$\begin{aligned} \mathbf{f}^i &= -\frac{1}{2}\left\{f_c'\left[A_{ij}^* - b_{ij}B_{ij}^*\right] + f_c(r^{ij})\left[\frac{dA_{ij}^*}{dr^{ij}} - b_{ij}\frac{dB_{ij}^*}{dr^{ij}}\right]\right\}\frac{\mathbf{r}^{ij}}{r_{ij}} + \frac{1}{2}c_{ij}f_c(r^{ij})B_{ij}^*\frac{\partial \eta_{ijk}}{\partial \mathbf{r}^i} \\ \mathbf{f}^j &= +\frac{1}{2}\left\{f_c'\left[A_{ij}^* - b_{ij}B_{ij}^*\right] + f_c(r^{ij})\left[\frac{dA_{ij}^*}{dr^{ij}} - b_{ij}\frac{dB_{ij}^*}{dr^{ij}}\right]\right\}\frac{\mathbf{r}^{ij}}{r_{ij}} + \frac{1}{2}c_{ij}f_c(r^{ij})B_{ij}^*\frac{\partial \eta_{ijk}}{\partial \mathbf{r}^j} \\ \mathbf{f}^k &= \frac{1}{2}c_{ij}f_c(r^{ij})B_{ij}^*\frac{\partial \eta_{ijk}}{\partial \mathbf{r}^k} \end{aligned} \tag{8.41}$$

where

$$f_c' \equiv \frac{df(r^{ij})}{dr^{ij}} = \begin{cases} 0 & : \quad r^{ij} \le R_{ij}^{\min} \\ -\dfrac{\pi}{2\left(R_{ij}^{\max} - R_{ij}^{\min}\right)} \sin\left(\dfrac{\pi\left(r^{ij} - R_{ij}^{\min}\right)}{R_{ij}^{\max} - R_{ij}^{\min}}\right) & : \quad R_{ij}^{\min} \le r^{ij} \le R_{ij}^{\max} \\ 0 & : \quad r^{ij} \ge R_{ij}^{\max} \end{cases} \tag{8.42}$$

$$\begin{aligned} \eta_{ijk} &\equiv f_c(r^{ik}) G^*[\cos(\theta_{ijk})] \\ A_{ij}^* &\equiv \left(1 + \frac{Q}{r^{ij}}\right) A e^{-\alpha r^{ij}}, \quad B_{ij}^* \equiv B e^{-\beta r^{ij}} \\ \frac{dA^*}{dr^{ij}} &= -\left[\alpha\left(1 + \frac{Q}{r^{ij}}\right) + \frac{Q}{(r^{ij})^2}\right] A e^{-\alpha r^{ij}}, \quad \frac{dB^*}{dr^{ij}} = -\beta B e^{-\beta r^{ij}} \\ c_{ij} &\equiv -\frac{1}{2}\left\{1 + P(N_i) + \xi_{ij}\right\}^{-3/2} \end{aligned} \tag{8.43}$$

$$\begin{aligned} \frac{\partial \eta_{ijk}}{\partial \mathbf{r}^i} &= \left\{ G^*\left[\cos(\theta_{ijk})\right] f_c'(r^{ik}) \frac{\mathbf{r}^{ik}}{r^{ik}} + f_c(r^{ik}) \frac{dG^*\left[\cos(\theta_{ijk})\right]}{d\cos(\theta_{ijk})} (\cos\theta)_{,i} \right\} \\ \frac{\partial \eta_{ijk}}{\partial \mathbf{r}^j} &= f_c(r^{ik}) \frac{dG^*\left[\cos(\theta_{ijk})\right]}{d\cos(\theta_{ijk})} (\cos\theta)_{,j} \\ \frac{\partial \eta_{ijk}}{\partial \mathbf{r}^k} &= \left\{ -G^*\left[\cos(\theta_{ijk})\right] f_c'(r^{ik}) \frac{\mathbf{r}^{ik}}{r^{ik}} + f_c(r^{ik}) \frac{dG^*\left[\cos(\theta_{ijk})\right]}{d\cos(\theta_{ijk})} (\cos\theta)_{,k} \right\} \end{aligned} \tag{8.44}$$

$$\begin{aligned} (\cos\theta_{ijk})_{,i} &= \left\{ \frac{1}{r^{ik}} - \frac{\cos\theta_{ijk}}{r^{ij}} \right\} \frac{\mathbf{r}^{ij}}{r^{ij}} + \left\{ \frac{1}{r^{ij}} - \frac{\cos\theta_{ijk}}{r^{ik}} \right\} \frac{\mathbf{r}^{ik}}{r^{ik}} \\ (\cos\theta_{ijk})_{,j} &= \frac{\cos\theta_{ijk}}{r^{ij}} \frac{\mathbf{r}^{ij}}{r^{ij}} - \frac{1}{r^{ij}} \frac{\mathbf{r}^{ik}}{r^{ik}} \\ (\cos\theta_{ijk})_{,k} &= -\frac{1}{r^{ik}} \frac{\mathbf{r}^{ij}}{r^{ij}} + \frac{\cos\theta_{ijk}}{r^{ik}} \frac{\mathbf{r}^{ik}}{r^{ik}} \end{aligned} \tag{8.45}$$

Again, it is seen that, among any three atoms, i, j, and k, AIREBO potential yields $\mathbf{f}^i + \mathbf{f}^j + \mathbf{f}^k = 0$, which means Newton's third law is also automatically satisfied.

8.5 OBJECTIVITY IN MD

The principle of objectivity, sometimes referred to as principle of material frame-indifference, addresses the invariance attributes of physical quantities and material properties under change of reference frame. This concept of objectivity has been introduced in almost every textbook of CM, for example, in *Mechanics of Continua* (Eringen, 1980) and *Microcontinuum Field Theories* (Eringen, 1999). The idea is very simple: material properties measured in different reference frames are required to be objectively equivalent, i.e., not subject to the motion of the observers. However, in MD, sharing the common ground with Newtonian mechanics, objectivity was rarely discussed. Recently, in-depth discussions about objectivity in MD have been given by Yang et al. (2016a, 2016b) and Liu and Lee (2016). Here, we briefly describe the concepts, end results, and implications of objectivity. We first recall the two definitions on objectivity given by Eringen (1980) as follows.

Definition 1. Two motions $x_k(\mathbf{X}, t)$ and $x_k^*(\mathbf{X}, t^*)$ are called objectively equivalent if and only if

$$\mathbf{x}^*(\mathbf{X}, t^*) = \mathbf{Q}(t)\mathbf{x}(X, t) + \mathbf{b}(t), \qquad t^* = t + a \tag{8.46}$$

where a is a constant time shift, $\mathbf{b}(t)$ is a time-dependent translation, and $\mathbf{Q}(t)$ is time-dependent orthogonal transformations, i.e.,

$$\mathrm{QQ}^T = \mathrm{Q}^T\mathrm{Q} = \mathrm{I}, \qquad \det(\mathrm{Q}) = 1 \tag{8.47}$$

It is seen that Q consists of all rigid rotations (det $\mathbf{Q}$ = +1). Two objectively equivalent motions differ only in relative frame and time.

Definition 2. Any tensorial quantity that obeys the following tensor transformation law is said to be objective or material frame-indifferent.

$$A^*_{abcd\ldots}(\mathbf{X}, t^*) = Q_{a\alpha}(t)Q_{b\beta}(t)Q_{c\gamma}(t)Q_{d\delta}(t)\ldots.A_{\alpha\beta\gamma\delta\ldots}(\mathbf{X}, t) \tag{8.48}$$

Following Equation 8.46, one may obtain

$$\mathrm{v}^* = \mathrm{Qv} + \dot{\mathrm{Q}}\mathrm{x} + \dot{\mathrm{b}}, \qquad \mathrm{a}^* = \mathrm{Qa} + \ddot{\mathrm{Q}}\mathrm{x} + 2\dot{\mathrm{Q}}\mathrm{v} + \ddot{\mathrm{b}} \tag{8.49}$$

From Equations 8.46 and 8.49, one may immediately conclude that time t^*, position $\mathbf{x}^*$, velocity $\mathbf{v}^*$, and acceleration $\mathbf{a}^*$ are not objective, i.e., simply because they do not obey the tensor transformation law, Equation 8.48. This may be regarded as common sense. However, it should be viewed as the beauty of theoretical work—because it has generality, as we can see in the later development.

Among N atoms, if one picks *atom i* as the reference point, the interatomic potential can be expressed as

$$U = U(\mathbf{r}^{i1}, \mathbf{r}^{i2}, \mathbf{r}^{i3}, \ldots.) \tag{8.50}$$

Now it is noticed that all the relative position vectors $\mathbf{r}^{ij}$ $j \in \{1, 2, 3, \ldots.\}$ are objective. From the viewpoint of objectivity, U is a scalar-valued isotropic function of many objective vectors. Following Wang's representative theorem for isotropic functions (Wang, 1970a, 1970b), one may rewrite U as a function of lengths and angles

$$r^{ij} = \left\|\mathbf{r}^{ij}\right\|, \qquad r^{ik} = \left\|\mathbf{r}^{ik}\right\|, \qquad \theta^{ijk} \equiv \cos^{-1}\left(\frac{\mathbf{r}^{ij}\cdot\mathbf{r}^{ik}}{r^{ij}r^{ik}}\right) \tag{8.51}$$

In fact, for two-body potential, e.g., Lennard-Jones potential, Buckingham potential, it takes the form

$$U_2 = \sum_{j>i}^{N} U^{ij}(r^{ij}) \tag{8.52}$$

for three-body potential, e.g., Tersoff potential (Tersoff, 1988, 1989), SW potential (Stillinger and Weber, 1985),

$$U_3 = \sum_{k>j>i}^{N} U^{ijk}(r^{ij},\ r^{ik},\ \theta^{ijk}) \tag{8.53}$$

for four-body potential, e.g., in Universal Force Field (UFF) (Rappe et al., 1992),

$$U_4 = \sum_{l>k>j>i}^{N} U^{ijkl}(r^{ij},\ r^{ik},\ r^{il}, \theta^{ijk},\ \phi,\ \varphi) \tag{8.54}$$

where ϕ and φ are the torsion angle and inversion angle, respectively.

Also, the interatomic force, for example $\mathbf{f}^i$, can be obtained as (*cf.* Equations 8.13 and 8.35)

$$\begin{aligned}
\mathbf{f}^i &= -\frac{\partial U(\mathbf{r}^{ij},\ \mathbf{r}^{ik},\ \mathbf{r}^{il},\ldots.)}{\partial \mathbf{r}^i} \\
&= -\frac{\partial U(\mathbf{r}^{ij},\ \mathbf{r}^{ik},\ \mathbf{r}^{il},\ldots.,\cos\theta^{ijk},\ \cos\theta^{ijl},\ \cos\theta^{ikl},\ldots.)}{\partial \mathbf{r}^i} \\
&= -\sum_j \frac{\partial U}{\partial r^{ij}}\frac{\partial r^{ij}}{\partial \mathbf{r}^i} - \sum_{j\neq k}\frac{\partial U}{\partial \cos\theta^{ijk}}\frac{\partial \cos\theta^{ijk}}{\partial \mathbf{r}^i} \\
&= \sum_j \frac{\partial U}{\partial r^{ij}}\frac{\mathbf{r}^{ij}}{r^{ij}} - \sum_{j\neq k}\frac{\partial U}{\partial \cos\theta^{ijk}}\left\{\left[\frac{1}{r^{ik}} - \frac{\cos\theta^{ijk}}{r^{ij}}\right]\frac{\mathbf{r}^{ij}}{r^{ij}} + \left[\frac{1}{r^{ij}} - \frac{\cos\theta^{ijk}}{r^{ik}}\right]\frac{\mathbf{r}^{ik}}{r^{ik}}\right\}
\end{aligned} \tag{8.55}$$

Theorem 1: $\mathbf{r}^{ij}$, r^{ij}, θ^{ijk}, U, and $\mathbf{f}^i$ are objective.

Proof:

From Equation 8.46, one has

$$\mathbf{r}^{i*} = \mathbf{Q}\mathbf{r}^i + \mathbf{b} \text{ and } \mathbf{r}^{j*} = \mathbf{Q}\mathbf{r}^j + \mathbf{b} \tag{8.56}$$

which means both $\mathbf{r}^i$ and $\mathbf{r}^j$ are not objective. However, we see that

$$\mathbf{r}^{ij*} = \mathbf{r}^{i*} - \mathbf{r}^{j*} = \mathbf{Q}\mathbf{r}^i + \mathbf{b} - (\mathbf{Q}\mathbf{r}^j + \mathbf{b}) = \mathbf{Q}\mathbf{r}^i - \mathbf{Q}\mathbf{r}^j = \mathbf{Q}\mathbf{r}^{ij} \tag{8.57}$$

$$r^{ij*} = \sqrt{\mathbf{r}^{ij*} \cdot \mathbf{r}^{ij*}} = \sqrt{r_\alpha^{ij*} r_\alpha^{ij*}} = \sqrt{Q_{\alpha\beta} r_\beta^{ij} Q_{\alpha\gamma} r_\gamma^{ij}} = \sqrt{\delta_{\beta\gamma} r_\beta^{ij} r_\gamma^{ij}} = \sqrt{r_\beta^{ij} r_\beta^{ij}} = r^{ij} \tag{8.58}$$

$$\theta^{ijk*} = \cos^{-1} \frac{\mathbf{r}^{ij*} \cdot \mathbf{r}^{ik*}}{r^{ij*} r^{ik*}} = \cos^{-1} \frac{Q_{\alpha\beta} r_\beta^{ij} Q_{\alpha\gamma} r_\gamma^{ik}}{r^{ij} r^{ik}} = \cos^{-1} \frac{r_\beta^{ij} r_\beta^{ik}}{r^{ij} r^{ik}} = \cos^{-1} \frac{\mathbf{r}^{ij} \cdot \mathbf{r}^{ik}}{r^{ij*} r^{ik*}} = \theta^{ijk} \tag{8.59}$$

$$\begin{aligned} U^* &= U(r^{ij*},\ r^{ik*},\ r^{il*}, \ldots, \cos\theta^{ijk*},\ \cos\theta^{ijl*},\ \cos\theta^{ikl*}, \ldots) \\ &= U(r^{ij},\ r^{ik},\ r^{il}, \ldots, \cos\theta^{ijk},\ \cos\theta^{ijl},\ \cos\theta^{ikl}, \ldots) = U \end{aligned} \tag{8.60}$$

Then, from Equation 8.55, one has

$$\begin{aligned} \mathbf{f}^{i*} &= \sum_j \frac{\partial U^*}{\partial r^{ij*}} \frac{\mathbf{r}^{ij*}}{r^{ij*}} - \sum_{j \neq k} \frac{\partial U^*}{\partial \cos\theta^{ijk*}} \left\{ \left[\frac{1}{r^{ik*}} - \frac{\cos\theta^{ijk*}}{r^{ij*}} \right] \frac{\mathbf{r}^{ij*}}{r^{ij*}} + \left[\frac{1}{r^{ij*}} - \frac{\cos\theta^{ijk*}}{r^{ik*}} \right] \frac{\mathbf{r}^{ik*}}{r^{ik*}} \right\} \\ &= \sum_j \frac{\partial U}{\partial r^{ij}} \frac{\mathbf{Q}\mathbf{r}^{ij}}{r^{ij}} - \sum_{j \neq k} \frac{\partial U}{\partial \cos\theta^{ijk}} \left\{ \left[\frac{1}{r^{ik}} - \frac{\cos\theta^{ijk}}{r^{ij}} \right] \frac{\mathbf{Q}\mathbf{r}^{ij}}{r^{ij}} + \left[\frac{1}{r^{ij}} - \frac{\cos\theta^{ijk}}{r^{ik}} \right] \frac{\mathbf{Q}\mathbf{r}^{ik}}{r^{ik}} \right\} \\ &= \mathbf{Q}\mathbf{f}^i \end{aligned} \tag{8.61}$$

This means r^{ij}, θ^{ijk}, and U are objective scalars; $\mathbf{r}^{ij}$ and $\mathbf{f}^i$ are objective vectors.

To capture the full motion of a material body with respect to a noninertial reference frame (a reference frame accelerates and rotates), the governing equation must account for the motion of the reference frame itself; thus, fictitious forces are introduced. *Rectilinear acceleration force*, *centrifugal force*, *Coriolis force*, *Euler force*, etc., are several commonly noticed fictitious forces. Moreover, Einstein pointed out that gravity, too, is a form of fictitious force, which leads to the birth of *general relativity*. It is regarded that the appearance of fictitious force resolves the discrepancy in different noninertial reference frames so that motion observed in one frame can be converted to the motion observed in another. As one has already observed that acceleration $\mathbf{a} \equiv \dot{\mathbf{v}}$ is not objective (cf. Equation 8.49), let $\mathbf{i}$ be the acceleration induced by the fictitious force and principle of objectivity imposes a requirement

$$\mathbf{a}^* - \mathbf{i}^* = \mathbf{Q}(\mathbf{a} - \mathbf{i}) \tag{8.62}$$

For simplicity we assume that there is no fictitious force in the inertia reference frame, which at most moves with constant speed, then Equation 8.62 simply means

$$\mathbf{i}^* = \ddot{\mathbf{Q}}\mathbf{x} + 2\dot{\mathbf{Q}}\mathbf{v} + \ddot{\mathbf{b}} \tag{8.63}$$

Yang et al. (2016a, 2016b) further developed the expressions for **a*** and **i*** as

$$\mathbf{a}^* = \mathbf{Q}\mathbf{a} + [-2\boldsymbol{\omega}\times\mathbf{v}^* - \boldsymbol{\omega}\times(\boldsymbol{\omega}\times\mathbf{x}^*) - \dot{\boldsymbol{\omega}}\times\mathbf{x}^*] + [2\boldsymbol{\omega}\times\dot{\mathbf{b}} + \boldsymbol{\omega}\times(\boldsymbol{\omega}\times\mathbf{b}) + \dot{\boldsymbol{\omega}}\times\mathbf{b}] + \ddot{\mathbf{b}} \tag{8.64}$$

and

$$\mathbf{i}^* = [-2\boldsymbol{\omega}\times\mathbf{v}^* - \boldsymbol{\omega}\times(\boldsymbol{\omega}\times\mathbf{x}^*) - \dot{\boldsymbol{\omega}}\times\mathbf{x}^*] + [2\boldsymbol{\omega}\times\dot{\mathbf{b}} + \boldsymbol{\omega}\times(\boldsymbol{\omega}\times\mathbf{b}) + \dot{\boldsymbol{\omega}}\times\mathbf{b}] + \ddot{\mathbf{b}} \tag{8.65}$$

where $\Omega_{ij} \equiv \dot{Q}_{ik}Q_{jk}$ and $\omega_k \equiv \frac{1}{2}e_{ijk}\Omega_{ij}$

We will impose principle of objectivity on constitutive relations in MD in the next section.

8.6 REFORMULATION OF NOSÉ–HOOVER THERMOSTAT

It is noticed that, in MD simulation, temperature is a dependent variable. Usually, but not correctly, it was expressed as

$$T = \frac{\sum_{i=1}^{N} m^i(\mathbf{v}^i\cdot\mathbf{v}^i)}{N_{dof}k_B} \tag{8.66}$$

where N_{dof} is the number of degrees of freedom of the system and k_B is the Boltzmann constant.

The revolutionary Nosé–Hoover dynamics, originally introduced by Nosé (1984a, 1984b) and developed further by Hoover (1985), modified Newtonian dynamics so as to reproduce canonical and isobaric–isothermal ensemble equilibrium systems. However, there is an increasing interest in conducting MD simulations that do not fall within the classification of these classical ensembles. A typical example is a nano-material system whose temperature varies spatially and temporally during the simulation with the imposition of a temperature gradient. Clearly, this is a nanoscale heat conduction problem and requires nonequilibrium MD (NEMD) with suitable algorithmic thermostat for the local temperature regulation. Li and Lee (2014a) pointed out the need to reformulate the Nosé–Hoover thermostat to locally regulate the temperatures at many distinct regions without introducing the unphysical linear and angular momenta. In this way, the trajectories of atoms and molecules can be generated more rigorously and accurately by NEMD simulations.

Here, for readers' convenience, we briefly review the procedures and results of the reformulation of Nosé–Hoover thermostat. Let us consider that the whole specimen is divided into N_G groups; in *group g*, there are n_g atoms, not necessarily of the same kind. Then it is straightforward to calculate the mass, the position of the centroid, and the average velocity of *group g* as follows:

$$m^g \equiv \sum_{i=1}^{n_g} m^i, \quad \overline{\mathbf{r}}^g \equiv \left(\sum_{i=1}^{n_g} m^i\mathbf{r}^i\right)/m^g, \quad \overline{\mathbf{v}}^g \equiv \left(\sum_{i=1}^{n_g} m^i\mathbf{v}^i\right)/m^g \tag{8.67}$$

Then the relative position and relative velocity can be obtained as

$$\overline{\mathbf{r}}^i \equiv \mathbf{r}^i - \overline{\mathbf{r}}^g, \qquad \overline{\mathbf{v}}^i \equiv \mathbf{v}^i - \overline{\mathbf{v}}^g \tag{8.68}$$

It is noticed that

$$\sum_i m^i \bar{\mathbf{r}}^i = 0, \quad \sum_i m^i \bar{\mathbf{v}}^i = 0 \tag{8.69}$$

From now on, if there is no ambiguity, we use the abbreviation: $\sum_i \equiv \sum_{i=1}^{n_g}$, $\bar{\mathbf{r}} \equiv \bar{\mathbf{r}}^g$, and $\bar{\mathbf{v}} \equiv \bar{\mathbf{v}}^g$. The angular momentum (with respect to the centroid) is calculated as

$$\mathbf{H} = \sum_i m^i \bar{\mathbf{r}}^i \times \bar{\mathbf{v}}^i \tag{8.70}$$

If, in general, a system of n particles has a rigid-body rotation about the centroid with an angular velocity $\boldsymbol{\omega}$, then the relative velocity of the ith particle $\boldsymbol{\eta}^i$ with respect to the centroid can be calculated as

$$\boldsymbol{\eta}^i = \boldsymbol{\omega} \times \bar{\mathbf{r}}^i \tag{8.71}$$

The angular momentum due to the rigid-body rotation can now be calculated as

$$\bar{\mathbf{H}} = \sum_i m^i \bar{\mathbf{r}}^i \times \boldsymbol{\eta}^i = \sum_i m^i \bar{\mathbf{r}}^i \times (\boldsymbol{\omega} \times \bar{\mathbf{r}}^i) = \mathbf{J}\boldsymbol{\omega} \tag{8.72}$$

where the moment of inertia tensor $\mathbf{J}$ is defined as

$$\mathbf{J} \equiv \sum_i m^i \left\{ (\bar{\mathbf{r}}^i \cdot \bar{\mathbf{r}}^i)\mathbf{I} - \bar{\mathbf{r}}^i \otimes \bar{\mathbf{r}}^i \right\} \tag{8.73}$$

By equating $\mathbf{H}$ and $\bar{\mathbf{H}}$, one may find the angular velocity of the system

$$\mathbf{H} = \bar{\mathbf{H}} = \mathbf{J}\boldsymbol{\omega} \quad \Rightarrow \quad \boldsymbol{\omega} = \mathbf{J}^{-1}\mathbf{H} \tag{8.74}$$

Now we define the *Nosé–Hoover thermal velocity* as

$$\tilde{\mathbf{v}}^i \equiv \mathbf{v}^i - \bar{\mathbf{v}} - \boldsymbol{\eta}^i = \mathbf{v}^i - \bar{\mathbf{v}} - \boldsymbol{\omega} \times \bar{\mathbf{r}}^i = \mathbf{v}^i - \bar{\mathbf{v}} - (\mathbf{J}^{-1}\mathbf{H}) \times \bar{\mathbf{r}}^i \tag{8.75}$$

In other words, the Nosé–Hoover thermal velocity is the velocity beyond rigid-body translation and rotation. Now one may prove the following theorem.

Theorem 2: The total linear momentum and angular momentum caused by the Nosé–Hoover thermal velocity are vanishing.

Proof:

$$\begin{aligned} \sum_i m^i \tilde{\mathbf{v}}^i &= \sum_i m^i (\mathbf{v}^i - \bar{\mathbf{v}} - \boldsymbol{\eta}^i) \\ &= \sum_i m^i \mathbf{v}^i - m\bar{\mathbf{v}} - \sum_i m^i \boldsymbol{\omega} \times \bar{\mathbf{r}}^i \\ &= -\boldsymbol{\omega} \times \sum_i m^i \bar{\mathbf{r}}^i = 0 \end{aligned} \tag{8.76}$$

$$\begin{aligned}\sum_i m^i \mathbf{r}^i \times \tilde{\mathbf{v}}^i &= \sum_i m^i (\overline{\mathbf{r}}^i + \overline{\mathbf{r}}) \times (\mathbf{v}^i - \overline{\mathbf{v}} - \boldsymbol{\eta}^i) \\ &= \sum_i m^i \overline{\mathbf{r}}^i \times (\mathbf{v}^i - \overline{\mathbf{v}} - \boldsymbol{\eta}^i) + \overline{\mathbf{r}} \times \sum_i m^i (\mathbf{v}^i - \overline{\mathbf{v}} - \boldsymbol{\eta}^i) \\ &= \sum_i m^i \overline{\mathbf{r}}^i \times (\mathbf{v}^i - \overline{\mathbf{v}} - \boldsymbol{\omega} \times \overline{\mathbf{r}}^i) \\ &= \sum_i m^i \overline{\mathbf{r}}^i \times \overline{\mathbf{v}}^i - \sum_i m^i \overline{\mathbf{r}}^i \times (\boldsymbol{\omega} \times \overline{\mathbf{r}}^i) \\ &= \mathbf{H} - \overline{\mathbf{H}} = 0\end{aligned} \tag{8.77}$$

Now the temperature of *group g* is calculated as

$$T_g = \frac{1}{N_g^{dof} k_B} \sum_{i=1}^{n_g} m^i \tilde{\mathbf{v}}^i \cdot \tilde{\mathbf{v}}^i \tag{8.78}$$

where $N_g^{dof} = 3n_g - 6$ is the number of degrees of freedom of *group g*. It is noticed that, according to Equations 8.76 and 8.77, rigid-body translation and rotation have no contribution to the calculation of temperature. The subtraction of 6 from $3n_g$ is due to the elimination of linear and angular momenta from the velocity field in the calculation of the Nosé–Hoover thermal velocity.

The governing equations for a material system with *upgraded Nosé–Hoover Thermostats* should now be expressed as

$$m^i \dot{\mathbf{v}}^i = \mathbf{f}^i + \boldsymbol{\varphi}^i - \chi_g m^i \tilde{\mathbf{v}}^i, \qquad i \in group\ g \tag{8.79}$$

where $-\chi_g m^i \tilde{\mathbf{v}}^i$ is named as *Nosé–Hoover temperature force*. The role of χ_g is similar to damping coefficient, except that χ_g is not a constant—instead it is governed by

$$\dot{\chi}_g = \frac{1}{\tau_g^2 T_g^o} \left(T_g - T_g^o \right) \tag{8.80}$$

where τ_g is a specified time constant associated with *group g* and T_g^o is the target temperature of the Nosé–Hoover thermostat. It is noticed that if *group g* does not have a thermostat, then it is a special case with $\chi_g = \dot{\chi}_g = 0$. From now on, if there is no ambiguity, we use the general case for description and derivation. Now one can readily prove the following theorem.

Theorem 3: The total force and total moment caused by the *Nosé–Hoover temperature forces* are vanishing.

Proof:

$$\sum_{i \in S_g} (-\chi_g m^i \tilde{\mathbf{v}}^i) = -\chi_g \sum_{i \in S_g} m^i \tilde{\mathbf{v}}^i = 0 \tag{8.81}$$

$$\sum_{i\in S_g} \mathbf{r}^i \times (-\chi_g m^i \tilde{\mathbf{v}}^i) = -\chi_g \sum_{i\in S_g} m^i \mathbf{r}^i \times \tilde{\mathbf{v}}^i = 0 \tag{8.82}$$

Together with Theorem 2, it means that the Nosé–Hoover thermal velocity does not introduce extra linear and angular momenta; the Nosé–Hoover temperature force does not introduce extra force and moment to the temperature-controlled group either. Actually, these conditions must be imposed whenever a thermostat is applied. Also, it is noticed that, according to Equation 8.78, (1) the temperature of *group g* can be calculated irrespective of whether that group have a thermostat or not and (2) the temperature depends on the Nosé–Hoover thermal velocity, but not on linear or angular momentum.

Although velocity $\mathbf{v}$ is not objective, Yang et al. (2016a, 2016b) verified numerically and Liu and Lee (2016) proved analytically that the Nosé–Hoover thermal velocity $\tilde{\mathbf{v}}$ is objective, i.e.,

$$\tilde{\mathbf{v}}^* = \mathbf{Q}\tilde{\mathbf{v}} \tag{8.83}$$

Then it is straightforward to conclude that temperature expressed in Equation 8.78 is objective, but the one expressed in Equation 8.66 is not. From now on, we name the temperature calculated by Equation 8.78 as *atomistic temperature*.

Also, we express the virial stress tensor $\mathbf{S}$, which is the counter part of Cauchy stress tensor in CM (Subramaniyan and Sun, 2008), as

$$\mathbf{S} = -\frac{1}{V_\Omega}\sum_{i\in\Omega}\left(m^i \tilde{\mathbf{v}}^i \otimes \tilde{\mathbf{v}}^i + \sum_j \mathbf{r}^i \otimes \mathbf{f}^i + \mathbf{r}^i \otimes (-m^i \chi \tilde{\mathbf{v}}^i)\right) \tag{8.84}$$

where Ω is the region of which one seeks to find the virial stress; V_Ω is the volume of Ω; $\mathbf{r}^i$ is the position vector of *atom i*; $\mathbf{f}^i$ is all the interatomic force acting on *atom i*; the summation on j means all the interactions between *atom i* and *atom j* should be counted, where it does not matter whether *atom j* is in Ω or not. Now let us examine whether the virial stress expressed in Equation 8.84 is objective. Since the kinetic part is obviously objective now, let us first take a look at the potential part

$$\begin{aligned}
\tilde{S}^*_{\alpha\beta} &= \sum_{i\in\Omega}\sum_j \tilde{\mathbf{r}}^{i*}_\alpha \tilde{\mathbf{f}}^{i*}_\beta = \sum_{i\in\Omega}\sum_j \left(Q_{\alpha\gamma}\mathbf{r}^i_\gamma + b_\alpha\right)Q_{\beta\delta}\mathbf{f}^i_\delta \\
&= \sum_{i\in\Omega}\sum_j Q_{\alpha\gamma}\mathbf{r}^i_\gamma Q_{\beta\delta}\mathbf{f}^i_\delta + b_\alpha Q_{\beta\delta}\sum_{i\in\Omega}\sum_j \mathbf{f}^i_\delta \\
&= Q_{\alpha\gamma}Q_{\beta\delta}\tilde{S}_{\gamma\delta} + b_\alpha Q_{\beta\delta}\sum_{i\in\Omega}\sum_j \mathbf{f}^i_\delta \\
&= Q_{ik}Q_{jl}\tilde{S}_{kl}
\end{aligned} \tag{8.85}$$

which means the potential part of the virial stress tensor expressed in Equation 8.84 is objective. It is noticed that the proof indicated in Equation 8.85 depends on two key points: (1) interatomic force fi is objective (which has been proved in Theorem 1) and (2) the total interatomic forces in a close system is zero, i.e.,

$$\sum_{i\in\Omega}\sum_{j}\mathbf{f}^{i}=0 \tag{8.86}$$

What if not all *atom j* are in Ω? For example, there is a *n*-body potential and there are *n* atoms among which one has

$$\sum_{i=1}^{n}\mathbf{f}^{i}=0 \;\Rightarrow\; \sum_{i=1}^{n}\mathbf{r}^{i}\otimes\mathbf{f}^{i} \text{ is objective} \tag{8.87}$$

If *atom i* is one of the *n* atoms, then consider $\frac{1}{n}\sum_{i=1}^{n}\mathbf{r}^{i}\otimes\mathbf{f}^{i}$ is the contribution of *atom i* to the virial stress of Ω. Now let us look at

$$\begin{aligned}\tilde{S}^{*}_{\alpha\beta} &= \sum_{i\in\Omega} r^{i*}_{\alpha}\otimes\left(-m^{i}\chi^{*}\tilde{v}^{i*}_{\beta}\right) = -\sum_{i\in\Omega} m^{i}\chi r^{i*}_{\alpha}\otimes\tilde{v}^{i*}_{\beta})\\ &= -\sum_{i\in\Omega} m^{i}\chi(Q_{\alpha\gamma}r^{i}_{\gamma}+b_{\alpha})Q_{\beta\xi}\tilde{v}^{i}_{\xi}\\ &= -\sum_{i\in\Omega} m^{i}\chi Q_{\alpha\gamma}r^{i}_{\gamma}Q_{\beta\xi}\tilde{v}^{i}_{\xi} - b_{\alpha}\chi Q_{\beta\xi}\sum_{i\in\Omega} m^{i}r^{i}_{\xi}\\ &= Q_{\alpha\gamma}Q_{\beta\xi}\tilde{S}_{\gamma\xi}\end{aligned} \tag{8.88}$$

which means the part of the virial stress tensor due to the Nosé–Hoover thermal velocity $\tilde{\mathbf{v}}$ is objective. It is noticed that the proof indicated in Equation 8.88 depends on three key points: (1) Nosé–Hoover thermal velocity $\tilde{\mathbf{v}}$ itself is objective, (2) the position vector r in Equation 8.84 is the relative position with respect to the group of atoms in Ω, and (3) χ is objective because atomistic temperature (cf. Equation 8.78) and $\dot{\chi}$ (cf. Equation 8.80) are objective.

8.6.1 Hamiltonian of the material system

The *Hamiltonian* of the entire system can be expressed as

$$\begin{aligned}H = &\sum_{g=1}^{N_g}\sum_{i\in g}\left\{\frac{1}{2}m^{i}\mathbf{v}^{i}\cdot\mathbf{v}^{i} - \int_{0}^{t}\boldsymbol{\varphi}^{i}(s)\cdot\mathbf{v}^{i}(s)\,ds\right\} + \sum_{g=1}^{N_g}\left\{\frac{1}{2}Q_g\chi_g^2+\frac{Q_g}{\tau_g^2}\int_{0}^{t}\chi_g(s)\,ds\right\}\\ &+U(\mathbf{r}^{1},\mathbf{r}^{2},\mathbf{r}^{3},\ldots,\mathbf{r}^{N})\end{aligned} \tag{8.89}$$

where *U* is the total interatomic potential energy of the entire system and

$$Q_g \equiv N_g^{dof}k_BT_g^{o}\tau_g^2 \tag{8.90a}$$

It is seen that the Hamiltonian, *H*, consists of four parts: (1) the kinetic energy, which is the sum of kinetic energies of all atoms, i.e., $\frac{1}{2}\,m^{i}\mathbf{v}^{i}\cdot\mathbf{v}^{i}(i=1,2,3,\ldots,N)$, (2) the potential energy $U(\mathbf{r}^1, \mathbf{r}^2, \mathbf{r}^3,\ldots, \mathbf{r}^N)$, which in principle cannot be divided into a summation of subsets, (3) the work done by force $\boldsymbol{\varphi}^i$ [i = 1,2,3,…,N], and (4) the sum of thermal energy E_g [g = 1,2,3,…,N_g] of all groups, where

$$E_g \equiv \left\{ \frac{1}{2} Q_g \chi_g^2 + \frac{Q_g}{\tau_g^2} \int_0^t \chi_g(s)\, ds \right\} \tag{8.90b}$$

Theorem 4: Hamiltonian is a constant.
Proof:

If there is no ambiguity, we use the abbreviation $\sum_{i=1}^{N} \equiv \sum_{g=1}^{N_g} \sum_{i \in g}$. Now we differentiate Equation 8.89 term by term with respect to time as follows

$$\frac{d}{dt} \sum_{i=1}^{N} \frac{1}{2} m^i \mathbf{v}^i \cdot \mathbf{v}^i = \sum_{i=1}^{N} m^i \dot{\mathbf{v}}^i \cdot \mathbf{v}^i \tag{8.91}$$

$$-\frac{d}{dt} \sum_{i=1}^{N} \int_0^t \boldsymbol{\varphi}^i(s) \cdot \mathbf{v}^i(s)\, ds = -\sum_{i=1}^{N} \boldsymbol{\varphi}^i \cdot \mathbf{v}^i \tag{8.92}$$

$$\begin{aligned} \frac{d}{dt} \sum_{g=1}^{N_g} \left\{ \frac{1}{2} Q_g \chi_g^2 + \frac{Q_g}{\tau_g^2} \int_0^t \chi_g(s)\, ds \right\} &= \sum_{g=1}^{N_g} \left\{ Q_g \chi_g \dot{\chi}_g + \frac{Q_g \chi_g}{\tau_g^2} \right\} \\ &= \sum_{g=1}^{N_g} N_g^{dof} k_B \chi_g T_g = \sum_{g=1}^{N_g} \sum_{i \in g} \chi_g m^i \tilde{\mathbf{v}}^i \cdot \tilde{\mathbf{v}}^i \\ &= \sum_{g=1}^{N_g} \sum_{i \in g} \chi_g m^i \tilde{\mathbf{v}}^i \cdot \mathbf{v}^i = \sum_{i=1}^{N_g} \chi_g m^i \tilde{\mathbf{v}}^i \cdot \mathbf{v}^i \end{aligned} \tag{8.93}$$

$$\frac{dU}{dt} = \sum_{i=1}^{N} \frac{\partial U}{\partial \mathbf{r}^i} \cdot \mathbf{v}^i = -\sum_{i=1}^{N} \mathbf{f}^i \cdot \mathbf{v}^i \tag{8.94}$$

Now we have

$$\dot{H} = \sum_{i=1}^{N} \{ m^i \dot{\mathbf{v}}^i - \mathbf{f}^i - \boldsymbol{\varphi}^i + \chi_g m^i \tilde{\mathbf{v}}^i \} \cdot \mathbf{v}^i \tag{8.95}$$

We recall that $m^i \dot{\mathbf{v}}^i - \mathbf{f}^i - \boldsymbol{\varphi}^i + \chi_g m^i \tilde{\mathbf{v}}^i = 0 \quad [i = 1,2,3,\ldots,N]$ is the governing equation for every atom in the system. Thus, we have proved *Theorem 4*. We also notice that

$$\dot{E}_g = N_g^{dof} k_B \chi_g T_g \tag{8.96}$$

Actually $\dot{E}_g$ is the flow of energy per unit time out of *group* g due to the action of Nosé–Hoover thermostat.

8.7 MICROSCOPIC MAXWELL'S EQUATIONS AND LORENTZ FORCE

The well-known macroscopic Maxwell's equations and Lorentz force equation in the Gaussian system are expressed as (Jackson, 1962):

$$\nabla \cdot \mathbf{D} = 4\pi\rho \tag{8.97}$$

$$\nabla \times \mathbf{H} = \frac{1}{c}\left\{4\pi \mathbf{J} + \frac{\partial \mathbf{D}}{\partial t}\right\} \tag{8.98}$$

$$\nabla \cdot \mathbf{B} = 0 \tag{8.99}$$

$$\nabla \times \mathbf{E} + \frac{1}{c}\frac{\partial \mathbf{B}}{\partial t} = 0 \tag{8.100}$$

$$\mathbf{F} = q\left\{\mathbf{E} + \frac{1}{c}\mathbf{v} \times \mathbf{B}\right\} \tag{8.101}$$

8.7.1 Microscopic Maxwell's equations

At atomic level, the electric and magnetic fields $\mathbf{e}(\mathbf{r}, t)$and $\mathbf{b}(\mathbf{r}, t)$ at the point with coordinates r and time t, generated by a collection of point particles located at $\mathbf{r}^i$ with charges q^i and velocities $\dot{\mathbf{r}}^i$ $\{i = 1,2,3,\ldots\}$, satisfy the following microscopic Maxwell field equations (in the Gaussian system) (de Groot and Suttorp, 1972):

$$\nabla \cdot \mathbf{e} = 4\pi \sum_i q^i \delta\left(\left|\mathbf{r} - \mathbf{r}^i\right|\right) \tag{8.102}$$

$$-\partial_o \mathbf{e} + \nabla \times \mathbf{b} = c^{-1} 4\pi \sum_i q^i \dot{\mathbf{r}}^i \delta\left(\left|\mathbf{r} - r^i\right|\right) \tag{8.103}$$

$$\nabla \cdot \mathbf{b} = 0 \tag{8.104}$$

$$\partial_o \mathbf{b} + \nabla \times \mathbf{e} = 0 \tag{8.105}$$

where $\nabla \equiv \frac{\partial}{\partial \mathbf{r}}$; $\partial_o \equiv \frac{1}{c}\frac{\partial}{\partial t}$; δ is the delta function.

It is noticed that, at atomic level, each particle is a "mathematical point"; i.e., there is no inner structure. Therefore, there are no differences between **E** and **D** and between **B** and **H**; charge density ρ is expressed as summation of products between charge and delta function; and current **J** is calculated as summation of products of charge, velocity, and delta function. In other words, the macroscopic Maxwell's equations and the microscopic Maxwell's equations are consistent.

8.7.2 Scalar and vector potentials

For convenience, a scalar potential ϕ and a vector potential **a** are introduced such that

$$\mathbf{b} = \nabla \times \mathbf{a} \tag{8.106}$$

and

$$\mathbf{e} = -\nabla\phi - \partial_o \mathbf{a} \tag{8.107}$$

We now substitute Equations 8.106 and 8.107 into Equations 8.102–8.105 and obtain, respectively,

$$\nabla \cdot (-\nabla\phi - \partial_o \mathbf{a}) = -\nabla^2\phi - \partial_o \nabla \cdot \mathbf{a} = 4\pi \sum_i q^i \delta\left(\left|\mathbf{r} - \mathbf{r}^i\right|\right) \tag{8.108}$$

$$\begin{aligned} &-\partial_o\{-\nabla\phi - \partial_o \mathbf{a}\} + \nabla \times (\nabla \times \mathbf{a}) \\ &= -\partial_o\{-\nabla\phi - \partial_o \mathbf{a}\} + \nabla(\nabla \cdot \mathbf{a}) - \nabla^2 \mathbf{a} \\ &= -\nabla^2 \mathbf{a} + \nabla\{\partial_o \phi + \nabla \cdot \mathbf{a}\} + \partial_o^2 \mathbf{a} \\ &= c^{-1} 4\pi \sum_i q^i \dot{\mathbf{r}}^i \delta\left(\left|\mathbf{r} - \mathbf{r}^i\right|\right) \end{aligned} \tag{8.109}$$

$$\nabla \cdot \mathbf{b} = \nabla \cdot (\nabla \times \mathbf{a}) = 0 \tag{8.110}$$

and

$$\partial_o \mathbf{b} + \nabla \times \mathbf{e} = \partial_o\{\nabla \times \mathbf{a}\} + \nabla \times \{-\partial_o \mathbf{a} - \nabla\phi\} = 0 \tag{8.111}$$

It is noticed that Equations 8.104 and 8.105 are identically satisfied. The scalar and vector potentials are not fixed uniquely by definitions, Equations 8.106 and 8.107; i.e., the same EM fields can be described by another set of potentials $\{\phi', \mathbf{a}'\}$ that are related to $\{\phi, \mathbf{a}'\}$ through a gauge transformation

$$\phi' = \phi - \partial_o \psi \tag{8.112}$$

and

$$\mathbf{a}' = \mathbf{a} + \nabla\psi \tag{8.113}$$

with a gauge function ψ. To prove this property, one may substitute Equations 8.112 and 8.113 into Equations 8.108 and 8.109 and see the following:

$$\begin{aligned}&\nabla^2\phi' + \partial_o \nabla\cdot \mathbf{a}'\\&= \nabla^2\phi + \partial_o\nabla\cdot\mathbf{a} + \nabla^2(-\partial_o\psi) + \partial_o(\nabla\cdot\nabla\psi)\\&= \nabla^2\phi + \partial_o\nabla\cdot\mathbf{a}\end{aligned} \tag{8.114}$$

and

$$\begin{aligned}&-\nabla^2\mathbf{a}' + \nabla\{\partial_o\phi' + \nabla\cdot\mathbf{a}'\} + \partial_o^2\mathbf{a}'\\&= -\nabla^2\mathbf{a} + \nabla\{\partial_o\phi + \nabla\cdot\mathbf{a}\} + \partial_o^2\mathbf{a} - \nabla^2(\nabla\psi) - \nabla\left(\partial_o^2\psi\right) + \nabla(\nabla\cdot\nabla\psi) + \partial_o^2(\nabla\psi)\\&= -\nabla^2\mathbf{a} + \nabla\{\partial_o\phi + \nabla\cdot\mathbf{a}\} + \partial_o^2\mathbf{a}\end{aligned} \tag{8.115}$$

This property is utilized to choose the potentials in such a way that they satisfy the Lorentz condition

$$\partial_o\phi + \nabla\cdot\mathbf{a} = 0 \tag{8.116}$$

It is seen that

$$\partial_o\phi' + \nabla\cdot\mathbf{a}' = \partial_o\phi + \nabla\cdot\mathbf{a} - \partial_o^2\psi + \nabla\cdot(\nabla\psi) \tag{8.117}$$

This means if {ϕ, **a**} satisfy the Lorentz condition, Equation 8.116, and a gauge function ψ can be found to satisfy a wave equation

$$\partial_o^2\psi = \nabla^2\psi \tag{8.118}$$

then any {ϕ′, **a**′} related to {ϕ, **a**} through Equations 8.112 and 8.113 satisfy the Lorentz condition. Now, Equations 8.108 and 8.109 can be rewritten as two uncoupled wave equations with forcing terms

$$\nabla^2\phi - \partial_o^2\phi = -4\pi\sum_i q^i\delta\left(\left|\mathbf{r} - \mathbf{r}^i\right|\right) \tag{8.119}$$

and

$$\nabla^2\mathbf{a} - \partial_o^2\mathbf{a} = -c^{-1}4\pi\sum_i q^i\dot{\mathbf{r}}^i\delta\left(\left|\mathbf{r} - \mathbf{r}^i\right|\right) \tag{8.120}$$

8.7.3 Nonrelativistic EM fields

From now on, we focus our attention to nonrelativistic electromagnetics; i.e., we are interested in solutions of equations involving terms up to c^{-1}. Now Equations 8.119 and 8.120 become two uncoupled Poisson equations:

$$\nabla^2\phi = -4\pi\sum_i q^i\delta\left(\left|\mathbf{r} - \mathbf{r}^i\right|\right) \tag{8.121}$$

and

$$\nabla^2 \mathbf{a} = -c^{-1} 4\pi \sum_i q^i \dot{\mathbf{r}}^i \delta(|\mathbf{r} - \mathbf{r}^i|) \tag{8.122}$$

We recall one of the properties of delta function as

$$\nabla^2 \left(\frac{1}{|\mathbf{r} - \mathbf{r}^i|} \right) = -4\pi\delta(|\mathbf{r} - \mathbf{r}^i|) \tag{8.123}$$

This leads to

$$\phi = \sum_i \frac{q^i}{|\mathbf{r} - \mathbf{r}^i|} \tag{8.124}$$

$$\mathbf{a} = c^{-1} \sum_i \frac{q^i \dot{\mathbf{r}}^i}{|\mathbf{r} - \mathbf{r}^i|} \tag{8.125}$$

$$\mathbf{e}^i = -\nabla \frac{q^i}{|\mathbf{r} - \mathbf{r}^i|}, \qquad \mathbf{e} = \sum_i \mathbf{e}^i \tag{8.126}$$

$$\mathbf{b}^i = c^{-1} \nabla \times \left\{ \frac{q^i \dot{\mathbf{r}}^i}{|\mathbf{r} - \mathbf{r}^i|} \right\}, \qquad \mathbf{b} = \sum_i \mathbf{b}^i \tag{8.127}$$

It is noticed that the nonrelativistic electric field is of order c^0, while the nonrelativistic magnetic field is of order c^{-1}.

8.7.4 Equation of motion of a point particle

Consider a material system, consisting of N point particles each with mass m^i, charge q^i, position $\mathbf{r}^i = \mathbf{u}^i + \mathbf{R}^i$, velocity $\dot{\mathbf{r}}^i = \dot{\mathbf{u}}^i = \mathbf{v}^i$, and acceleration $\ddot{\mathbf{r}}^i = \ddot{\mathbf{u}}^i = \dot{\mathbf{v}}^i$, subjected to an external EM fields $\{\mathbf{E}^e, \mathbf{B}^e\}$. The equation of motion, following Newton's law, is obtained as

$$m^i \ddot{\mathbf{r}}^i = q^i \{\mathbf{e}^t(\mathbf{r}^i, t) + c^{-1} \mathbf{v}^i \times \mathbf{b}^t(\mathbf{r}^i, t)\} \tag{8.128}$$

Notice that the Lorentz force appears at the right-hand side of Equation 8.128, where the total EM fields $\{\mathbf{e}^t, \mathbf{b}^t\}$ are the summation of the external EM fields and those generated by all the other point particles, i.e.,

$$\begin{aligned} \mathbf{e}^t(\mathbf{r}^i, t) &= \mathbf{E}^e(\mathbf{r}^i, t) + \sum_{j=1, j\neq i}^{N} \mathbf{e}^j(\mathbf{r}^i, t) \\ &= \mathbf{E}^e(\mathbf{r}^i, t) - \sum_{j=1, j\neq i}^{N} q^j \nabla_i \left\{ \frac{1}{|\mathbf{r}^i - \mathbf{r}^j|} \right\} \end{aligned} \tag{8.129}$$

and

$$\mathbf{b}^{t}(\mathbf{r}^{i},t)=\mathbf{B}^{e}(\mathbf{r}^{i},t)+c^{-1}\sum_{j=1,j\neq i}^{N}q^{j}\nabla_{i}\times\frac{\dot{\mathbf{r}}^{j}}{\left|\mathbf{r}^{i}-\mathbf{r}^{j}\right|}$$
$$\approx\mathbf{B}^{e}(\mathbf{r}^{i},t) \tag{8.130}$$

Now Equation 8.128 can be rewritten as

$$m^{i}\ddot{\mathbf{r}}^{i}=q^{i}\{\mathbf{E}^{e}(\mathbf{r}^{i},t)+c^{-1}\mathbf{v}^{i}\times\mathbf{B}^{e}(\mathbf{r}^{i},t)\}+\sum_{j=1,j\neq i}^{N}q^{i}q^{j}\frac{\mathbf{r}^{i}-\mathbf{r}^{j}}{\left|\mathbf{r}^{i}-\mathbf{r}^{j}\right|^{3}} \tag{8.131}$$

It is recognized that Equation 8.131 is the foundation of our theory on multiphysics of which the governing equation is expressed as

$$m^{i}\ddot{\mathbf{r}}^{i}=q^{i}\{\mathbf{E}^{e}(\mathbf{r}^{i},t)+c^{-1}\mathbf{v}^{i}\times\mathbf{B}^{e}(\mathbf{r}^{i},t)\}+\mathbf{f}^{i}-\chi^{i}m^{i}\tilde{\mathbf{v}}^{i} \tag{8.132}$$

If we take the Coulomb–Buckingham potential as an example, Equation 8.132 can be further derived to be

$$m^{i}\ddot{\mathbf{r}}^{i}=q^{i}\left\{\mathbf{E}^{e}(\mathbf{r}^{i},t)+c^{-1}\mathbf{v}^{i}\times\mathbf{B}^{e}(\mathbf{r}^{i},t)\right\}+\sum_{j=1,j\neq i}\frac{q^{i}q^{j}}{\left|\mathbf{r}^{i}-\mathbf{r}^{j}\right|^{3}}(\mathbf{r}^{i}-\mathbf{r}^{j})$$
$$-\frac{\partial}{\partial\mathbf{r}^{i}}\left\{\sum_{j=1,j\neq i}^{N}\left[A^{ij}e^{-B^{ij}r^{ij}}-C^{ij}(r^{ij})^{-6}+D^{ij}(r^{ij})^{-12}\right]\right\} \tag{8.133}$$
$$-\chi^{i}m^{i}\tilde{\mathbf{v}}^{i}$$

where, in the right-hand side of Equation 8.133, the first, second, third, and fourth terms are the Lorentz force due to the external EM fields, Coulomb force between *atom i* and *atom j*, the force due to the Buckingham potential, and the force due to the upgraded Nosé–Hoover Thermostat, respectively. At this moment, one may wonder whether the Lorentz force is objective or not because it depends on the velocity of the charged particle. To answer this question directly, first we notice that principle of objectivity is a concept in Newtonian physics; for electromagnetism and special relativity, the counterpart is Lorentz transformation. The Dutch physicist Hendrik Lorentz tried to explain how the speed of light was observed to be independent of the reference frame and understand the symmetries of the laws of electromagnetism. He derived the Lorentz transformation before special relativity was derived. We put a brief introduction of Lorentz transformation in the Appendix.

8.8 HEUN'S METHOD

Now we may rewrite the governing equations for the thermomechanical– EM coupling problem in MD simulation as (cf. Equations 8.78–8.80)

$$m^i\dot{\mathbf{v}}^i = \mathbf{f}^i - \chi_g m^i\tilde{\mathbf{v}}^i + q^i(\mathbf{E}^e + c^{-1}\mathbf{v}^i \times \mathbf{B}^e) + \hat{\mathbf{f}}^i(t), \qquad i \in group\ g \tag{8.134}$$

$$\dot{\chi}_g = \frac{1}{\tau_g^2 T_g^o}\left(\frac{1}{N_g^{dof} k_B}\sum_{i=1}^{n_g} m^i\tilde{\mathbf{v}}^i \cdot \tilde{\mathbf{v}}^i - T_g^o\right) \tag{8.135}$$

with the understanding that the whole specimen is divided into N_G groups; in *group g*, there are n_g atoms; N_g^{dof} is the number of degrees of freedom of *group g*; T_g^o is the target temperature of the Nosé–Hoover thermostat; τ_g is a specified time constant associated with *group g*; k_B is the Boltzmann constant; $\mathbf{f}^i$ is the interatomic force acting on *atom i*; $\hat{\mathbf{f}}^i(t)$ is the specified applied force on *atom i* at time t; q^i is the electric charge of *atom i*; and $\mathbf{E}^e$ and $\mathbf{B}^e$ are specified external electric and magnetic fields—functions of positions and time, respectively. The displacement-specified boundary conditions are given as

$$u_\alpha^i(t) = \hat{u}_\alpha^i(t) \tag{8.136}$$

which means the αth component of the displacement of *atom i* is specified as a function of time, $\hat{u}_\alpha^i(t)$. Symbolically, we express Equations 8.134–8.136 as

$$\mathbf{U} = \mathbf{A}(\mathbf{U}, \mathbf{V}, \boldsymbol{\chi}, t) \tag*{(8.134)*}$$

$$\dot{\boldsymbol{\chi}} = \mathbf{B}(\mathbf{U}, \mathbf{V}, t) \tag*{(8.135)*}$$

and

$$\mathbf{U}_I = \hat{\mathbf{U}}_I(t), \qquad I \in I_{BC} \tag*{(8.136)*}$$

One may differentiate Equation 8.136* and obtain

$$\mathbf{V}_I = \frac{d\hat{\mathbf{U}}_I(t)}{dt}, \qquad I \in I_{BC}. \tag{8.137}$$

To solve the governing equations, Equations 8.134 and 8.135*, numerically, one may employ Heun's method, which may be referred to as the improved Euler's method or two-stage Runge–Kutta method (Suli and Mayers, 2003). The numerical procedures of Heun's method can be outlined as follows:

Step 1:
Suppose at $t = t^n$ we know

$$\mathbf{U}^n = \mathbf{U}(t^n), \mathbf{V}^n = \mathbf{V}(t^n), \chi^n = \chi(t^n) \tag{8.138}$$

and $\mathbf{U}^n$ and $\mathbf{V}^n$ satisfy the boundary conditions, i.e.,

$$\mathbf{U}_I = \hat{\mathbf{U}}_I(t^n), \mathbf{V}_I = \hat{\mathbf{V}}_I(t^n) \qquad I \in I_{BC} \tag{8.139}$$

Calculate

$$\ddot{\mathbf{U}}^n = \mathbf{A}(\mathbf{U}^n, \mathbf{V}^n, \chi^n, t^n) \equiv \mathbf{A}^n$$
$$\dot{\chi}^n = \mathbf{B}(\mathbf{U}^n, \mathbf{V}^n, t^n) \equiv \mathbf{B}^n \tag{8.140}$$

Step 2:

$$\mathbf{U}^{*n+1} = \mathbf{U}^n + \Delta t \mathbf{V}^n$$
$$\mathbf{V}^{*n+1} = \mathbf{V}^n + \Delta t \mathbf{A}^n \tag{8.141}$$
$$\chi^{*n+1} = \chi^n + \Delta t \mathbf{B}^n$$

Impose the boundary conditions, i.e.,

$$\mathbf{U}_I^{*n+1} = \hat{\mathbf{U}}_I(t^{n+1}),\ \mathbf{V}_I^{*n+1} = \hat{\mathbf{V}}_I(t^{n+1}) \qquad I \in I_{BC} \tag{8.142}$$

where $t^{n+1} = t^n + \Delta t$; Δt is the time step.

Step 3:

Calculate

$$\ddot{\mathbf{U}}^{*n+1} = \mathbf{A}\left(\mathbf{U}^{*n+1}, \mathbf{V}^{*n+1}, \chi^{*n+1}, t^{n+1}\right) \equiv \mathbf{A}^{*n+1}$$
$$\dot{\chi}^{*n+1} = \mathbf{B}(\mathbf{U}^{*n+1}, \mathbf{V}^{*n+1}, t^{n+1}) \equiv \mathbf{B}^{*n+1} \tag{8.143}$$

Step 4:

Calculate

$$\bar{\mathbf{A}} = (\mathbf{A}^n + \mathbf{A}^{*n+1})/2$$
$$\bar{\mathbf{B}} = (\mathbf{B}^n + \mathbf{B}^{*n+1})/2 \tag{8.144}$$

Step 5:

$$\mathbf{U}^{n+1} = \mathbf{U}^n + \Delta t \bar{\mathbf{V}}$$
$$\mathbf{V}^{n+1} = \mathbf{V}^n + \Delta t \bar{\mathbf{A}} \tag{8.145}$$
$$\boldsymbol{\chi}^{n+1} = \boldsymbol{\chi}^n + \Delta t \bar{\mathbf{B}}$$

Impose the boundary conditions, i.e.,

$$\mathbf{U}_I^{n+1} = \hat{\mathbf{U}}_I(t^{n+1}), \qquad \mathbf{V}_I^{n+1} = \hat{\mathbf{V}}_I(t^{n+1}) \qquad I \in I_{BC} \tag{8.146}$$

Step 6:

Calculate

$$\ddot{\mathbf{U}}^{n+1} = \mathbf{A}(\mathbf{U}^{n+1}, \mathbf{V}^{n+1}, \boldsymbol{\chi}^{n+1}, t^{n+1}) \equiv \mathbf{A}^{n+1}$$
$$\dot{\boldsymbol{\chi}}^{n+1} = \mathbf{B}(\mathbf{U}^{n+1}, \mathbf{V}^{n+1}, t^{n+1}) \equiv \mathbf{B}^{n+1} \tag{8.147}$$

Step 7:

Update $n \leftarrow n + 1$ and go to Step 2.

8.9 NUMERICAL RESULTS OF A SAMPLE PROBLEM

EM field–induced motion has been the focus of several research efforts (Aichinger et al., 2010; Lemeshko et al., 2013). For example, polymeric materials can interact with a magnetic field through the diamagnetic anisotropy of the constituent chemical units (Al-Haik et al., 2006a, 2006b). For another example, oxide materials, such as perovskite, can interact with a magnetic field through metal ions and oxygen ions (Wang and Lee, 2010).

Since one of the goals of this study is to simulate the environmental effects on a material system, the external EM field is considered as part of the environment. It is worthwhile to note that, due to the nonrelativistic approximation, i.e., $\|\mathbf{v}_i\| \ll c$, the Lorentz force exists only between charged atom and the external $\mathbf{E}^e$ and $\mathbf{B}^e$ fields; between charged atoms themselves, only Coulomb forces exist, which is usually included in the interatomic force field.

In a simulation, one of the following three situations can be chosen:

(1) There is no external EM field: $\mathbf{E}^e = \mathbf{B}^e = 0$.
(2) External EM field is constant in space and in time: $\mathbf{E}^e = \mathbf{c}^1$ and $\mathbf{B}^e = \mathbf{c}^2$.
(3) External EM field is specified as function of space and time: $\mathbf{E}^e = \mathbf{E}(\mathbf{x}, t)$ and $\mathbf{B}^e = \mathbf{B}(\mathbf{x},t)$.

However, it is emphasized that when the EM field is set up in the simulation, it has to satisfy Maxwell's equations; otherwise, the simulation result is not physical and cannot be validated by experiments.

From the expression of Lorentz force, it is clear that the magnetic field affects atom motion via the cross product $\mathbf{v} \times \mathbf{B}$. At zero temperature, all atoms stand still and have zero velocity. Therefore, magnetic field has no effect on atom motion when the temperature is at absolute zero. At any nonzero temperature, atoms do move and are influenced by the magnetic field. It is interesting to see the atomistic effect due to external magnetic field and temperature.

A magnesium oxide (MgO) specimen consisting of 16,848 atoms is used in the simulations. All simulations run 20,000 time steps with $\Delta t = 40$ atomic time units ($\tau = 2.4188433 \times 10^{-17}$ seconds). The Coulomb–Buckingham potential used in this simulation, with the remedy term neglected, is expressed as (cf. Equation 8.16)

$$U(r_{ij}) = \frac{q_i q_j}{r_{ij}} + A_{ij} \exp\left(-\frac{r_{ij}}{\rho_{ij}}\right) - \frac{C_{ij}}{r_{ij}^6}$$

where, in atomic units, the numerical values of A, ρ, C are given in Table 8.1.

In this simulation, Heun's method is employed with an external EM field and a thermostat, where the external EM field is set at $\mathbf{E} = 0$ and $\mathbf{B} = (1, 1, 1)$, and an upgrade Nosé–Hoover thermostat is used to heat up the specimen from 400 K to 1,200 K. Figure 8.4 shows the target temperature and the actual temperature as obtained from the

Table 8.1 Material parameters for MgO

Type	*A(Hartree)*	*ρ(Bohr)*	*C(Hartree Bohr⁶)*
Mg-Mg	0	0	0
Mg-O	52.4964	0.5565	0
O-O	836.5611	0.2816	466.9241

simulation. It is seen that the thermostat successfully controls the temperature, and the temperature fluctuation can be averaged out via time averaging. More specifically, the fluctuation appears to be smaller compared to that obtained by the Velocity Verlet Method or Inversion Method (Yang et al., 2017). It is noted that, in Heun's method, the Nosé–Hoover temperature force and the Lorentz force are calculated based on the velocities at same time step. Therefore, the results are much more accurate. This is seen in Figure 8.4, in which the red line is the target temperature and the green line is the actual temperature.

Figures 8.5 and 8.6 show the linear momentum and angular momentum of the material body. The result demonstrates that the linear momentum and angular momentum maintain at a stable level and do not drift away. The red, green, and blue lines are for x, y, and z components, respectively.

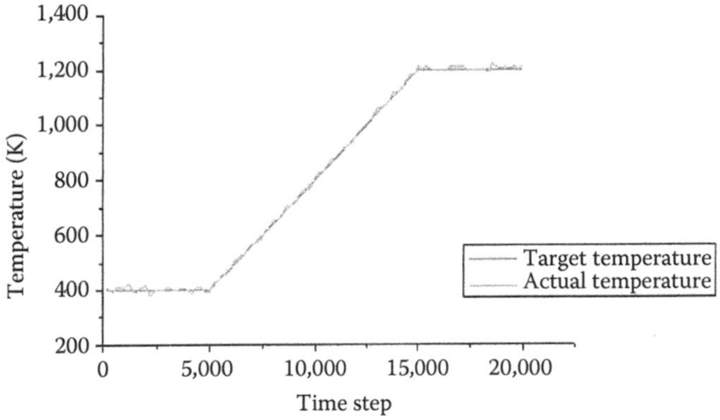

Figure 8.4 **(See color insert.)** Target and actual temperature under a thermostat that raises temperature from T = 400 K to T = 1,200 K.

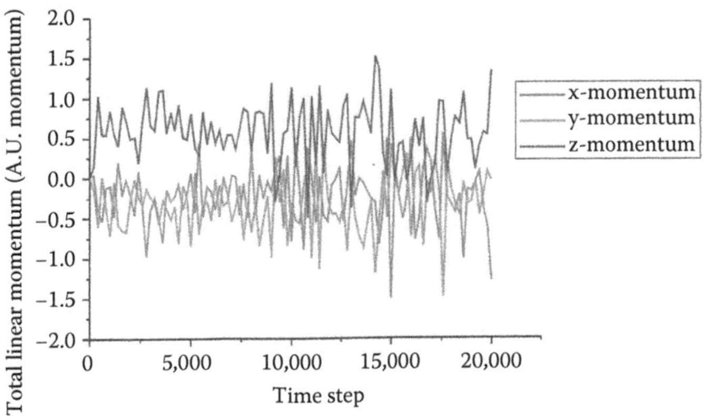

Figure 8.5 **(See color insert.)** History of linear momentum of a material body under the effect of a magnetic field B = (1, 1, 1), with thermostat that raises temperature from T = 400 K to T = 1,200 K.

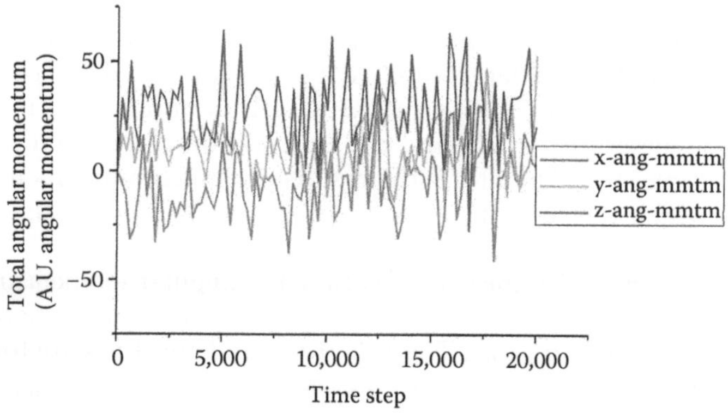

Figure 8.6 **(See color insert.)** History of angular momentum of a material body under the effect of a magnetic field B = (1, 1, 1), with thermostat that raises temperature from T = 400 K to T = 1,200 K.

8.10 CLOSING REMARKS

This chapter systematically analyzes multiphysics in MD simulations. From basic differential equations to numerical methods, from interatomic potentials to advanced thermostat and Lorentz force, from principle of objectivity at the continuum level to principle of objectivity in atomic scale, we try to show you the concepts, methodologies, and usage. This chapter is intended to help one to investigate a multiphysics material system using MD simulations. Moreover, based on the understanding of basic and advanced concepts in this chapter, one may be able to analyze new methodologies and even more complicated material systems.

APPENDIX: LORENTZ TRANSFORMATION

The fundamental problem is to obtain a set of transformation equations connecting the variables of two systems, which will enable one to transform the description of any given kinematical occurrence from the variables of one system to those of the other. In other words, if some given kinematical occurrence has been measured by an observer moving with system S' and described in terms of quantities x', y', z', and t', we desire a set of expressions of these quantities that, on substitution, will give a correct description of the same occurrence in terms of the variables x, y, z, and t, used by an observer moving with system S. The correct expressions for this purpose were first obtained by Lorentz and hence are usually referred to as Lorentz transformation, which may be expressed as (Tolman, 1934)

$$
\begin{aligned}
x' &= \frac{x - Vt}{\beta} \\
y' &= y \\
z' &= z \\
t' &= \frac{t - xV/c^2}{\beta}
\end{aligned}
\tag{8.148}
$$

where V is the relative velocity of the two systems, c is the speed of light, and the Lorentz contraction factor is defined as

$$\beta \equiv \sqrt{1 - V^2/c^2} \tag{8.149}$$

One may readily rewrite Equation 8.148 as

$$\begin{aligned} x &= \frac{x' + Vt'}{\beta} \\ y &= y' \\ z &= z' \\ t &= \frac{t' + x'V/c^2}{\beta} \end{aligned} \tag{8.150}$$

With nonrelativistic assumption, i.e.,

$$\beta \approx 1, \qquad V/c \approx 0 \tag{8.151}$$

Equation 8.148 is reduced to

$$\begin{aligned} x' &= x - Vt \\ y' &= y \\ z' &= z \\ t' &= t \end{aligned} \tag{8.152}$$

This is called Galilean transformation, maybe considered as common sense in classical physics.

Also, one may readily verify that

$$\begin{aligned} dx^2 + dy^2 + dz^2 - c^2dt^2 &= \frac{(dx' + Vdt')^2}{\beta^2} + dy'^2 + dz'^2 - c^2\frac{(dt' + (Vdx'/c^2)^2}{\beta^2} \\ &= dx'^2 + dy'^2 + dz'^2 - c^2dt'^2 \end{aligned} \tag{8.153}$$

which says the quantity $dx^2 + dy^2 + dz^2 - c^2dt^2$ is invariant in Lorentz transformation and also it means the speed of light will measure the same in both systems. Notice that in Galilean transformation, $ds^2 = dx^2 + dy^2 + dz^2$ and dt are two invariant quantities.

Now we write Equation 8.148 as

$$
\begin{aligned}
dx' &= \frac{dx - Vdt}{\beta} \\
dy' &= dy \\
dz' &= dz \\
dt' &= \frac{dt - dxV/c^2}{\beta}
\end{aligned}
\tag{8.154}
$$

Then it is straightforward to obtain

$$
\begin{aligned}
v'_x &= \frac{dx'}{dt'} = \frac{dx - Vdt}{dt - dxV/c^2} = \frac{\frac{dx}{dt} - V}{1 - \frac{dx}{dt}V/c^2} = \frac{v_x - V}{1 - v_x V/c^2} \\
v'_y &= \frac{dy'}{dt'} = \frac{dy}{dt}\frac{dt}{dt'} = \frac{dy}{dt}\frac{\beta}{1 - v_x V/c^2} = v_y \frac{\beta}{1 - v_x V/c^2} \\
v'_z &= \frac{dz'}{dt'} = \frac{dz}{dt}\frac{dt}{dt'} = \frac{dz}{dt}\frac{\beta}{1 - v_x V/c^2} = v_z \frac{\beta}{1 - v_x V/c^2}
\end{aligned}
\tag{8.155}
$$

The transformation of electric and magnetic fields are (Tolman, 1934)

$$
\begin{aligned}
E'_x &= E_x \\
E'_y &= \frac{E_y - B_z V/c}{\beta} \\
E'_z &= \frac{E_z + B_y V/c}{\beta}
\end{aligned}
\tag{8.156}
$$

and

$$
\begin{aligned}
B'_x &= B_x \\
B'_y &= \frac{B_y + E_z V/c}{\beta} \\
B'_z &= \frac{B_z - E_y V/c}{\beta}
\end{aligned}
\tag{8.157}
$$

Remark 1: The electric charge, e, is also invariant in Lorentz transformation. The electric charge is essentially a quantity having discrete magnitude to be determined by counting numbers of electrons and protons and hence necessarily invariant for different observers in different systems.

Remark 2: Lorentz was the first to show that Maxwell's equations are invariant with respect to the change of variables defined by Equation 8.148, but not invariant under the Galilean transformation, Equation 8.152 (Stratton, 1941).

The Lorentz force acting on a electric charge, e, moving with velocity $\mathbf{v}$ can be expressed as

$$\mathbf{F} = e\{\mathbf{E} + \mathbf{v} \times \mathbf{B} / c\} \tag{8.158}$$

or

$$\begin{aligned} F_x &= e\{E_x + [v_y B_z - v_z B_y] / c\} \\ F_y &= e\{E_y + [v_z B_x - v_x B_z] / c\} \\ F_z &= e\{E_z + [v_x B_y - v_y B_x] / c\} \end{aligned} \tag{8.158)*}$$

With the Lorentz transformation of $\mathbf{v}'$, $\mathbf{E}'$, and $\mathbf{B}'$ obtained in Equations 8.155–8.157, the transformation of Lorentz force is obtained as

$$\begin{aligned} F'_x &= e\left\{E_x + \left[v_y \frac{\beta}{1 - v_x V/c^2} \frac{B_z - E_y V/c}{\beta} - v_z \frac{\beta}{1 - v_x V/c^2} \frac{B_y + E_z V/c}{\beta}\right] \Big/ c\right\} \\ &= e\left\{E_x + \left[v_y \frac{B_z - E_y V/c}{1 - v_x V/c^2} - v_z \frac{B_y + E_z V/c}{1 - v_x V/c^2}\right] \Big/ c\right\} \end{aligned} \tag{8.159}$$

$$F'_y = e\left\{\frac{E_y - B_z V/c}{\beta} + \left[v_z \frac{\beta}{1 - v_x V/c^2} B_x - \frac{v_x - V}{1 - v_x V/c^2} \frac{B_z - E_y V/c}{\beta}\right] \Big/ c\right\} \tag{8.160}$$

$$F'_z = e\left\{\frac{E_z + B_y V/c}{\beta} + \left[\frac{v_x - V}{1 - v_x V/c^2} \frac{B_y + E_z V/c}{\beta} - v_y \frac{\beta}{1 - v_x V/c^2} B_x\right] \Big/ c\right\} \tag{8.161}$$

With nonrelativistic assumption, Equation 8.151, it is seen that

$$\begin{aligned} F'_x &= e\left\{E_x + \left[v_y \frac{B_z - E_y V/c}{1 - v_x V/c^2} - v_z \frac{B_y + E_z V/c}{1 - v_x V/c^2}\right] \Big/ c\right\} \\ &= e\left\{E_x + \left[\frac{1}{1 - v_x V/c^2}(v_y B_z - v_z B_y)\right] \Big/ c - E_y \frac{v_y V/c^2}{1 - v_x V/c^2} - E_z \frac{v_z V/c^2}{1 - v_x V/c^2}\right\} \\ &\approx e\{E_x + [v_y B_z - v_z B_y] / c\} \end{aligned} \tag{8.162}$$

$$\begin{aligned}
F'_y &= e\left\{\frac{E_y - B_z V/c}{\beta} + \left[v_z \frac{\beta}{1-v_x V/c^2} B_x - \frac{v_x - V}{1-v_x V/c^2}\frac{B_z - E_y V/c}{\beta}\right]\bigg/ c\right\} \\
&= e\left\{\frac{E_y}{\beta} + \frac{v_z B_x}{c}\frac{\beta}{1-v_x V/c^2} - \frac{v_x B_z}{c}\frac{1}{(1-v_x V/c^2)\beta}\right. \\
&\left. + E_y\left[\frac{v_x V/c^2}{(1-v_x V/c^2)\beta} - \frac{V^2/c^2}{(1-v_x V/c^2)\beta}\right] + B_z\left[\frac{V/c}{(1-v_x V/c^2)\beta} - \frac{V/c}{\beta}\right]\right\} \\
&\approx e\{E_y + [v_z B_x - v_x B_z]/c\}
\end{aligned} \tag{8.163}$$

$$\begin{aligned}
F'_z &= e\left\{\frac{E_z + B_y V/c}{\beta} + \left[\frac{v_x - V}{1-v_x V/c^2}\frac{B_y + E_z V/c}{\beta} - v_y \frac{\beta}{1-v_x V/c^2} B_x\right]\bigg/ c\right\} \\
&= e\left\{\frac{E_z}{\beta} + \left[v_x B_y \frac{1}{(1-v_x V/c^2)\beta} - v_y B_x \frac{\beta}{1-v_x V/c^2}\right]\bigg/ c\right. \\
&\left. + E_z\left[\frac{v_x V/c^2}{(1-v_x V/c^2)\beta} - \frac{V^2/c^2}{(1-v_x V/c^2)\beta}\right] + B_y\left[\frac{V/c}{\beta} - \frac{V/c}{(1-v_x V/c^2)\beta}\right]\right\} \\
&\approx e\{E_z + [v_x B_y - v_y B_x]/c\}
\end{aligned} \tag{8.164}$$

This simply says, under nonrelativistic condition, that the Lorentz force remains unchanged. This is the justification that the governing equation for multiphysics in atomistic scale can be expressed in Equation 8.132.

REFERENCES

Aichinger, M., Janecek, S., and Räsänen, E. 2010. Billiards in magnetic fields: A molecular dynamics approach, *Phys. Rev. E*, **81**(1), 016703.

Alder, B. J., and Wainwright, T. E. 1959. Studies in molecular dynamics. I. General method, *J. Chem. Phys.*, **31**(2), 459–466.

Al-Haik, M., and Hussaini, M. 2006a. Molecular dynamics simulation of reorientation of polyethylene chains under a high magnetic field, *Mol Simul.*, **32**(8), 601–608.

Al-Haik, M., and Hussaini, M. 2006b. Molecular dynamics simulation of magnetic field induced orientation of nanotube-polymer composite, *Japan. J. Appl. Phys.*, **45**(11R), 8984.

Andersen, H. 1980. Molecular dynamics simulations at constant pressure and/or temperature, *J. Chem. Phys.*, **72**(4), 2384–2393.

Berendsen, H., Postma, J., van Gunsteren, W., DiNola, A., and Haak, J. 1984. Molecular dynamics with coupling to an external bath, *J. Chem. Phys.*, **81**(8), 3684–3690.

Brenner, D., Shenderova, O., and Areshkin, D. 1998. Quantum-based analytic interatomic forces and materials simulation, *Rev. Computat. Chem.*, **12**, 207–240.

Bi, Y., Porras, A., and Li, T. 2016a. Free energy landscape and molecular pathways of gas hydrate nucleation, *J. Chem. Phys.*, **145**, 211909.

Bi, Y., Cabriolu, R., and Li, T. 2016b. Heterogeneous ice nucleation controlled by the coupling of surface crystallinity and surface hydrophilicity, *J. Phys. Chem.*, **120**, 1507–1514.

Cabriolu, R., and Li, T. 2015. Ice nucleation on carbon surface supports the classical theory for heterogeneous nucleation, *Phys. Rev. E*, **91**, 052402.
Chen, J., and Lee, J. D. 2011. The Buckingham catastrophe in multiscale modelling of fracture, *Int. J. Theor. Appl. Multiscale Mech.*, **2**(1), 3–11.
English, N., and MacElroy, J. 2003. Molecular dynamics simulations of microwave heating of water, *J. Chem. Phys.*, **118(4)**, 1589–1592.
English, N., and Mooney, D. 2007. Denaturation of hen egg white lysozyme in electromagnetic fields: a molecular dynamics study, *J. Chem. Phys.*, **126**(9), 091105.
English, N., Mooney, D., and O'Brien, S. 2011. Ionic liquids in external electric and electromagnetic fields: a molecular dynamics study, *Mol. Phys.*, **109**(4), 625–638.
English, N., Sorescu, D., and Johnson, J. 2006. Effects of an external electromagnetic field on rutile TiO2: a molecular dynamics study, *J. Phys. Chem. Solids*, **67**(7), 1399–1409.
Eringen, A. C. 1980. *Mechanics of Continua*, R. E. Krieger Pub. Co., Melbourne, Australia.
Eringen, A. C. 1999. *Microcontinuum Field Theories I: Foundation and Solids*, Springer, New York.
Erkoc, Ş. 1997. Empirical many-body potential energy functions used in computer simulations of condensed matter properties, *Phys. Rep.*, **278**(2), 79–105.
Filla, N., Zhang, L., Becton, M., and Wang, X. 2017. Failure mechanism of nanoscale graphene kirigami, *Int. J. Terraspace Sci. Eng.*, 9(1), 63–67.
Garrison, B., and Srivastava, D. 1995. Potential energy surfaces for chemical reactions at solid surfaces, *Ann. Rev. Phys. Chem.*, **46**(1), 373–396.
Geim, A. K., and Novoselov, K. S. 2007. The rise of graphene, *Nat. Mater.*, **6**, 183–191.
de Groot, S. R., and Suttorp, L. G. 1972. *Foundations of Electrodynamics*, North-Holland Pub. Co., Amsterdam.
Hoover, W. G. 1985. Canonical dynamics: Equilibrium phase-space distributions, *Phys. Rev. A*, **31**(3), 1695.
Jackson, J. D. 1962. *Classical Electrodynamics*, Wiley, New York.
Jiang, J., Park, H., and Rabczuk, T. 2013. Molecular dynamics simulations of single-layer molybdenum disulphide (MoS_2): Stillinger–Weber parametrization, mechanical properties, and thermal conductivity, *J. Appl. Phys.*, **114**(6), 064307.
Kittel, C. 2005. *Introduction to Solid State Physics*, Wiley, New York.
Landau, L. D. 1937. Theory of phase changes I, *Phys. Z. Sowjetunion*, **11**, 26–47.
Lemeshko, M., Krems, R. V., Doyle, J. M., and Kais, S. 2013. Manipulation of molecules with electromagnetic fields, *Mol. Phys.*, **111**(12–13), 1648–1682.
Lemons, D., and Gythiel, A. 1997. Paul Langevin's 1908 paper "on the theory of Brownian motion" ["Sur la théorie du mouvement brownien," *CR Acad. Sci. (Paris)* 146, 530–533 (1908)], *Am. J. Phys.*, **65**(11), 1079–1081.
Lennard-Jones, J. E. 1924. On the determination of molecular fields, *Proc. R. Soc. Lond. A*, **106**(738), 463–477.
Li, J., and Lee, J. D. 2014a. Reformulation of the Nosé–Hoover thermostat for heat conduction simulation at nanoscale. *Acta Mech.*, **225**, 1223–1233.
Li, J., and Lee, J. D. 2014b. Stiffness-based course-grained molecular dynamics, *J. Nanomech. Micromech.*, **4**, 1–8.
Liang, T., Phillpot, S. R., and Sinnott, S. B. 2009. Parametrization of a reactive many-body potential for Mo-S systems, *Phys. Rev. B*, **79**, 245110.
Liang, T., Phillpot, S. R., and Sinnott, S. B. 2012. Erratum: Parametrization of a reactive many-body potential for Mo-S systems, *Phys. Rev. B*, **85**, 199903.
Lifshitz, E. M., and Pitaevskii, L. P. 1980. *Statistical Physics Part I*, Pergamon Press, Oxford, UK.
Liu, I. S. 2002. *Continuum Mechanics*. Springer Science & Business Media, Berlin.
Liu, I. S., and Lee, J. D. 2016. On material objectivity of intermolecular force in molecular dynamics, *Acta Mech.*, **228**(2), 731–738.
Martyna, G., Klein, M., and Tuckerman, M. 1992. Nosé–Hoover chains: The canonical ensemble via continuous dynamics, *J. Chem. Phys.*, **97**(4), 2635–2643.
Metaxas, A., and Meredith, R. 1983. *Industrial Microwave Heating*, Peter Peregrinus, Ltd., London.

Nosé, S. 1984a. A unified formulation of the constant temperature molecular dynamics methods, *J. Chem. Phys.*, **81**, 511–519.

Nosé, S. 1984b. A molecular dynamics method for simulations in the canonical ensemble, *Mol. Phys.*, **53**, 255–268.

Novoselov, K. S., Geim, A. K., Morozov, S. V., Jiang, D., Zhang, Y., Dubonos, S. V. et al. 2004. Electric field effect in atomically thin carbon films, *Science*, **306**, 666–669.

Peierls, R. E. 1935. Quelques proprietes typiques des corpses solides, *Annales de l'Institut Henri Poincaré*, **5**, 177–222.

Qin, Z., Jung, G. S., Kang, M. J., and Buehler, M. J. 2017. The mechanics and design of a lightweight three-dimensional graphene assembly, *Sci. Adv.*, 3(1), p.e1601536.

Rahman, A. 1964. Correlations in the motion of atoms in liquid argon, *Phys. Rev.*, **136**(2A), A405–A411.

Rappe, A. K., Casewit, C. J., Colwell, K. S., Goddard, W. A., Skiff, W. M. 1992. UFF, a full period table force field for molecular mechanics and molecular dynamics simulations, *J. Am. Chem. Soc.*, **114**, 10024–10035.

Roussy, G., and Pearce, J. 1995. *Foundations and Industrial Applications of Microwaves and Radio Frequency Fields. Physical and Chemical Processes*, Wiley, Chichester, England.

Sinnott, S., and Brenner, D. 2012. Three decades of many-body potentials in materials research, *Mrs Bull.*, **37**(5), 469–473.

Srolovitz, D., and Vitek, V. E. 2012. *Atomistic Simulation of Materials: Beyond Pair Potentials*, Springer Science & Business Media, Berlin.

Stewart, J., and Spearot, D. 2013. Atomistic simulations of nanoindentation on the basal plane of crystalline molybdenum disulfide (MoS2), *Model. Simul. Mater. Sci. Eng.*, **21**(4), 045003.

Stillinger, F. H., and Weber, T. A. 1985. Computer simulation of local order in condensed phases of silicon, *Phys. Rev. B*, **31**, 5262.

Stratton, J. A. 1941. *Electromagnetic Theory*, McGraw-Hill, New York.

Subramaniyan, A. K., and Sun, C. T. 2008. Continuum interpretation of virial stress in molecular simulations, *Int. J. Solids Struct.*, **45**(14–15), 4340–4346.

Suli, E., and Mayers, D. 2003. *An Introduction to Numerical Analysis*, Cambridge University Press, Cambridge, UK.

Tersoff, J. 1988. New empirical approach for the structure and energy of covalent systems, *Phys. Rev. B*, **37**, 6991–7000.

Tersoff, J. 1989. Modeling solid-state chemistry: Interatomic potentials for multicomponent systems, *Phys. Rev. B*, **39**, 5566–5568.

Tolman, R. C. 1934. *Relativity Thermodynamics and Cosmology*, Clarendon Press, Oxford.

Wang, C. C. 1970a. A new representation theorem for isotropic functions. Part I: An answer to Professor GF Smith's criticism of my papers on representations for isotropic functions, *Arch. Rational Mech. Anal.*, **36**(3), 166–197.

Wang, C. C. 1970b. A new representation theorem for isotropic functions. Part II: An answer to Professor GF Smith's criticism of my papers on representations for isotropic functions, *Arch. Rational Mech. Anal.*, **36**(3), 198–223.

Wang, X., and Lee, J. D. 2010. Atomistic simulation of MgO nanowires subject to electromagnetic wave, *Model. Simul. Mater. Sci. Eng.*, **18**, 085010.

Wood, M., van Duin, A., and Strachan, A. 2014. Coupled thermal and electromagnetic induced decomposition in the molecular explosive αHMX; A reactive molecular dynamics study, *J. Phys. Chem. A*, **118**(5), 885–895.

Yang, L., Huang, K., and Yang, X. 2009. Dielectric properties of *N*, *N*-dimethylformamide aqueous solutions in external electromagnetic fields by molecular dynamics simulation, *J. Phys. Chem. A*, **114**(2), 1185–1190.

Yang, Z., Lee, J. D., and Eskandarian, A. 2016a. Objectivity in molecular dynamics simulations, *Int. J. Terraspace Sci. Eng.*, **8**(1), 79–92.

Yang, Z., Lee, J. D., and Eskandarian, A. 2017. Non-equilibrium molecular dynamics in electromagnetic field, *Int. J. Terraspace Sci. Eng.*, 9(1), 77–85.
Yang, Z., Lee, J. D., Liu, I. S., and Eskandarian, A. 2016b. On non-equilibrium molecular dynamics with Euclidean objectivity, *Acta Mech.*, **228**(2), 693–710.

BIBLIOGRAPHY

Allen, M. P., and Tildesley, D. J. 1987. *Computer Simulation of Liquids*, Clarendon Press, Oxford.
Boresi, A. P., Chong, K. P., and Lee, J. D. 2011. *Elasticity in Engineering Mechanics*, Wiley, New York.
Frank, I., Tanenbaum, D., Van der Zande, A., and McEuen, P. 2007. Mechanical property of suspended graphene sheets, *J. Vacuum Sci. Tech. B*, **25**(6), 2558–2561.
Fried, L., and Howard, W. 2000. Explicit Gibbs free energy equation of state applied to the carbon phase diagram, *Phys. Rev. B*, **61**(13), 8734.
Guo, Z., Ding, J., and Gong, X. 2012. Substrate effects on the thermal conductivity of epitaxial graphene nanoribbons, *Phys. Rev. B*, **85**(23), 235429.
Guo, Z., Zhang, D., and Gong, X. 2009. Thermal conductivity of graphene nanoribbons, *Appl. Phys. Lett.*, **95**(16), 163103.
Haile, J. 1992. *Molecular Dynamics Simulation*, Wiley, New York.
Jiang, J., Lan, J., Wang, J., and Li, B. 2010. Isotopic effects on the thermal conductivity of graphene nanoribbons: localization mechanism, *J. Appl. Phys.*, **107**(5), 054314.
Jiang, J., Wang, J., and Li, B. 2010. Elastic and nonlinear stiffness of graphene: A simple approach, *Phys. Rev. B*, **81**(7), 073405.
Jing, N., Xue, Q., Ling, C., Shan, M., Zhang, T., Zhou, X., and Jiao, Z. 2012. Effect of defects on Young's modulus of graphene sheets: A molecular dynamics simulation, *Rsc Adv.*, **2**(24), 9124–9129.
Lee, C., Wei, X., Kysar, J., and Hone, J. 2008. Measurement of the elastic properties and intrinsic strength of monolayer graphene, *Science,* **321**(5887), 385–388.
Li, J., Wang X., and Lee J. D. 2012. Multiple time scale algorithm for multiscale modeling, *Comput. Model. Eng. Sci.*, **85**, 463–480.
Lindsay, L., Broido, D., and Mingo, N. 2011. Flexural phonons and thermal transport in multilayer graphene and graphite, *Phys. Rev. B*, **83**(23), 235428.
Liu, W. K., Karpov, E. G., and Park, H. S. 2006. *Nano Mechanics and Materials-Theory, Multiscale Methods and Applications*, Wiley, New York.
Neek-Amal, M., and Peeters, F. 2010. Nanoindentation of a circular sheet of bilayer graphene., *Phys. Rev. B*, **81**(23), 235421.
Oden, J. T. et al. 2006. Simulation-based engineering science, NSF Blue Ribbon Panel Report, http://www.nsf.gov/pubs/reports/sbes_final_report.pdf
Pei, Q., Zhang, Y., and Shenoy, V. 2012. Mechanical properties of graphynes under tension: A molecular dynamics study, *Appl. Phys. Lett.*, **101**(8), 081909.
Roldan, R., Castellanos-Gomez, A., Cappelluti, E., and Guinea, F. 2015. Strain engineering in semiconducting two-dimensional crystals, *J. Phys. Condens. Matter*, **27**, 333201.
Shaina, P., George, L., Yadav, V., and Jaiswal, M. 2016. Estimating the thermal expansion coefficient of graphene: The role of graphene–substrate interactions, *J. Phys. Cond. Matter*, **28**(8), 085301.
Shen, Y., and Wu, H. 2012. Interlayer shear effect on multilayer graphene subjected to bending, *Appl. Phys. Lett.*, **100**(10), 101909.
Tohei, T., Kuwabara, A., Oba, F., and Tanaka, I. 2006. Debye temperature and stiffness of carbon and boron nitride polymorphs from first principles calculations, *Phys. Rev. B*, **73**(6), 064304.
Wei, Z., Ni, Z., Bi, K., Chen, M., and Chen, Y. 2011. In-plane lattice thermal conductivities of multilayer graphene films, *Carbon,* **49**(8), 2653–2658.

Wei, N., Xu, L., Wang, H., and Zheng, J. 2011. Strain engineering of thermal conductivity in graphene sheets and nanoribbons: A demonstration of magic flexibility, *Nanotechnology*, **22**(10), 105705.

Yoon, D., Son, Y., and Cheong, H. 2011. Negative thermal expansion coefficient of graphene measured by Raman spectroscopy, *Nano Lett.*, **11**(8), 3227–3231.

Zhang, Y., Pei, Q., and Wang, C. 2012. Mechanical properties of graphynes under tension: A molecular dynamics study, *Appl. Phys. Lett.*, **101**(8), 081909.

Zhao, H., Min, K., and Aluru, N. 2009. Size and chirality dependent elastic properties of graphene nanoribbons under uniaxial tension, *Nano Lett.*, **9**(8), 3012–3015.

Zheng, Y., Wei, N., Fan, Z., Xu, L., and Huang, Z. 2011. Mechanical properties of grafold: A demonstration of strengthened graphene, *Nanotechnology*, **22**(40), 405701.

Chapter 9

Multiscale modeling from atoms to genuine continuum

James D. Lee[1], *Jiaoyan Li*[2], *Zhen Zhang*[1], *and Kerlin P. Robert*[1]

[1]Department of Mechanical and Aerospace Engineering, The George Washington University, Washington, DC

[2]School of Engineering, Brown University, Providence, Rhode Island

9.1 INTRODUCTION

Basically, there exist two fundamental physical models that provide foundations for almost all material behavioral simulations: microscopic discrete atomistic models and macroscopic continuum models. In the range of nanometer or below, the material body is viewed as a collection of discrete particles moving under the influence of their mutual interaction forces. On the other hand, the results of centuries of experimental work at larger length scale are usually formalized into well-structured continuum theories (Chen et al., 2006; Eringen, 1980; Liu, 2002; Truesdell and Noll, 2004; Truesdell and Toupin, 1960) with the purpose to inscribe all objects of our perception into the familiar framework of a four-dimensional space-time. Although the underlying physical concept may be the same, the descriptions by the two distinct models are as different as day and night, whereas the success of both models has been demonstrated and tested throughout the history of science in explaining and predicting various physical phenomena.

Among all microscopic discrete atomistic models, all-atom molecular dynamics (AAMD), as described in Chapter 8, has become routine in numerous fields, including nanoscience, molecular biology, medicinal chemistry, and so on. Unfortunately, the extension of molecular dynamics (MD) into computational science over a realistic range of length and time is limited, due to the large number of particles involved as well as the complex nature of their interactions. The limitations are also imposed by the requirement of smallness of the time step, even though one may be primarily interested in events that occur over a much longer time scale. To overcome this difficulty, basically, there are two approaches: (1) reduce the atomistic model directly with coarser descriptions, referred to as coarse-grained molecular dynamics (CGMD) and (2) link the atomistic model with the continuum model through sequential and concurrent multiscale modeling.

Usually, the basic idea of CGMD is to treat several atoms as a virtual particle and even to make approximations regarding interatomic potentials, leading to the significant reduction of degrees of freedom and complexity of the force field of complex systems. This idea is widely employed in dynamical simulation of biomolecule or long chain polymer at the atomic level. Gu et al. (2012) designed 20 coarse-grained (CG) beads based on the structures of amino acid residues, with which an amino acid can be represented by one or two beads, and adopted a CG solvent model with five water molecules to ensure the consistency with the protein CG beads. Two classes of biological coarse-graining, namely, residue-based (Shih et al., 2006, 2007) and shape-based (Arkhipov et al., 2006, 2008) coarse-graining, have been developed. Despite the enormous success of these approaches, the fundamental questions still remain: is there a rigorous way to obtain the generic force field instead of matching the experimental data for parameterization of CG models? Rudd and Broughton (1998, 2005) proposed their CGMD model, which avoids the previously mentioned problem. They derived the equation of motion directly from finite temperature MD through a statistical coarse-graining procedure and the dimension reduction is done in the partition function by integrating out the excess atomic degrees of freedom. Li (2010) reformulated the CGMD model by incorporating the history-dependent and random noise terms based on the Mori-Zwanzig formalism (Mori, 1965; Zwanzig, 1973). In this chapter, we start from the AAMD model and rigorously derive a different kind CGMD model based on a kinematic constraint and Taylor series expansion (Li and Lee, 2014a). The formulation procedure is straightforward and easy to understand and implement, as shown in Section 9.2.

The past several years have witnessed the explosive growth of interest in multiple length scale theories and simulations. There have been several reviews on multiscale modeling and simulation in the literature, focusing on different aspects. Generally speaking, there are two categories of multiscale modeling methods: sequential and concurrent. The sequential multiscale methodology separates calculation at each scale and passes the results between scales. For instance, almost all conventional MD simulations rely on interatomic potentials obtained from first principle calculations. Material constants in constitutive relations for finite element (FE) analysis could be predicted from MD simulation. The idea of sequential multiscale modeling and simulation is straightforward and has shown applicability for systems that are weakly coupled at different scales. The other one, concurrent multiscale methodology, integrates calculations and solves the multiscale problem simultaneously, and therefore, it is more efficient and more challenge than the sequential one. In a concurrent simulation, the system is often decomposed into several subdomains characterized by different scales and physics. Different theories are applied to different domains to simulate the material behaviors. A successful concurrent multiscale model seeks a smooth coupling between these subdomains, interpreted as the construction of interfacial conditions. In this chapter, a concurrent multiple length/time scale modeling is presented with rigorously derived interfacial conditions. This concurrent approach is more desirable for systems in which system behavior at each scale depends strongly on what happens at the higher or lower scales. The emergence of multiple length scale and time scale approach, along with the development of massively parallel computers, remarkably expands the realm of modeling and simulation from nanoscale to microscale.

9.2 FORCE-BASED CGMD

Consider a crystalline material system, which is built on periodic arrays of atoms or groups of atoms called *unit cells*. The unit cell can embrace just one atom, a group of same kind of atoms, or even a group of different kinds of atoms. The position of a unit cell is considered as the mass center of all atoms in it. The solution region of CGMD in this work is modeled on a FE mesh whose size may vary depending on the expected resolution (cf. the atom-based continuum region in Figure 9.1). A unit cell is shown in Figure 9.1; the unit cell in a circle is also a node in the FE mesh; the one in a square is a unit cell, but not a node; each of the colored squares in the atomic region is just a single atom—not necessarily the same kind of atom, not necessarily arranged in a pattern. The FE mesh with smallest element size is the one where each node coincides with each unit cell; in other words, there is no extra unit cell in the interior of elements.

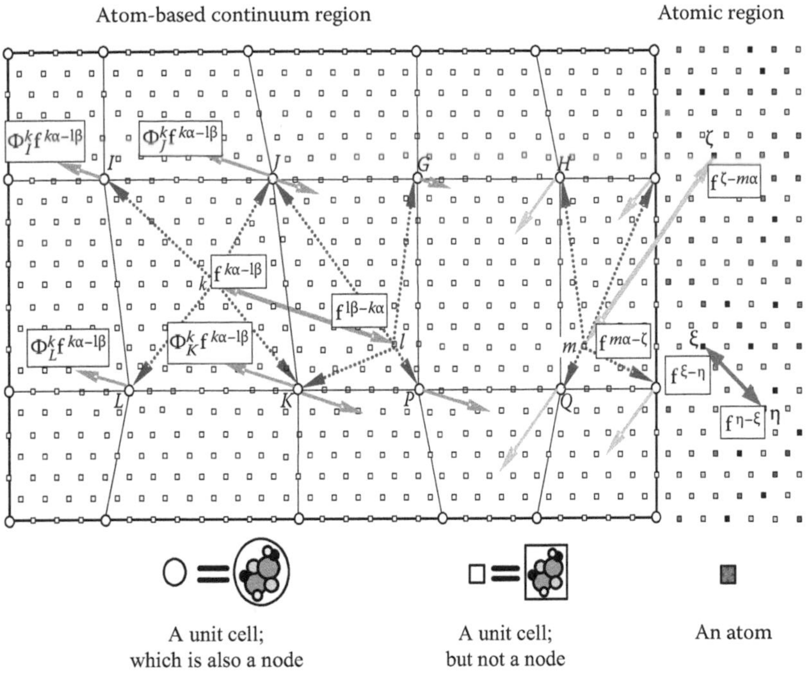

Figure 9.1 **(See color insert.)** Illustrations of atom-based continuum region and atomic region. Circle = finite element node, square = unit cell, but not node, colored square = atom.

Now, one may express the equation of motion of *atom* α in *k*th unit cell (abbreviated as *atom k*α) as

$$m^{\alpha}\ddot{\mathbf{u}}^{k\alpha} = \mathbf{f}^{k\alpha} + \boldsymbol{\varphi}^{k\alpha}, \qquad \alpha \in [1, N_a], \qquad k \in [1, N_c] \tag{9.1}$$

where N_a is the number of atoms within each unit cell; N_c is the total number of unit cells in the material system; $\mathbf{u}^{k\alpha}$ is the displacement of *atom k*α; and $\mathbf{f}^{k\alpha}$ and $\boldsymbol{\varphi}^{k\alpha}$ are interatomic force and body force (may include electromagnetic force, Nosé–Hoover temperature force, applied force) acting on *atom* α, respectively. Borrowing the idea from classical FE method, an assumption is made, called a *kinematic constraint* (Li et al., 2012), with which the motion of atoms in unit cells are connected to that of the corresponding nodes

$$\mathbf{u}^{k\alpha} = \sum_{I=1}^{N_s} N_I(k)\mathbf{U}_I^{\alpha} \tag{9.2}$$

where N_s is the number of nodes of the FE mesh; $N_I(k)$ is the *I*th shape function evaluated at the centroid of the *k*th unit cell; $\mathbf{U}_I^{\alpha}$ is the nodal value of the displacement vector of *atom* α in the *I*th node of the element where the *k*th unit cell resides. In this work, eight-node three-dimensional (3D) solid element is employed; therefore, $N_s = 8$. Equation 9.2 indicates that *atom k*α is just a *follower*—the motion of *atom k*α depends on the motion of the nodes $\mathbf{U}_I^{\alpha}$.

The virtual work done by $\mathbf{f}^{k\alpha}$ is

$$\begin{aligned} \mathbf{f}^{k\alpha}\cdot\delta\mathbf{u}^{k\alpha} &= \mathbf{f}^{k\alpha}\cdot\delta\left[\sum_{I=1}^{8} N_I(k)\mathbf{U}_I^{\alpha}\right] \\ &= \sum_{I=1}^{8}[N_I(k)\mathbf{f}^{k\alpha}]\cdot\delta\mathbf{U}_I^{\alpha} \\ &\equiv \sum_{I=1}^{8}\mathbf{F}_I^{\alpha}\cdot\delta\mathbf{U}_I^{\alpha} \end{aligned} \tag{9.3}$$

This means the nodal force contributed by the atomic force can be calculated as

$$\mathbf{F}_I^{\alpha}(k) = N_I(k)\mathbf{f}^{k\alpha} \tag{9.4}$$

One may say *atom kα* now serves as a messenger that conveys the atomic force to the nodal force. Now, one may look at Figure 9.1 and notice that there is an interaction between *atom ς* in the atomic region and *atom mα* in the ABC region, i.e., $\mathbf{f}^{m\alpha-\varsigma} = -\mathbf{f}^{\varsigma-m\alpha}$. The force acting on *atom ς* due to the interaction with *atom mα* simply emerges in Equation 9.1. However, the force acting on *atom mα* due to the interaction with *atom ς* needs to be distributed to the nodes of the element where *unit cell m* resides according to Equation 9.4.

It is worthwhile to emphasize that the pattern of Equation 9.4 also applies to body force $\varphi^{k\alpha}$, i.e.,

$$\mathbf{\Phi}_I^{\alpha}(k) = N_I(k)\boldsymbol{\varphi}^{k\alpha} \tag{9.5}$$

One may also define

$$M_I^{\alpha}(k) \equiv m^{\alpha}N_I(k) \tag{9.6}$$

This means the nodal mass of the Ith node, M_I^{α}, contributed by the atomic mass, m^{α}, located at kth unit cell can be calculated through Equation 9.6. Then the atomic inertia force, $m^{\alpha}\ddot{\mathbf{u}}^{k\alpha}$, can be further derived as

$$m^{\alpha}\ddot{\mathbf{u}}^{k\alpha} = \sum_{I=1}^{8} m^{\alpha}N_I(k)\ddot{\mathbf{U}}_I^{\alpha} = \sum_{I=1}^{8} M_I^{\alpha}\ddot{\mathbf{U}}_I^{\alpha} \tag{9.7}$$

which means the atomic inertia force, $m^{\alpha}\ddot{\mathbf{u}}^{k\alpha}$, is distributed to the nodal inertia forces—the product of M_I^{α} and $\ddot{\mathbf{U}}_I^{\alpha}$. The well-known *partition of unity* dictates

$$\sum_{I=1}^{8} N_I(k) = 1 \tag{9.8}$$

which leads to the following expected

$$\sum_{I=1}^{8} \mathbf{F}_I^{\alpha} = \mathbf{f}^{k\alpha}, \quad \sum_{I=1}^{8} \mathbf{\Phi}_I^{\alpha} = \boldsymbol{\varphi}^{k\alpha}, \quad \sum_{I=1}^{8} M_I^{\alpha} = m^{\alpha} \tag{9.9}$$

Therefore, the governing equations for all nodal displacements are obtained as

$$M_{\underline{I}}^{\alpha} \ddot{\mathbf{U}}_{\underline{I}}^{\alpha} = \mathbf{F}_I^{\alpha} + \mathbf{\Phi}_I^{\alpha} \tag{9.10}$$

where the bar under the index I indicates that the summation is suspended. In other words, in a natural way, we get a set of governing equations for lumped-mass system. Also, it should be mentioned that the index I now extends for all nodes in the material system. The reduction of degrees of freedom from all-atoms MD (cf. Equation 9.1) to CGMD (cf. Equation 9.10) is tremendous—the ratio is number of atoms vs. number of nodes.

9.3 STIFFNESS-BASED CGMD

In this section, we are going to derive the governing equations for stiffness-based CGMD (SB-CGMD). Recall that the interatomic force $\mathbf{f}^{k\alpha}(\mathbf{r})$ is a function of atomic positions. Let $\mathbf{r}$ and $\mathbf{u}$ represent the totality of the atomic positions and displacements, respectively, at time t:

$$\mathbf{r}(t) = \{\mathbf{r}^{k\alpha} \mid k = 1,2,3,\ldots,N_c \; ; \; \alpha = 1,2,3,\ldots,N_a\} \tag{9.11}$$

and

$$\mathbf{u}(t) = \{\mathbf{u}^{k\alpha} \mid k = 1,2,3,\ldots,N_c \; ; \; \alpha = 1,2,3,\ldots,N_a\} \equiv \mathbf{r} - \mathbf{r}_o \tag{9.12}$$

where $\mathbf{r}_o = \mathbf{r}(0)$ are the initial atomic positions. Now we perform a Taylor series expansion as follows:

$$\mathbf{f}^{k\alpha}(\mathbf{r}) = \mathbf{f}^{k\alpha}(\mathbf{r}_o) + \sum_{l}^{N_c} \sum_{\beta}^{N_a} \frac{\partial \mathbf{f}^{k\alpha}(\mathbf{r}_o)}{\partial \mathbf{r}^{l\beta}} \left(\mathbf{r}^{l\beta} - \mathbf{r}_o^{l\beta}\right) + \ldots \tag{9.13}$$

Define

$$\mathbf{f}_o^{k\alpha} \equiv \mathbf{f}^{k\alpha}(\mathbf{r}_o), \; \mathbf{s}_{l\beta}^{k\alpha} \equiv -\sum_{l}^{N_c} \sum_{\beta}^{N_a} \frac{\partial \mathbf{f}^{k\alpha}(\mathbf{r}_o)}{\partial \mathbf{r}^{l\beta}} \tag{9.14}$$

where $\mathbf{s}_{l\beta}^{k\alpha}$ is often referred to as dynamical matrix. One may now express Equation 9.13 as

$$\mathbf{f}^{k\alpha}(\mathbf{r}) \approx \mathbf{f}_o^{k\alpha} - \sum_{l}^{N_c} \sum_{\beta}^{N_a} \mathbf{s}_{l\beta}^{k\alpha} \mathbf{u}^{l\beta} \tag{9.15}$$

Substituting Equation 9.15 into Equation 9.4 and applying Equation 9.2 results in

$$\begin{aligned}\mathbf{F}_I^{\alpha} &= N_I(k)\mathbf{f}^{k\alpha}(\mathbf{r}) = N_I(k)\mathbf{f}_o^{k\alpha} - \sum_{l}^{N_c}\sum_{\beta}^{N_a} N_I(k)\mathbf{s}_{l\beta}^{k\alpha}\mathbf{u}^{l\beta} \\ &= N_I(k)\mathbf{f}_o^{k\alpha} - \sum_{J=1}^{8}\sum_{l}^{N_c}\sum_{\beta}^{N_a} N_I(k)N_J(l)s_{l\beta}^{k\alpha}\mathbf{U}_J^{\beta}\end{aligned} \tag{9.16}$$

Define

$$\tilde{\mathbf{F}}_I^{\alpha}(k) \equiv N_I(k)\mathbf{f}_o^{k\alpha}, \qquad \mathbf{S}_{J\beta}^{I\alpha}(k,l) \equiv N_I(k)N_J(l)s_{l\beta}^{k\alpha} \tag{9.17}$$

Then Equation 9.16 can be rewritten as

$$\mathbf{F}_I^{\alpha}(k) = \tilde{\mathbf{F}}_I^{\alpha}(k) - \sum_{J=1}^{8}\sum_{l}^{N_c}\sum_{\beta}^{N_a} \mathbf{S}_{J\beta}^{I\alpha}(k,l)\mathbf{U}_J^{\beta} \tag{9.18}$$

Now the governing equations of the force-based CGMD, Equation 9.10, is reduced to

$$M_{\underline{I}}^{\alpha}\ddot{\mathbf{U}}_{\underline{I}}^{\alpha} + \sum_{J}\sum_{\beta}^{N_a} \mathbf{K}_{J\beta}^{I\alpha}\mathbf{U}_J^{\beta} = \tilde{\mathbf{F}}_I^{\alpha} + \mathbf{\Phi}_I^{\alpha}, \qquad \alpha \in [1, N_a] \tag{9.19}$$

with the understanding that $M_I^{\alpha}, \tilde{\mathbf{F}}_I^{\alpha}$, and $\mathbf{\Phi}_I^{\alpha}$ are the assembly of $M_I^{\alpha}(k), \tilde{\mathbf{F}}_I^{\alpha}(k)$, and $\mathbf{\Phi}_I^{\alpha}(k)$, respectively, over all unit cells ($k = 1, 2, 3, \ldots, N_c$) and $\mathbf{K}_{J\beta}^{I\alpha}$ is the assembly of $\mathbf{S}_{J\beta}^{I\alpha}(k,l)$ over all unit cells ($k \in [1, N_c]$ and $l \in [1, N_c]$). In abbreviation, Equation 9.19 can be written symbolically as

$$\mathbf{M}\ddot{\mathbf{U}} + \mathbf{K}\mathbf{U} = \hat{\mathbf{F}}(t) \tag{9.19*}$$

Assume the system is initially at rest, i.e., $\mathbf{U}(0) = \dot{\mathbf{U}}(0) = 0$, then the conservation of energy can be represented by

$$\frac{1}{2}(\dot{\mathbf{U}}^T\mathbf{M}\dot{\mathbf{U}} + \mathbf{U}^T\mathbf{K}\mathbf{U}) = \int_0^t \hat{\mathbf{F}}(\tau)\cdot\dot{\mathbf{U}}(\tau)\,d\tau \tag{9.20}$$

To solve the governing equations, Equation 9.19, the nodal displacements $\mathbf{U}_I^{\alpha}$ and accelerations $\ddot{\mathbf{U}}_I^{\alpha}$ need to be updated at each time step. However, it is stressed that the initial nodal force $\tilde{\mathbf{F}}_I^{\alpha}$ and the stiffness matrix $\mathbf{K}_{J\beta}^{I\alpha}$ need to be calculated only once. In other words, once they are obtained, they can be used for the entire time-marching process. It is noticed that the SB-CGMD is limited to small atomic displacements as seen in Equation 9.15.

9.3.1 General dynamical matrix

The most general interatomic potential energy in MD simulation can be expressed as

$$V(\mathbf{r})=\frac{1}{2!}\sum_i\sum_j V^{ij}(\mathbf{r}^i,\mathbf{r}^j)+\frac{1}{3!}\sum_i\sum_j\sum_k V^{ijk}(\mathbf{r}^i,\mathbf{r}^j,\mathbf{r}^k)+\frac{1}{4!}\sum_i\sum_j\sum_k\sum_l V^{ijkl}(\mathbf{r}^i,\mathbf{r}^j,\mathbf{r}^k,\mathbf{r}^l)+\dots. \tag{9.21}$$

with the understanding that the atoms involved in any of the many-body potentials must be distinct (e.g., $i \neq j \neq k \neq l$ *in* V^{ijkl}). The potential must be form-invariant with respect to rigid motions of the spatial frame of reference—coined as axiom of objectivity in *continuum mechanics* (CM) (Eringen, 1980, 1999; Liu, 2002). Principle of objectivity recently has been extended to MD by Yang et al. (2016a, 2016b) and Liu and Lee (2016). From a physical viewpoint, this means material properties should not depend on the motion of the observer. If two frames $\bar{\mathbf{r}}$ and $\mathbf{r}$ can be made to coincide with each other through a rigid-body motion, then they must be related to each other by

$$\bar{\mathbf{r}}=\mathbf{Q}(t)\mathbf{r}+\mathbf{b}(t) \tag{9.22}$$

where $\mathbf{Q}(t)$ belong to full group of orthogonal transformation, i.e.,

$$\mathbf{Q}^T(t)\mathbf{Q}(t)=\mathbf{Q}(t)\mathbf{Q}^T(t)=\mathbf{I}, \qquad \det(\mathbf{Q})=\pm 1 \tag{9.23}$$

Then, according to the *principle of objectivity*, one must have

$$V^{ijkl\dots}(\bar{\mathbf{r}}^i,\bar{\mathbf{r}}^j,\bar{\mathbf{r}}^k,\bar{\mathbf{r}}^l,\dots)=V^{ijkl\dots}(\mathbf{r}^i,\mathbf{r}^j,\mathbf{r}^k,\mathbf{r}^l,\dots) \tag{9.24}$$

Now, let $\mathbf{Q}(t) = \mathbf{I}$ and $\mathbf{b}(t) = -\mathbf{r}^i(t)$. Then Equation 9.24 reads as

$$V^{ijkl\dots}(0,\mathbf{r}^j-\mathbf{r}^i,\mathbf{r}^k-\mathbf{r}^i,\mathbf{r}^l-\mathbf{r}^i,\dots)=V^{ijkl\dots}(\mathbf{r}^i,\mathbf{r}^j,\mathbf{r}^k,\mathbf{r}^l,\dots) \tag{9.25}$$

which simply says, from now on, we write the many-body potentials as

$$V^{ijkl\dots}=V^{ijkl\dots}(\mathbf{r}^{ij},\mathbf{r}^{ik},\mathbf{r}^{il},\dots) \tag{9.26}$$

and it obeys

$$V^{ijkl\dots}=V^{ijkl\dots}(\mathbf{r}^{ij},\mathbf{r}^{ik},\mathbf{r}^{il},\dots)=V^{ijkl\dots}(\mathbf{Q}\mathbf{r}^{ij},\mathbf{Q}\mathbf{r}^{ik},\mathbf{Q}\mathbf{r}^{il},\dots) \tag{9.27}$$

This means $V^{ijkl\dots}$ is an isotropic scalar-valued functions of vectors $\mathbf{r}^{ij}$, $\mathbf{r}^{ik}$, $\mathbf{r}^{il}$,... and according to Wang's representation theorem for isotropic functions (Wang, 1970a, 1970b, 1971), $V^{ijkl\dots}$ should be expressed as a function of all the inner products between those vectors, i.e.,

$$V^{ijkl\dots}=V^{ijkl\dots}(r^{ij},r^{ik},r^{il},\dots,\cos\theta_{ijk},\cos\theta_{ijl},\cos\theta_{ikl},\dots) \tag{9.28}$$

where r^{ij} is the length of $\mathbf{r}^{ij}$ and θ_{ijk} is the angle between $\mathbf{r}^{ij}$ and $\mathbf{r}^{ik}$. It is noticed that the n-body interatomic potential is a scalar-valued function of $n(n-1)/2$ scalars. It can be readily derived that

$$\frac{\partial r^{ij}}{\partial \mathbf{r}^{i}} = \frac{r^{ij}}{\mathbf{r}^{ij}}, \quad \frac{\partial r^{ij}}{\partial \mathbf{r}^{j}} = -\frac{r^{ij}}{\mathbf{r}^{ij}} \tag{9.29}$$

$$\begin{aligned}
\frac{\partial \cos\theta_{ijk}}{\partial \mathbf{r}^{i}} &= \left(\frac{1}{r^{ik}} - \frac{\cos\theta_{ijk}}{r^{ij}}\right)\frac{\mathbf{r}^{ij}}{r^{ij}} + \left(\frac{1}{r^{ij}} - \frac{\cos\theta_{ijk}}{r^{ik}}\right)\frac{\mathbf{r}^{ik}}{r^{ik}} \\
\frac{\partial \cos\theta_{ijk}}{\partial \mathbf{r}^{j}} &= \frac{\cos\theta_{ijk}}{r^{ij}}\frac{\mathbf{r}^{ij}}{r^{ij}} - \frac{1}{r^{ij}}\frac{\mathbf{r}^{ik}}{r^{ik}} \\
\frac{\partial \cos\theta_{ijk}}{\partial \mathbf{r}^{k}} &= \frac{\cos\theta_{ijk}}{r^{ik}}\frac{\mathbf{r}^{ik}}{r^{ik}} - \frac{1}{r^{ik}}\frac{\mathbf{r}^{ij}}{r^{ij}}
\end{aligned} \tag{9.30}$$

In general, for a given n-body potential, one may derive the interatomic forces as

$$\begin{aligned}
\mathbf{f}^{i} &= -\frac{\partial V^{ijkl\ldots}}{\partial \mathbf{r}^{i}} \\
&= -\frac{\partial V^{ijkl\ldots}}{\partial r^{ij}}\frac{\partial r^{ij}}{\partial \mathbf{r}^{i}} - \frac{\partial V^{ijkl\ldots}}{\partial r^{ik}}\frac{\partial r^{ik}}{\partial \mathbf{r}^{i}} - \frac{\partial V^{ijkl\ldots}}{\partial r^{il}}\frac{\partial r^{il}}{\partial \mathbf{r}^{i}} - \ldots \\
&\quad - \frac{\partial V^{ijkl\ldots}}{\partial \cos\theta_{ijk}}\frac{\partial \cos\theta_{ijk}}{\partial \mathbf{r}^{i}} - \frac{\partial V^{ijkl\ldots}}{\partial \cos\theta_{ijl}}\frac{\partial \cos\theta_{ijl}}{\partial \mathbf{r}^{i}} - \frac{\partial V^{ijkl\ldots}}{\partial \cos\theta_{ikl}}\frac{\partial \cos\theta_{ikl}}{\partial \mathbf{r}^{i}} - \ldots
\end{aligned} \tag{9.31}$$

and

$$\begin{aligned}
\mathbf{f}^{j} &= -\frac{\partial V^{ijkl\ldots}}{\partial \mathbf{r}^{j}} \\
&= -\frac{\partial V^{ijkl\ldots}}{\partial r^{ij}}\frac{\partial r^{ij}}{\partial \mathbf{r}^{j}} - \frac{\partial V^{ijkl\ldots}}{\partial \cos\theta_{ijk}}\frac{\partial \cos\theta_{ijk}}{\partial \mathbf{r}^{j}} - \frac{\partial V^{ijkl\ldots}}{\partial \cos\theta_{ijl}}\frac{\partial \cos\theta_{ijl}}{\partial \mathbf{r}^{j}} - \ldots
\end{aligned} \tag{9.32}$$

and similarly, following the same pattern as in Equation 9.32, one may find $\mathbf{f}^{k}$, $\mathbf{f}^{l}$,.... Now, for three-body potential case, one finds

$$\begin{aligned}
\mathbf{f}^{i} &= -\frac{\partial V^{ijk}}{\partial r^{ij}}\frac{\partial r^{ij}}{\partial \mathbf{r}^{i}} - \frac{\partial V^{ijk}}{\partial r^{ik}}\frac{\partial r^{ik}}{\partial \mathbf{r}^{i}} - \frac{\partial V^{ijk}}{\partial \cos\theta_{ijk}}\frac{\partial \cos\theta_{ijk}}{\partial \mathbf{r}^{i}} \\
&= -\frac{\partial V^{ijk}}{\partial r^{ij}}\frac{\mathbf{r}^{ij}}{r^{ij}} - \frac{\partial V^{ijk}}{\partial r^{ik}}\frac{\mathbf{r}^{ik}}{r^{ik}} - \frac{\partial V^{ijk}}{\partial \cos\theta_{ijk}}\left\{\left(\frac{1}{r^{ik}} - \frac{\cos\theta_{ijk}}{r^{ij}}\right)\frac{\mathbf{r}^{ij}}{r^{ij}} + \left(\frac{1}{r^{ij}} - \frac{\cos\theta_{ijk}}{r^{ik}}\right)\frac{\mathbf{r}^{ik}}{r^{ik}}\right\}
\end{aligned} \tag{9.33}$$

$$\mathbf{f}^j = -\frac{\partial V^{ijk}}{\partial \mathbf{r}^{ij}}\frac{\partial r^{ij}}{\partial \mathbf{r}^j} - \frac{\partial V^{ijk}}{\partial \cos\theta_{ijk}}\frac{\partial \cos\theta_{ijk}}{\partial \mathbf{r}^j}$$
$$= \frac{\partial V^{ijk}}{\partial \mathbf{r}^{ij}}\frac{\mathbf{r}^{ij}}{r^{ij}} - \frac{\partial V^{ijk}}{\partial \cos\theta_{ijk}}\left\{\frac{\cos\theta_{ijk}}{r^{ij}}\frac{\mathbf{r}^{ij}}{r^{ij}} - \frac{1}{r^{ij}}\frac{\mathbf{r}^{ik}}{r^{ik}}\right\} \tag{9.34}$$

and

$$\mathbf{f}^k = -\frac{\partial V^{ijk}}{\partial \mathbf{r}^{ik}}\frac{\partial r^{ik}}{\partial \mathbf{r}^k} - \frac{\partial V^{ijk}}{\partial \cos\theta_{ijk}}\frac{\partial \cos\theta_{ijk}}{\partial \mathbf{r}^k}$$
$$= \frac{\partial V^{ijk}}{\partial \mathbf{r}^{ik}}\frac{\mathbf{r}^{ik}}{r^{ik}} - \frac{\partial V^{ijk}}{\partial \cos\theta_{ijk}}\left\{\frac{\cos\theta_{ijk}}{r^{ik}}\frac{\mathbf{r}^{ik}}{r^{ik}} - \frac{1}{r^{ik}}\frac{\mathbf{r}^{ij}}{r^{ij}}\right\} \tag{9.35}$$

It is seen that, in general many-body potential, the interatomic forces have a general form

$$\begin{aligned}
\mathbf{f}^i &= A^{ij}\mathbf{r}^{ij} + A^{ik}\mathbf{r}^{ik} + A^{il}\mathbf{r}^{il} + \ldots \\
\mathbf{f}^j &= B^{ij}\mathbf{r}^{ij} + B^{ik}\mathbf{r}^{ik} + B^{il}\mathbf{r}^{il} + \ldots \\
\mathbf{f}^k &= C^{ij}\mathbf{r}^{ij} + C^{ik}\mathbf{r}^{ik} + C^{il}\mathbf{r}^{il} + \ldots \\
&\ldots\ldots\ldots
\end{aligned} \tag{9.36}$$

The general dynamical matrix can be obtained as

$$\mathbf{s}^{\alpha}_{\beta} = -\frac{\partial \mathbf{f}^{\alpha}}{\partial \mathbf{r}^{\beta}} \tag{9.37}$$

From Equation 9.36, one may find

$$\mathbf{s}^i_i = -\left\{\frac{\partial A^{ij}}{\partial \mathbf{r}^i}\otimes \mathbf{r}^{ij} + A^{ij}\mathbf{I}\right\} - \left\{\frac{\partial A^{ik}}{\partial \mathbf{r}^i}\otimes \mathbf{r}^{ik} + A^{ik}\mathbf{I}\right\} - \left\{\frac{\partial A^{il}}{\partial \mathbf{r}^i}\otimes \mathbf{r}^{il} + A^{il}\mathbf{I}\right\} - \ldots \tag{9.38}$$

$$\mathbf{s}^i_j = -\left\{\frac{\partial A^{ij}}{\partial \mathbf{r}^j}\otimes \mathbf{r}^{ij} - A^{ij}\mathbf{I}\right\} - \left\{\frac{\partial A^{ik}}{\partial \mathbf{r}^j}\otimes \mathbf{r}^{ik}\right\} - \left\{\frac{\partial A^{il}}{\partial \mathbf{r}^j}\otimes \mathbf{r}^{il}\right\} - \ldots \tag{9.39}$$

$$\mathbf{s}^j_i = -\left\{\frac{\partial B^{ij}}{\partial \mathbf{r}^i}\otimes \mathbf{r}^{ij} + B^{ij}\mathbf{I}\right\} - \left\{\frac{\partial B^{ik}}{\partial \mathbf{r}^i}\otimes \mathbf{r}^{ik} + B^{ik}\mathbf{I}\right\} - \left\{\frac{\partial B^{il}}{\partial \mathbf{r}^i}\otimes \mathbf{r}^{il} + B^{il}\mathbf{I}\right\} - \ldots \tag{9.40}$$

and

$$\mathbf{s}^j_j = -\left\{\frac{\partial B^{ij}}{\partial \mathbf{r}^j}\otimes \mathbf{r}^{ij} - B^{ij}\mathbf{I}\right\} - \left\{\frac{\partial B^{ik}}{\partial \mathbf{r}^j}\otimes \mathbf{r}^{ik}\right\} - \left\{\frac{\partial B^{il}}{\partial \mathbf{r}^j}\otimes \mathbf{r}^{il}\right\} - \ldots \tag{9.41}$$

Following the same pattern, one may find other components of the dynamical matrix. Actually, we may prove analytically and verify numerically that, for the most general interatomic potential energy, the summation of interatomic forces is zero and the dynamical matrix is symmetric.

9.4 FROM AAMD TO CGMD

In this section, we are going to construct a theory of multiscale modeling from AAMD to CGMD. First, we divide the atomic region, modeled by MD, into *ng* groups: in group *g*, $g \in [1, ng]$, there are n_g atoms, not necessarily of the same kind, not necessarily being a single crystal either. Now we recall the governing equations, Equation 8.79, from Chapter 8 as

$$m^i \dot{\mathbf{v}}^i = \mathbf{f}^i + \boldsymbol{\varphi}^i - \chi_g m^i \tilde{\mathbf{v}}^i, \qquad i \in group\ g \tag{9.42}$$

where $-\chi_g m^i \tilde{\mathbf{v}}^i$ is the *Nosé–Hoover temperature force*, $\mathbf{f}^i$ is the interatomic force acting on *atom i* from interaction with all other atoms in the atomic region, and $\boldsymbol{\varphi}^i$ now stands for all other kinds of forces acting on *atom i*. Let the atom-based-continuum (ABC) region, modeled by CGMD, be divided into *NG* groups: in group *G*, $G \in [1, NG]$, there are N_G nodes. In a similar way, we are going to show that the governing equation of the *atom* α atom in the *I*th node of group *G* in the ABC region can be derived to be

$$M_{\underline{I}}^{\alpha} \ddot{\mathbf{U}}_{\underline{I}}^{\alpha} = \mathbf{F}_I^{\alpha} + \boldsymbol{\Phi}_I^{\alpha} - \chi_G M_{\underline{I}}^{\alpha} \tilde{\mathbf{V}}_I^{\alpha}, \qquad I \in \text{group } G, \qquad \alpha \in [1, N_a] \tag{9.43}$$

where $\mathbf{F}_I^{\alpha}$ is the interatomic force acting on *atom* α in the *I*th node of the group *G* due to the interaction between this atom and atoms in other part of the ABC region; $-\chi_G M_{\underline{I}}^{\alpha} \tilde{\mathbf{V}}_I^{\alpha}$ is the Nosé–Hoover temperature force, and $\boldsymbol{\Phi}_I^{\alpha}$ now stands for all other kinds of forces acting on *atom* α. The interaction between atom in the atomic region and atom in the ABC region is reflected in $\boldsymbol{\varphi}^i$ and $\boldsymbol{\Phi}_I^{\alpha}$. To be precise, assume that there is a pair potential between *atom i* in the atomic region and *atom* α, which is located in the *unit cell k* of the ABC region. The force, $\boldsymbol{\varphi}^i$, acting on *atom i* will be incorporated into Equation 9.42; the force acting on *atom* α, $\boldsymbol{\varphi}^{k\alpha} = -\boldsymbol{\varphi}^i$, will be distributed to nodes according to

$$\boldsymbol{\Phi}_I^{\alpha}(k) = N_I(k)\boldsymbol{\varphi}^{k\alpha} = -N_I(k)\boldsymbol{\varphi}^i \tag{9.44}$$

9.4.1 Nosé–Hoover thermostat in the ABC region

Similar to the situation in AAMD simulation (cf. Chapter 8), if group *G* in the ABC region has a Nosé–Hoover thermostat, then

$$\dot{\chi}_G = \frac{1}{\tau_G^2} \frac{T_G - T_G^*}{T_G^*} \tag{9.45}$$

If group G has no Nosé–Hoover thermostat, then simply

$$\chi_G = \dot{\chi}_G = 0 \tag{9.46}$$

In the following, we outline the procedures to formulate the Nosé–Hoover thermostat in CGMD. From now on, if there is no ambiguity, we use the following abbreviations

$$\sum_I \equiv \sum_{I \in G}, \quad \sum_\alpha \equiv \sum_{\alpha=1}^{N_a} \tag{9.47}$$

Position and velocity of the centroid of group G:

$$\begin{aligned} M_G &\equiv \sum_I \sum_\alpha M_I^\alpha \\ \bar{\mathbf{R}} &\equiv \frac{\sum_I \sum_\alpha M_I^\alpha \mathbf{R}_I^\alpha}{M_G} \\ \bar{\mathbf{V}} &\equiv \frac{\sum_I \sum_\alpha M_I^\alpha \mathbf{V}_I^\alpha}{M_G} \end{aligned} \tag{9.48}$$

Relative position and relative velocity:

$$\bar{\mathbf{R}}_I^\alpha \equiv \mathbf{R}_I^\alpha - \bar{\mathbf{R}}, \qquad \bar{\mathbf{V}}_I^\alpha \equiv \mathbf{V}_I^\alpha - \bar{\mathbf{V}} \tag{9.49}$$

It is seen that $\sum_I \sum_\alpha M_I^\alpha \bar{\mathbf{R}}_I^\alpha = 0$ and $\sum_I \sum_\alpha M_I^\alpha \bar{\mathbf{V}}_I^\alpha = 0$. That is why $\bar{\mathbf{R}}_I^\alpha$ and $\bar{\mathbf{V}}_I^\alpha$ are named as relative position and relative velocity, respectively.

Moment of inertia:

$$\mathbf{J} = \sum_I \sum_\alpha \left\{ M_I^\alpha [(\bar{\mathbf{R}}_I^\alpha \cdot \bar{\mathbf{R}}_I^\alpha)] \mathbf{I} - \bar{\mathbf{R}}_I^\alpha \otimes \bar{\mathbf{R}}_I^\alpha \right\} \tag{9.50}$$

Angular momentum:

$$\mathbf{H} = \sum_I \sum_\alpha M_I^\alpha \bar{\mathbf{R}}_I^\alpha \times \bar{\mathbf{V}}_I^\alpha \tag{9.51}$$

Angular velocity:

$$\boldsymbol{\omega} \equiv \mathbf{J}^{-1}\mathbf{H}, \qquad \hat{\mathbf{V}}_I^{\alpha} \equiv \boldsymbol{\omega} \times \bar{\mathbf{R}}_I^{\alpha} \tag{9.52}$$

Thermal velocity:

$$\tilde{\mathbf{V}}_I^{\alpha} \equiv \mathbf{V}_I^{\alpha} - \bar{\mathbf{V}} - \hat{\mathbf{V}}_I^{\alpha} \tag{9.53}$$

Atomistic temperature:

$$T = \frac{1}{\kappa_B N^{dof}} \sum_I \sum_{\alpha} M_I^{\alpha} \left(\tilde{\mathbf{V}}_I^{\alpha} \cdot \tilde{\mathbf{V}}_I^{\alpha} \right) \tag{9.54}$$

Theorem I: Total thermal force is zero, i.e.,

$$\sum_I \sum_{\alpha} \chi M_I^{\alpha} \tilde{\mathbf{V}}_I^{\alpha} = 0 \tag{9.55}$$

Proof:

$$\begin{aligned} \sum_I \sum_{\alpha} \chi M_I^{\alpha} \tilde{\mathbf{V}}_I^{\alpha} &= \sum_I \sum_{\alpha} \chi M_I^{\alpha} \left(\mathbf{V}_I^{\alpha} - \bar{\mathbf{V}} - \boldsymbol{\omega} \times \bar{\mathbf{R}}_I^{\alpha} \right) \\ &= -\boldsymbol{\omega} \times \sum_I \sum_{\alpha} \chi m_I^{\alpha} \bar{\mathbf{R}}_I^{\alpha} = 0 \end{aligned} \tag{9.56}$$

Theorem II: Total thermal moment is zero, i.e.,

$$\sum_I \sum_{\alpha} \chi M_I^{\alpha} \mathbf{R}_I^{\alpha} \times \tilde{\mathbf{V}}_I^{\alpha} = 0 \tag{9.57}$$

Proof:

$$\begin{aligned} &\sum_I \sum_{\alpha} \chi M_I^{\alpha} \mathbf{R}_I^{\alpha} \times \tilde{\mathbf{V}}_I^{\alpha} = \sum_I \sum_{\alpha} \chi M_I^{\alpha} \left(\bar{\mathbf{R}}_I^{\alpha} + \bar{\mathbf{R}} \right) \times \left(\bar{\mathbf{V}}_I^{\alpha} - \hat{\mathbf{V}}_I^{\alpha} \right) \\ &= \chi \sum_I \sum_{\alpha} M_I^{\alpha} \left\{ \bar{\mathbf{R}}_I^{\alpha} \times \bar{\mathbf{V}}_I^{\alpha} - \bar{\mathbf{R}}_I^{\alpha} \times \hat{\mathbf{V}}_I^{\alpha} + \bar{\mathbf{R}} \times \bar{\mathbf{V}}_I^{\alpha} - \bar{\mathbf{R}} \times \hat{\mathbf{V}}_I^{\alpha} \right\} \\ &= \chi \left\{ \mathbf{H} - \sum_I \sum_{\alpha} \left\{ M_I^{\alpha} \bar{\mathbf{R}}_I^{\alpha} \times \left(\boldsymbol{\omega} \times \bar{\mathbf{R}}_I^{\alpha} \right) - M_I^{\alpha} \bar{\mathbf{R}} \times \left(\boldsymbol{\omega} \times \bar{\mathbf{R}}_I^{\alpha} \right) \right\} \right\} \\ &= \chi \left\{ \mathbf{H} - \mathbf{J}\boldsymbol{\omega} - \bar{\mathbf{R}} \times \left(\boldsymbol{\omega} \times \sum_I \sum_{\alpha} M_I^{\alpha} \bar{\mathbf{R}}_I^{\alpha} \right) \right\} = 0 \end{aligned} \tag{9.58}$$

9.4.2 Hamiltonian of AAMD ⊕ CGMD

The *Hamiltonian* of the entire system can be expressed as

$$H=\sum_{G=1}^{NG}\sum_{I\in G}\sum_{\alpha=1}^{N_a}\left\{\frac{1}{2}M_I^{\alpha}\mathbf{V}_I^{\alpha}\cdot\mathbf{V}_I^{\alpha}-\int_0^t\mathbf{\Phi}_I^{\alpha}(s)\cdot\mathbf{V}_I^{\alpha}(s)\,ds\right\}+\sum_{g=1}^{ng}\sum_{i\in g}\left\{\frac{1}{2}m^i\mathbf{v}^i\cdot\mathbf{v}^i-\int_0^t\boldsymbol{\varphi}^i(s)\cdot\mathbf{v}^i(s)\,ds\right\}$$
$$+\sum_{G=1}^{NG}\left\{\frac{1}{2}Q_G\chi_G^2+\frac{Q_G}{\tau_G^2}\int_0^t\chi_G(s)\,ds\right\}+\sum_{g=1}^{ng}\left\{\frac{1}{2}Q_g\chi_g^2+\frac{Q_g}{\tau_g^2}\int_0^t\chi_g(s)\,ds\right\}+U(\mathbf{R},\mathbf{r}) \tag{9.59}$$

where U is the total interatomic potential energy of the entire system, which consists of atomic region and ABC region; $\mathbf{R}$ represents all the position vectors of *atom* α in the *unit cell* k in ABC region; and $\mathbf{r}$ represents all the position vectors of *atom* i in the atomic region.

Theorem III: Hamiltonian is constant in time.

Proof:

From now on, if there is no ambiguity, we use the following abbreviations:

$$\sum_{G,I,\alpha}\equiv\sum_{G=1}^{NG}\sum_{I\in G}\sum_{\alpha=1}^{N_a},\qquad \sum_{g,i}\equiv\sum_{g=1}^{ng}\sum_{i\in g} \tag{9.60}$$

We differentiate Equation 9.59 term by term with respect to time as follows:

$$\frac{d}{dt}\sum_{G,I,\alpha}\frac{1}{2}M_I^{\alpha}\mathbf{V}_I^{\alpha}\cdot\mathbf{V}_I^{\alpha}=\sum_{G,I,\alpha}M_I^{\alpha}\dot{\mathbf{V}}_I^{\alpha}\cdot\mathbf{V}_I^{\alpha} \tag{9.61}$$

$$\frac{d}{dt}\sum_{g,i}\frac{1}{2}m^i\mathbf{v}^i\cdot\mathbf{v}^i=\sum_{g,i}\frac{1}{2}m^i\dot{\mathbf{v}}^i\cdot\mathbf{v}^i \tag{9.62}$$

$$-\frac{d}{dt}\sum_{G,I,\alpha}\int_0^t\mathbf{\Phi}_I^{\alpha}(s)\cdot\mathbf{V}_I^{\alpha}(s)\,ds=-\sum_{G,I,\alpha}\mathbf{\Phi}_I^{\alpha}\cdot\mathbf{V}_I^{\alpha} \tag{9.63}$$

$$-\frac{d}{dt}\sum_{g,i}\int_0^t\boldsymbol{\varphi}^i(s)\cdot\mathbf{v}^i(s)\,ds=-\sum_{g,i}\boldsymbol{\varphi}^i\cdot\mathbf{v}^i \tag{9.64}$$

$$\begin{aligned}
&\frac{d}{dt}\sum_{G=1}^{NG}\left\{\frac{1}{2}Q_G\chi_G^2+\frac{Q_G}{\tau_G^2}\int_0^t\chi_G(s)\,ds\right\}=\sum_{G=1}^{NG}\left\{Q_G\chi_G\dot{\chi}_G+\frac{Q_G\chi_G}{\tau_G^2}\right\}\\
&=\sum_{G=1}^{NG}N_G^{dof}k_B\chi_G T_G=\sum_{G,I,\alpha}\chi_G M_I^\alpha\tilde{\mathbf{V}}_I^\alpha\cdot\tilde{\mathbf{V}}_I^\alpha\\
&=\sum_{G,I,\alpha}\chi_G M_I^\alpha\tilde{\mathbf{V}}_I^\alpha\cdot\mathbf{V}_I^\alpha
\end{aligned} \tag{9.65}$$

$$\begin{aligned}
&\frac{d}{dt}\sum_{g=1}^{ng}\left\{\frac{1}{2}Q_g\chi_g^2+\frac{Q_g}{\tau_g^2}\int_0^t\chi_g(s)\,ds\right\}=\sum_{g=1}^{ng}\left\{Q_g\chi_g\dot{\chi}_g+\frac{Q_g\chi_g}{\tau_g^2}\right\}\\
&=\sum_{g=1}^{ng}N_g^{dof}k_B\chi_g T_g=\sum_{g,i}\chi_g m^i\tilde{\mathbf{v}}^i\cdot\tilde{\mathbf{v}}^i\\
&=\sum_{g,i}\chi_g m^i\tilde{\mathbf{v}}^i\cdot\mathbf{v}^i
\end{aligned} \tag{9.66}$$

$$\begin{aligned}
\frac{dU}{dt}&=\sum_{G,I,\alpha}\frac{\partial U}{\partial\mathbf{R}_I^\alpha}\cdot\mathbf{V}^{I\alpha}+\sum_{g,i}\frac{\partial U}{\partial\mathbf{r}^i}\cdot\mathbf{v}^i\\
&=\sum_{G,I,\alpha}\sum_k\left\{\frac{\partial U}{\partial\mathbf{r}^{k\alpha}}\frac{\partial\mathbf{r}^{k\alpha}}{\partial\mathbf{R}_I^\alpha}\right\}\cdot\mathbf{V}_I^\alpha-\sum_{g,i}\mathbf{f}^i\cdot\mathbf{v}^i\\
&=-\sum_{G,I,\alpha}\sum_k\left\{\mathbf{f}^{k\alpha}\sum_{J=1}^{8}N_J(k)\delta_{IJ}\right\}\cdot\mathbf{V}_I^\alpha-\sum_{g,i}\mathbf{f}^i\cdot\mathbf{v}^i\\
&=-\sum_{G,I,\alpha}\sum_k\{\mathbf{f}^{k\alpha}N_I(k)\}\cdot\mathbf{V}_I^\alpha-\sum_{g,i}\mathbf{f}^i\cdot\mathbf{v}^i\\
&=-\sum_{G,I,\alpha}\mathbf{F}_I^\alpha\cdot\mathbf{V}_I^\alpha-\sum_{g,i}\mathbf{f}^i\cdot\mathbf{v}^i
\end{aligned} \tag{9.67}$$

Note: In deriving Equation 9.67, we utilized

$$\mathbf{r}^{k\alpha}=\sum_{J=1}^{8}N_J(k)\mathbf{R}_J^\alpha,\qquad \mathbf{F}_I^\alpha=\sum_k N_I(k)\mathbf{f}^{k\alpha} \tag{9.68}$$

which is actually the basic assumption in CGMD. Now we have

$$\dot{H}=\sum_{G,I,\alpha}\left\{M_I^\alpha\dot{\mathbf{V}}_I^\alpha-\mathbf{F}_I^\alpha-\mathbf{\Phi}_I^\alpha-\chi_G M_I^\alpha\tilde{\mathbf{V}}_I^\alpha\right\}\cdot\mathbf{V}_I^\alpha+\sum_{g,i}\{m^i\dot{\mathbf{v}}^i-\mathbf{f}^i-\boldsymbol{\varphi}^i-\chi_g m^i\tilde{\mathbf{v}}^i\}\cdot\mathbf{v}^i \tag{9.69}$$

Now, we recall

$$m^i\dot{\mathbf{v}}^i = \mathbf{f}^i + \boldsymbol{\varphi}^i - \chi_g m^i\tilde{\mathbf{v}}^i, \qquad i \in group\ g \tag{9.42)*}$$

and

$$M_{\underline{I}}^{\alpha}\ddot{\mathbf{U}}_{\underline{I}}^{\alpha} = \mathbf{F}_I^{\alpha} + \boldsymbol{\Phi}_I^{\alpha} - \chi_G M_{\underline{I}}^{\alpha}\tilde{\mathbf{V}}_I^{\alpha}, \qquad I \in group\ G, \qquad \alpha \in [1, N_a] \tag{9.43)*}$$

which are the governing equations for atoms in the atomic region and for nodes in the ABC region, respectively. Thus, Theorem III is proved.

9.5 SEQUENTIAL MULTISCALE MODELING FROM MD TO THERMOELASTICITY

9.5.1 Thermoelasticity and sequential multiscale modeling

In small-strain thermoelasticity (a branch of CM), the relevant balance laws and constitutive equations may be expressed as (Boresi et al., 2011; Chen et al., 2006; Eringen, 1980):

$$\rho^o\dot{\mathbf{v}} = \nabla\cdot\boldsymbol{\sigma} + \rho^o\boldsymbol{\varphi} \tag{9.70}$$

$$\rho^o\dot{e} - \boldsymbol{\sigma} : \nabla\mathbf{v} + \nabla\cdot\mathbf{q} - \rho^o h = 0 \tag{9.71}$$

$$\sigma_{ij} = -\beta_{ij}(T - T^{ref}) + A_{ijkl}e_{kl} \tag{9.72}$$

$$\rho^o\dot{e} = \rho^o\gamma\dot{T} + A_{ijkl}e_{ij}\dot{e}_{kl} \tag{9.73}$$

and

$$q_i = -\kappa_{ij}T_{,j} \tag{9.74}$$

where ρ^o is the mass density in the reference state; $\mathbf{u}$ is the displacement vector and $\mathbf{v} = \dot{\mathbf{u}}$ is the velocity vector; σ is the Cauchy stress tensor; e is the internal energy density; $\mathbf{q}$ is the heat flux; T is the absolute temperature; T^{ref} is the reference temperature; $\mathbf{e}$ is the strain tensor ($\mathbf{e} \approx (\nabla\mathbf{u} + (\nabla\mathbf{u})^T)/2$); $\boldsymbol{\beta}$ is named as the thermal expansion coefficients; $\mathbf{A}$ is the elastic constant; $\boldsymbol{\kappa}$ is the thermal conductivity; γ is the specific heat; and $\boldsymbol{\varphi}$ is the body force per unit mass. Notice that, in thermoelasticity, the Cauchy stress is derivable from a scalar-valued Helmholtz free energy density function; stress tensor and strain tensor are both symmetric. Therefore, one has

$$\beta_{ij} = \beta_{ji}, \qquad A_{ijkl} = A_{jikl} = A_{ijlk} = A_{klij} \tag{9.75}$$

It is emphasized that in CM, temperature is an independent variable. Therefore, an energy equation, Equation 9.73, is needed. On the contrary, Nosé–Hoover thermostat is not needed—all one needs to do is to set temperature-specified boundary conditions. Of course, one may choose to set heat-flux-specified boundary conditions instead. Also, the temperature field and displacement field in CM are functions of spatial and temporal variables—that is why we see terms such as $\nabla \cdot \sigma$, $\nabla \mathbf{u}$, $\nabla \mathbf{v}$, $\nabla \cdot \mathbf{q}$, and ∇T in Equations 9.70–9.74. From now on, we refer the temperature in CM as continuum temperature, to distinguish it from atomistic temperature in MD.

In FE analysis, relate the displacement and temperature fields with their nodal values as

$$\begin{aligned} u_i &= N_{i\alpha} U_\alpha, \qquad e_{ij} = B_{ij\alpha} U_\alpha \\ T &= N_\xi T_\xi, \qquad T_{,k} = C_{k\xi} T_\xi \end{aligned} \tag{9.76}$$

Then it is straightforward to obtain the dynamic FE equations as

$$\mathbf{M}\ddot{\mathbf{U}} + \mathbf{K}\mathbf{U} = \mathbf{P}\mathbf{T} + \mathbf{F}^1 + \mathbf{F}^2 \tag{9.77}$$

$$\mathbf{G}\dot{\mathbf{T}} + \mathbf{H}\mathbf{T} + T^{ref}\mathbf{P}^{\mathrm{T}}\dot{\mathbf{U}} = -\mathbf{Q}^1 + \mathbf{Q}^2 \tag{9.78}$$

where

$$M_{\alpha\beta} \equiv \int_v \rho^o N_{i\alpha} N_{i\beta}\, dv = M_{\beta\alpha} \tag{9.79}$$

$$K_{\alpha\beta} \equiv \int_v A_{ijkl} B_{kl\alpha} B_{ij\beta}\, dv = K_{\beta\alpha} \tag{9.80}$$

$$P_{\beta\eta} \equiv \int_v \beta_{ij} N_\eta B_{ij\beta}\, dv \tag{9.81}$$

$$G_{\xi\eta} \equiv \int_v \rho^o \gamma N_\xi N_\eta\, dv = G_{\eta\xi} \tag{9.82}$$

$$H_{\xi\eta} \equiv \int_v \kappa_{kl} C_{k\xi} C_{l\eta}\, dv = H_{\eta\xi} \tag{9.83}$$

$$F_\beta^1 \equiv \int_{s_\sigma} \hat{\sigma}_i N_{i\beta}\, ds \tag{9.84}$$

$$F_\beta^2 \equiv \int_{v} \rho^o \varphi_i N_{i\beta}\, dv \tag{9.85}$$

$$Q_\eta^1 \equiv \int_{s_q} \hat{q} N_\eta\, ds \tag{9.86}$$

and

$$Q_\eta^2 \equiv \int_{v} \rho^o h N_\eta\, dv \tag{9.87}$$

Now it is noticed that, if one has the values of **A**, **β**, **κ**, and γ, then one may proceed to solve Equations 9.77 and 9.78 for $\mathbf{U}(\mathbf{X}, t)$ and $T(\mathbf{X}, t)$. If, further, we can obtain those material properties from MD simulation, then this approach is named as a *sequential multiscale modeling.*

9.5.2 Material constants from MD simulation

In Voigt's convention, one may rewrite Equations 9.72 and 9.74 as

$$\begin{Bmatrix} \sigma_{11} \\ \sigma_{22} \\ \sigma_{33} \\ \sigma_{23} \\ \sigma_{31} \\ \sigma_{12} \end{Bmatrix} = -\begin{Bmatrix} \beta_{11} \\ \beta_{22} \\ \beta_{33} \\ \beta_{23} \\ \beta_{31} \\ \beta_{12} \end{Bmatrix}(T - T^{ref}) + \begin{bmatrix} A_{1111} & A_{1122} & A_{1133} & A_{1123} & A_{1131} & A_{1112} \\ A_{1122} & A_{2222} & A_{2233} & A_{2223} & A_{2231} & A_{2212} \\ A_{1133} & A_{2233} & A_{3333} & A_{3323} & A_{3331} & A_{3312} \\ A_{1123} & A_{2223} & A_{3323} & A_{2323} & A_{2331} & A_{2312} \\ A_{1131} & A_{2231} & A_{3331} & A_{2331} & A_{3131} & A_{3112} \\ A_{1121} & A_{2212} & A_{3312} & A_{2312} & A_{3112} & A_{1212} \end{bmatrix} \begin{Bmatrix} e_{11} \\ e_{22} \\ e_{33} \\ \gamma_{23} \\ \gamma_{31} \\ \gamma_{12} \end{Bmatrix} \tag{9.88}$$

and

$$\begin{Bmatrix} q_1 \\ q_2 \\ q_3 \end{Bmatrix} = -\begin{bmatrix} \kappa_{11} & \kappa_{12} & \kappa_{13} \\ \kappa_{21} & \kappa_{22} & \kappa_{23} \\ \kappa_{31} & \kappa_{32} & \kappa_{33} \end{bmatrix} \begin{Bmatrix} T_{,1} \\ T_{,2} \\ T_{,3} \end{Bmatrix} \tag{9.89}$$

where $\gamma_{23} = 2e_{23}$, $\gamma_{31} = 2e_{31}$, and $\gamma_{12} = 2e_{12}$.

9.5.2.1 Elastic constants

In simple strain problem, the deformation can be expressed as

$$\begin{Bmatrix} x \\ y \\ z \end{Bmatrix} = \begin{bmatrix} 1+e_1 & \gamma_{12} & \gamma_{13} \\ \gamma_{21} & 1+e_2 & \gamma_{23} \\ \gamma_{31} & \gamma_{32} & 1+e_3 \end{bmatrix} \begin{Bmatrix} X \\ Y \\ Z \end{Bmatrix} \quad \text{or} \quad x_k = F_{kK} X_K \tag{9.90}$$

which implies the Green deformation tensor, and the Lagrangian strain tensor can be calculated as

$$C_{KL} = F_{kK} F_{kL}, \qquad E_{KL} = (C_{KL} - \delta_{KL})/2 \tag{9.91}$$

One may obtain the Lagrangian strains and their small-strain counterparts as follows

$$E_{11} = 0.5\left\{(1+e_1)^2 + \gamma_{21}^2 + \gamma_{31}^2 - 1\right\} \approx e_1 \tag{9.92}$$

$$E_{22} = 0.5\left\{(1+e_2)^2 + \gamma_{12}^2 + \gamma_{32}^2 - 1\right\} \approx e_2 \tag{9.93}$$

$$E_{33} = 0.5\left\{(1+e_3)^2 + \gamma_{13}^2 + \gamma_{23}^2 - 1\right\} \approx e_3 \tag{9.94}$$

$$\begin{aligned} E_{12} = E_{21} &= 0.5\{(1+e_1)\gamma_{12} + (1+e_2)\gamma_{21} + \gamma_{31}\gamma_{32}\} \\ &\approx (\gamma_{12} + \gamma_{21})/2 \end{aligned} \tag{9.95}$$

$$\begin{aligned} E_{13} = E_{31} &= 0.5\{(1+e_1)\gamma_{13} + (1+e_3)\gamma_{31} + \gamma_{21}\gamma_{23}\} \\ &\approx (\gamma_{13} + \gamma_{31})/2 \end{aligned} \tag{9.96}$$

and

$$\begin{aligned} E_{23} = E_{32} &= 0.5\{(1+e_2)\gamma_{23} + (1+e_3)\gamma_{32} + \gamma_{12}\gamma_{13}\} \\ &\approx (\gamma_{23} + \gamma_{32})/2 \end{aligned} \tag{9.97}$$

In MD simulation, let the whole specimen have N atoms. There are six independent simple strains that can be created in the following six cases:

1. Simple tension along the x-axis

$$r_1^i \leftarrow (1+e_1) r_1^i, \qquad i \in [1,2,3,\ldots,N] \tag{9.98}$$

where $\mathbf{r}^i$ is the position vector of *atom i*; r_2^i and r_3^i remain unchanged.

2. Simple tension along the y-axis

$$r_2^i \leftarrow (1+e_2)r_2^i \tag{9.99}$$

3. Simple tension along the z-axis

$$r_3^i \leftarrow (1+e_3)r_3^i \tag{9.100}$$

4. Simple shear on the $x-y$ plane

$$r_1^i \leftarrow r_1^i + \gamma_{12}r_2^i, \qquad r_2^i \leftarrow r_2^i + \gamma_{21}r_1^i \tag{9.101}$$

with r_3^i unchanged.

5. Simple shear on the $y-z$ plane

$$r_2^i \leftarrow r_2^i + \gamma_{23}r_3^i, \qquad r_3^i \leftarrow r_3^i + \gamma_{32}r_2^i \tag{9.102}$$

with r_1^i unchanged.

6. Simple shear on the $z-x$ plane

$$r_1^i \leftarrow r_1^i + \gamma_{13}r_3^i, \quad r_3^i \leftarrow r_3^i + \gamma_{31}r_1^i \tag{9.103}$$

with r_2^i unchanged.

Because these six cases are static cases, the virial stress tensor is reduced to

$$\mathrm{S} = -\frac{1}{V}\sum_{i=1}^{N} \mathrm{r}^i \otimes \mathrm{f}^i \tag{9.104}$$

where V is the volume of the whole specimen. For each case, we obtain six components of the virial stress, from which the elastic constants are deduced as follows:

Case 1:

$$\{A_{1111}, A_{1122}, A_{1133}, A_{1123}, A_{1131}, A_{1112}\} = \{S_{11}, S_{22}, S_{33}, S_{23}, S_{31}, S_{12}\}/e_1 \tag{9.105}$$

Case 2:

$$\{A_{1122}, A_{2222}, A_{2233}, A_{2223}, A_{2231}, A_{2212}\} = \{S_{11}, S_{22}, S_{33}, S_{23}, S_{31}, S_{12}\}/e_2 \tag{9.106}$$

Case 3:

$$\{A_{1133}, A_{2233}, A_{3333}, A_{3323}, A_{3331}, A_{3312}\} = \{S_{11}, S_{22}, S_{33}, S_{23}, S_{31}, S_{12}\}/e_3 \tag{9.107}$$

Case 4:

$$\{A_{1112}, A_{2212}, A_{3312}, A_{2312}, A_{3112}, A_{1212}\} = \{S_{11}, S_{22}, S_{33}, S_{23}, S_{31}, S_{12}\}/\gamma_{12} \tag{9.108}$$

Case 5:

$$\{A_{1123}, A_{2223}, A_{3323}, A_{2323}, A_{2331}, A_{2312}\} = \{S_{11}, S_{22}, S_{33}, S_{23}, S_{31}, S_{12}\}/\gamma_{23} \tag{9.109}$$

Case 6:

$$\{A_{1131}, A_{2231}, A_{3331}, A_{2331}, A_{3131}, A_{3112}\} = \{S_{11}, S_{22}, S_{33}, S_{23}, S_{31}, S_{12}\}/\gamma_{31} \tag{9.110}$$

In general, the stress–strain relation is nonlinear, and we focus our attention to the small strain theory, therefore, to evaluate the elastic constants; those specified strains, namely, e_1, e_2, e_3, γ_{12}, γ_{23}, and γ_{31}, should be in the linear elastic range.

9.5.2.2 Thermal conductivity

Now suppose we have several atomic groups lined up in series, and let the first group be subject to Nosé–Hoover thermostat at atomistic temperature T_H and let the last group be subject to Nosé–Hoover thermostat at atomistic temperature T_L. Those groups in between do not have Nosé–Hoover thermostat (Figure 9.2). After the system reaches steady state, one may obtain the thermal energies E_H and E_L as

$$E_H \equiv \left\{\frac{1}{2}Q_H\chi_H^2 + \frac{Q_H}{\tau_H^2}\int_0^t \chi_H(s)\,ds\right\} \tag{9.111}$$

and

$$E_L \equiv \left\{\frac{1}{2}Q_L\chi_L^2 + \frac{Q_L}{\tau_L^2}\int_0^t \chi_L(s)\,ds\right\} \tag{9.112}$$

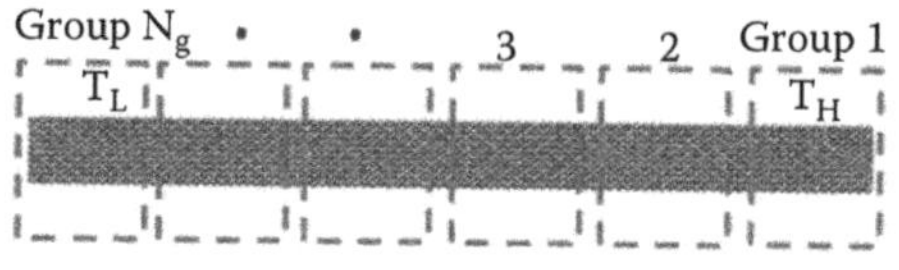

Figure 9.2 **(See color insert.)** Depiction of the groups may or may not be controlled by Nosé–Hoover thermostat.

Because in MD simulation, atomistic temperature is a statistical quantity and involves a significant amount of noise, one may plot the thermal energies E_H and E_L as functions of time, which can be approximated by two straight lines. Numerically, one should obtain

$$\frac{\Delta E_H}{\Delta t} \approx \dot{E}_H = N_H^{dof} k_B \chi_H T_H, \qquad \frac{\Delta E_L}{\Delta t} \approx \dot{E}_L = N_L^{dof} k_B \chi_L T_L \tag{9.113}$$

where $\dot{E}_H$ and $\dot{E}_L$ should be approximately equal in magnitude but opposite in sign. The heat flux in magnitude is equal to $\dot{E}_H$ and $\dot{E}_L$ divided by a cross-sectional area; the temperature gradient is equal to $\Delta T/\Delta L$, where $\Delta T = T_H - T_L$ and ΔL is the distance between the centers of mass of the two atomic groups controlled by Nosé–Hoover thermostats. Following this way, one may find the numerical values of thermal conductivities.

9.5.2.3 Specific heat and thermal expansion coefficients

We now recall two constitutive equations in small strain thermoelasticity:

$$\rho^o \dot{e} = \rho^o \gamma \dot{T} + A_{ijkl} e_{ij} \dot{e}_{kl} \tag{9.71)*$$

and

$$\sigma_{ij} = -\beta_{ij} T + A_{ijkl} e_{kl} \tag{9.72)*$$

where γ and $\boldsymbol{\beta}$ are the specific heat and thermal expansion coefficients, respectively. In MD simulation, one may consider a group of atoms in relaxed and idealized state, i.e., absolute zero temperature, vanishing interatomic forces, and vanishing virial stresses. Then imagine that this group of atoms is put into a rigid box and prohibited from moving out of the box. In this situation, the strain and the strain rate are zero, and Equations 9.71 and 9.72 are rewritten as

$$\dot{e} = \gamma \dot{T}, \qquad S_{ij} = -\beta_{ij} T \tag{9.114}$$

Here, we assume that the Cauchy stress tensor in CM is equivalent to the virial stress tensor in MD simulation (Subramaniyan and Sun, 2008). This may be considered as an approximation. But without it, it is very difficult, even impossible, to bridge the gap between atoms and genuine continuum (GC).

Now let the atomistic temperature of atoms in the box be raised to a specified temperature T by Nosé–Hoover thermostat, and after steady state is reached, one may calculate virial stress S_{ij} and internal energy density e (sum of potential energy and thermal energy divided by total mass of atoms). It results in

$$\beta_{ij} = -S_{ij}/T, \qquad \gamma = \frac{e}{T} \tag{9.115}$$

9.6 CONCURRENT MULTISCALE MODELING FROM MD TO CM

9.6.1 One specimen, two regions

The basic strategy for concurrent modeling may be described as follows: divide the entire solution domain into noncritical far field and several critical subregions, where stress concentrations, crack initiation and propagation, dislocations, and other critical physical phenomena may occur. We have successfully modeled the critical regions (or atomic regions) by MD simulation and the ABC region by CGMD simulation (Li et al., 2012; Li and Lee, 2014a, 2014b) in a single theoretical framework. One may extend this approach by dividing the entire solution domain into three parts: atomic region, ABC region, and GC region, modeled by CM.

To illustrate the idea, one may simply divide the entire solution domain into two regions: atomic region and GC region, and construct a concurrent multiscale theory and a corresponding computational methodology (Lee et al., 2017; Lee and Robert, 2017). As mentioned in Section 9.2, the atomic region is further divided into N_g groups; group g has n_g atoms; total number of atoms in the atomic region is N. Let the first M_g groups have Nosé–Hoover thermostats, $N_g > M_g$. For example, Figure 9.3b shows $N_g = 22$ and $M_g = 17$. These M_g groups are linked to the first M_g nodes in the GC region (cf. Figure 9.3a). It is noticed that there are $N_g - M_g = 5$ groups not controlled by thermostats because they are not linked to any node in the GC region. If one had viewed the N_g groups as nodes and inserted into the FE mesh (cf. Figure 9.3a) of GC, then an apparent FE mesh would have been created, as shown in Figure 9.3c. The basic idea is that there is an *interface* between the atomic region and the continuum region. In the *interface*, each node corresponds to a group of atoms, not just a single atom (cf. Figure 9.3a and 9.3b).

In this work, we make two major assumptions:

Assumption 1:

Each node in the interface is anchored at the mass center of its corresponding group.

Assumption 2:

The sum of heat fluxes into the node and its corresponding group is zero.

Now, the mathematical formulation proceeds as follows: First, for the atomic region, we recall the relevant governing equations as (cf. Chapter 8)

$$m^i\dot{\mathbf{v}}^i = \mathbf{f}^i + \boldsymbol{\varphi}^i + \mathbf{f}^i_{\text{int}} - \chi_g m^i\tilde{\mathbf{v}}^i \tag{9.116}$$

$$\dot{\chi}_g = \frac{1}{\tau_g^2 T_g^c}\left(T_g^a - T_g^c\right) \tag{9.117}$$

$$T_g^a = \frac{1}{N_g^{dof} k_B}\sum_{i=1}^{n_g} m^i\tilde{\mathbf{v}}^i\cdot\tilde{\mathbf{v}}^i \tag{9.118}$$

and

$$\dot{E}_g = N_g^{dof} k_B \chi_g T_g^a \tag{9.119}$$

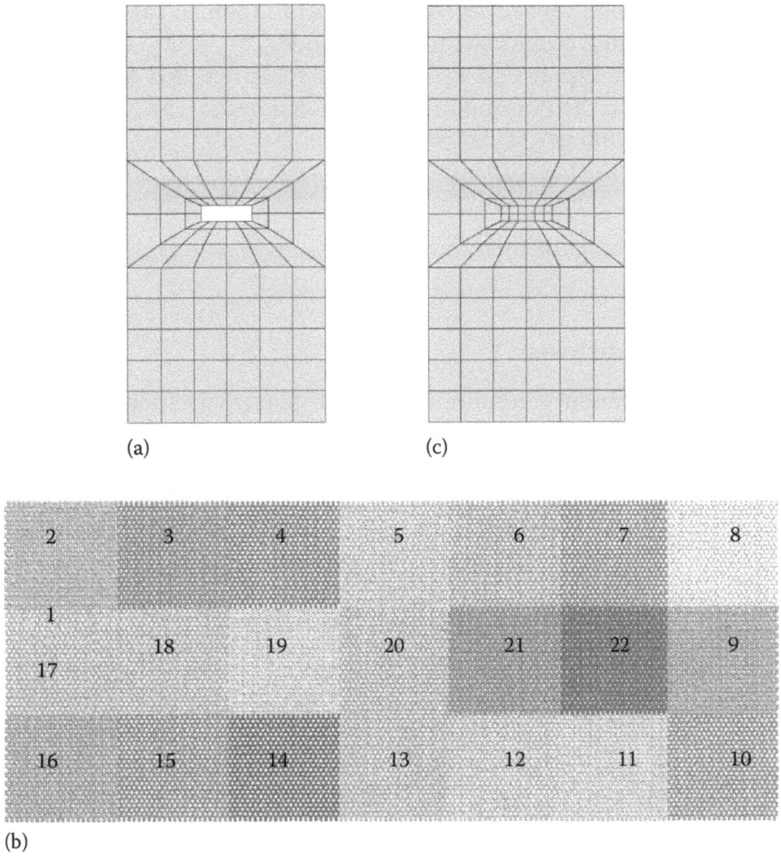

Figure 9.3 **(See color insert.)** (a) Finite element mesh in the genuine continuum region, (b) groups of atoms in the atomic region, and (c) apparent FE mesh.

However, we replace T_g by T_g^a to emphasize that it is the atomistic temperature of group g calculated through the Nosé–Hoover thermal velocity $\tilde{\mathbf{v}}^i$ in the atomic region; we replace T_g^o by T_g^c to indicate the target temperature of group g is no longer a constant, but the nodal temperature of the corresponding *node* g in the GC region, $g \in [1, 2, 3, \ldots, M_g]$. Especially, it is noticed that we add an interactive term $\mathbf{f}_{\text{int}}^i$ into Equation 9.116 to indicate the interaction between atoms in group g and its corresponding *node* g in the GC region. We employ the velocity Verlet method to solve Equations 9.116 and 9.117. Detailed formula will be derived later. The time step used in the atomic region is denoted by Δt^a.

For the GC region, we recall and rewrite the FE equations as

$$\mathbf{M}\ddot{\mathbf{U}} + \mathbf{K}\mathbf{U} = \mathbf{P}\mathbf{T} + \mathbf{F}^1 + \mathbf{F}^2 + \mathbf{F}_{\text{int}} \qquad (9.77)^*$$

and

$$\mathbf{G}\dot{\mathbf{T}} + \mathbf{H}\mathbf{T} + T^{ref}\mathbf{P}^T\dot{\mathbf{U}} = -\mathbf{Q}^1 + \mathbf{Q}^2 + \mathbf{Q}_{\text{int}} \qquad (9.78)^*$$

It is seen that, for those nodes anchored in groups of atoms in the atomic region, we add $\mathbf{F}_{\text{int}}$ and $\mathbf{Q}_{\text{int}}$ to the governing equations; for other nodes, $\mathbf{F}_{\text{int}} = \mathbf{Q}_{\text{int}} = 0$. We employ central difference method to solve Equations 9.77* and 9.78*. A detailed formula will be derived later. The time step used in the continuum region is denoted by Δt^c. The ratio of time steps N^{time}, which should be an integer, is defined as

$$N^{time} \equiv \Delta t^c \big/ \Delta t^a \quad \text{or} \quad \Delta t^c = N^{time} \Delta t^a \tag{9.120}$$

9.6.2 Interfacial conditions

One needs to impose a boundary condition, either displacement specified or force specified, at each component of each node, similarly, impose boundary condition, either temperature specified or heat flow specified, at each node, at the outer boundary of the GC region, but not at the inner boundary, which is the interface.

At each node/group in the interface, one should have the following two conditions:

$$\sum_{i \in g} \mathbf{f}^i_{\text{int}} + \mathbf{F}^g_{\text{int}} = 0 \tag{9.121}$$

and

$$Q^g_{\text{int}} = \dot{E}_g = N^{dof}_g k_B \chi_g T^a_g \tag{9.122}$$

The first condition, Equation 9.121, simply means the sum of interactive forces of the node/group pair should be vanishing. Recall that $\dot{E}_g$ is the flow of energy per unit time out of group g due to the action of Nosé–Hoover thermostat. Therefore, the second condition simply says that $\dot{E}_g$ should be the inward flow of energy per unit time to the corresponding *node g*.

We now rewrite Equation 9.116 and Equation 9.77*, respectively, as

$$m^i \ddot{\mathbf{u}}^i = m^i \dot{\mathbf{v}}^i = \mathbf{f}^i + \boldsymbol{\varphi}^i + \mathbf{f}^i_{\text{int}} - \chi_g m^i \tilde{\mathbf{v}}^i \equiv \overline{\mathbf{f}}^i + \mathbf{f}^i_{\text{int}} \tag{9.123}$$

and

$$\mathbf{M}\ddot{\mathbf{U}} = -\mathbf{K}\mathbf{U} + \mathbf{P}\mathbf{T} + \mathbf{F}^1 + \mathbf{F}^2 + \mathbf{F}_{\text{int}} \equiv \overline{\mathbf{F}} + \mathbf{F}_{\text{int}} \tag{9.124}$$

where $\mathbf{u}^i$ denotes the position vector of *atom i*. In this work, lumped-mass system is adopted; i.e., the mass matrix $\mathbf{M}$ is diagonalized, and hence, nodal mass is well defined. Similarly, the $\mathbf{G}$ matrix in Equation 9.98* is also diagonalized. Equation 9.121 leads to

$$M^g \ddot{\mathbf{U}}^g + \sum_{i \in g} m^i \ddot{\mathbf{u}}^i = \overline{\mathbf{F}}^g + \sum_{i \in g} \overline{\mathbf{f}}^i \tag{9.125}$$

Assumption 1 that we made says that *node g* is anchored at the mass center of group *g*. It implies that

$$\ddot{\mathbf{U}}^g = \frac{\sum_{i\in g} m^i \ddot{\mathbf{u}}^i}{\sum_{i\in g} m^i} \equiv \frac{\sum_{i\in g} m^i \ddot{\mathbf{u}}^i}{m^g} \equiv \ddot{\mathbf{u}}^g_{avg} \tag{9.126}$$

To ensure Equation 9.126, one needs

$$\ddot{\mathbf{u}}^i = \frac{\bar{\mathbf{f}}^i}{m^i} + \ddot{\mathbf{U}}^g - \frac{\sum_{i\in g} \bar{\mathbf{f}}^i}{m^g} \tag{9.127}$$

Substituting Equation 9.126 into Equation 9.125 results in

$$\ddot{\mathbf{U}}^g = \frac{1}{M^g + m^g}\left\{\bar{\mathbf{F}}^g + \sum_{i\in g} \bar{\mathbf{f}}^i\right\} \tag{9.128}$$

9.6.3 Multiple time scale algorithm

There are two regions, atomic and continuum, and therefore, there are two time scales, n and m. Then time t equals

$$t(n,m) = n\Delta t^c + m\Delta t^a \tag{9.129}$$

It is seen that $t(n+1, 0) = t(n, N^{time})$. First, suppose at $t(n, 0)$ we know $\mathbf{u}^{n,0}$, $\mathbf{a}^{n,0} \equiv \ddot{\mathbf{u}}^{n,0}$, $\chi^{n,0}$, $\mathbf{v}^{n,-\frac{1}{2}} = \dot{\mathbf{u}}^{n,-\frac{1}{2}}$, $\mathbf{U}^{n,0}$, $\mathbf{V}^{n,0} \equiv \dot{\mathbf{U}}^{n,0}$, $\mathbf{A}^{n,0} \equiv \ddot{\mathbf{U}}^{n,0}$, $\mathbf{T}^{n,0}$, and $\dot{\mathbf{T}}^{n,0}$. The numerical procedures to solve Equations 9.116, 9.117, 9.77*, and 9.78* are listed in the following.

Step 1:

Calculate $\mathbf{U}^{n,m}$ and $\mathbf{T}^{n,m}$ ($m = 1, 2, 3, \ldots, N^{time}$)

$$\mathbf{U}^{n,m} = \mathbf{U}^{n,0} + m\Delta t^a \mathbf{V}^{n,0} + 0.5(m\Delta t^a)^2 \mathbf{A}^{n,0} \tag{9.130}$$

$$\mathbf{T}^{n,m} = \mathbf{T}^{n,0} + m\Delta t^a \dot{\mathbf{T}}^{n,0} \tag{9.131}$$

Impose boundary conditions at $t(n, m)$ on $\mathbf{U}^{n,m}$ and $\mathbf{T}^{n,m}$. Notice that $\mathbf{U}^{n+1,0} = \mathbf{U}^{n,N^{time}}$ and $\mathbf{T}^{n+1,0} = \mathbf{T}^{n,N^{time}}$.

Step 2:

Based on the velocity Verlet method, let the velocity and position of each atom be updated as

$$\mathbf{v}^{n,m+1/2} = \mathbf{v}^{n,m-1/2} + \Delta t^a \mathbf{a}^{n,m} \tag{9.132}$$

and

$$\mathbf{u}^{n,m+1} = \mathbf{u}^{n,m} + \Delta t^a \mathbf{v}^{n,m+1/2} \tag{9.133}$$

For each node/group in the interface, identify the position of *node* g as $\mathbf{R}_g = \mathbf{R}_g^o + \mathbf{U}_g$ ($\mathbf{R}_g^o$ denotes the position of *node* g in the reference state and $\mathbf{U}_g$ is the displacement vector obtained from Step 1) and calculate the position of the mass center of group g as

$$\mathbf{u}_{avg}^g = \frac{\sum_{i \in g} m^i \mathbf{u}^i}{m^g} \tag{9.134}$$

Then let every atom in group g move according to

$$\mathbf{u}^i \leftarrow \mathbf{u}^i + \mathbf{R}_g - \mathbf{u}_{avg}^g \tag{9.135}$$

This simply means that Assumption 1 is enforced on the atomic positions at all time $t(n, m)$.

Step 3:

Calculate T_g^a according to Equation 9.118 based on $\mathbf{v}^{n,m+1/2}$ (cf. Equation 9.132). Notice that $\tilde{\mathbf{v}}^{n,m+1/2}$ can be obtained from $\mathbf{v}^{n,m+1/2}$ and $\mathbf{u}^{n,m+1}$. Then calculate

$$\dot{\chi}_g^{n,m+1/2} = \frac{1}{\tau_g^2 T_g^{c(n,m+1)}} \left(T_g^a - T_g^{c(n,m+1)} \right) \tag{9.136}$$

$$\chi_g^{n,m+1} = \chi_g^{n,m} + \Delta t^a \dot{\chi}_g^{n,m+1/2} \tag{9.137}$$

The force acting on all atoms, without counting the interactive force between *node* g and group g, can now be calculated as

$$\overline{\mathbf{f}}^{i(n,m+1)} = \mathbf{f}^{i(n,m+1)} + \boldsymbol{\varphi}^{i(n,m+1)} - \chi_g^{n,m+1} m^i \tilde{\mathbf{v}}^{i(n,m+1/2)} \tag{9.138}$$

Then the acceleration is updated as

$$\mathbf{a}^{i(n,m+1)} = \overline{\mathbf{f}}^{i(n,m+1)} / m^i \tag{9.139}$$

with the understanding that at $m = N^{time}$, a special treatment (cf. Equations 9.126–9.128) needs to be implemented. Repeat the cycle of Steps 2 and 3 for N^{time} times.

Step 4:

After m reaches N^{time}, calculate

$$\mathbf{V}^{n+1/2,0} = \mathbf{V}^{n,0} + \Delta t^c \mathbf{A}^{n,0} \tag{9.140}$$

$$\overline{\mathbf{F}}^{n+1,0} = -\mathbf{K}\mathbf{U}^{n+1,0} + \mathbf{P}\mathbf{T}^{n+1,0} + \mathbf{F}^{1(n+1,0)} + \mathbf{F}^{2(n+1,0)} \tag{9.141}$$

and

$$\mathbf{A}^{n+1.0} \equiv \ddot{\mathbf{U}}^{n+1,0} = \overline{\mathbf{F}}^{n+1,0}/\mathbf{M} \tag{9.142}$$

Here, we notice that (1) because we adopt a lumped-mass system, we can express $\ddot{\mathbf{U}}^{n+1,0} = \overline{\mathbf{F}}^{n+1,0}/\mathbf{M}$ and (2) only for the nodes in the interface, we have special treatment as follows.

$$\mathbf{A}^{n+1,0} \equiv \ddot{\mathbf{U}}^{g(n+1,0)} = \frac{1}{M^g + m^g}\left\{\overline{\mathbf{F}}^{g(n+1,0)} + \sum_{i\in g} \overline{\mathbf{f}}^{i(n+1,0)}\right\} \tag{9.143}$$

$$\mathbf{a}^{n+1,0} = \ddot{\mathbf{u}}^{i(n+1,0)} = \frac{\overline{\mathbf{f}}^{i(n+1,0)}}{m^i} + \ddot{\mathbf{U}}^{g(n+1,0)} - \frac{\sum_{i\in g} \overline{\mathbf{f}}^{i(n+1,0)}}{m^g} \tag{9.144}$$

Now, the velocity vector in GC region is updated as

$$\mathbf{V}^{n+1,0} = \mathbf{V}^{n+1/\frac{1}{2}} + \frac{1}{2}\Delta t^c \mathbf{A}^{n+1,0} \tag{9.145}$$

Step 5:
Following Equation 9.78*, calculate

$$\mathbf{Q}^{n+1,0} \equiv -\mathbf{H}\mathbf{T}^{n+1,0} - T^{ref}\mathbf{P}^T\mathbf{V}^{n+1,0} - \mathbf{Q}^{1(n+1,0)} + \mathbf{Q}^{2(n+1,0)} + \mathbf{Q}_{\text{int}}^{g(n+1,0)} \tag{9.146}$$

$$\dot{\mathbf{T}}^{n+1,0} = \mathbf{Q}^{n+1,0}/\mathbf{G} \tag{9.147}$$

where

$$\mathbf{Q}_{\text{int}}^{g(n+1,0)} = N_g^{dof} k_B \chi_g^{n+1,0} \mathbf{T}_g^{a(n+1,0)} \tag{9.122*}$$

It is seen that, after going from Step 1 to Step 5, all the unknown variables are updated as

$$\mathbf{U}^{n,0}, \mathbf{V}^{n,0}, \mathbf{T}^{n,0}, \mathbf{A}^{n,0},\ \mathbf{T}^{n,0} \rightarrow \mathbf{U}^{n+1,0}, \mathbf{V}^{n+1,0}, \mathbf{T}^{n+1,0}, \mathbf{A}^{n+1,0}, \mathbf{T}^{n+1,0} \tag{9.148}$$

and

$$\mathbf{u}^{n,m}, \mathbf{v}^{n,m-\frac{1}{2}}, \mathbf{a}^{n,m}, \boldsymbol{\chi}^{n,m} \rightarrow \mathbf{u}^{n,m+1}, \mathbf{v}^{n,m+\frac{1}{2}}, \mathbf{a}^{n,m+1}, \boldsymbol{\chi}^{n,m+1} \tag{9.149}$$

9.6.4 Sample problems and numerical results

For illustrative purposes, the material we studied in this work is graphene. Graphene belongs to a broad family of 2D materials, characterized by strong covalent in-plane bonds and weak interlayer van der Waals interactions, which give them a layered structure. One of the fascinating properties of 2D crystals is their high stretchability and the possibility to use external strain to manipulate their optical and electronic properties, coined as strain engineering. For interested readers, a review article by Roldan et al. (2015) is recommended.

9.6.4.1 Material constants obtained from MD simulations

The interatomic potential for graphene used in this work is Tersoff potential (Tersoff, 1988, 1989), which is a three-body potential. In the continuum region, we model the material as a 2D thermoelastic solid with its material properties obtained from MD simulation (cf. Section 9.5). Since graphene is anisotropic, we identify x_1*-axis* ≡ *x-axis* as the one perpendicular to the *armchair* edge and x_2*-axis* ≡ *y-axis* as the one perpendicular to the *zigzag* edge. The material properties may be summarized as follows:

$$\begin{Bmatrix} \sigma_{11} \\ \sigma_{22} \\ \sigma_{12} \end{Bmatrix} = -\begin{Bmatrix} 0.1723\times10^{-6} \\ 0.1710\times10^{-6} \\ 0 \end{Bmatrix}(T - T^{ref}) + \begin{bmatrix} 0.04496 & -0.00792 & 0 \\ -0.00792 & 0.04624 & 0 \\ 0 & 0 & 0.02646 \end{bmatrix}\begin{Bmatrix} e_{11} \\ e_{22} \\ \gamma_{12} \end{Bmatrix} \quad (9.150)$$

$$\begin{Bmatrix} q_1 \\ q_2 \end{Bmatrix} = -\begin{bmatrix} 0.6238\times10^{-7} & 0 \\ 0 & 0.4936\times10^{-7} \end{bmatrix}\begin{Bmatrix} T_{,1} \\ T_{,2} \end{Bmatrix} \quad (9.151)$$

$$\gamma = 0.2551\times10^{-9} \quad (9.152)$$

In this work, we use atomic units; i.e., the dimensions and units of stress, temperature, heat expansion coefficient, heat flux, thermal conductivity, and specific heat are

$$\begin{aligned} &[\sigma] = \frac{Hartree}{Bohr^3}, \quad [T] = Kelvin, \quad [\beta] = \frac{Hartree}{Bohr^3\ Kelvin} \\ &[q] = \frac{Hartree}{\tau\ Bohr^2}, \quad [\kappa] = \frac{Hartree}{\tau\ Bohr\ Kelvin}, \quad [\gamma] = \frac{Hartree}{m_e\ Kelvin} \end{aligned} \quad (9.153)$$

where

$$\begin{aligned} &Hartree = 4.3597482\times10^{-18}\ \text{Joule} \\ &Bohr = 5.29177249\times10^{-11}\ \text{meter} \\ &\tau = 2.418884326555\times10^{-17}\ \text{second} \\ &m_e = 9.10938291\times10^{-31}\ \text{kg} \end{aligned} \quad (9.154)$$

9.6.4.2 Case studies

The entire specimen is divided into two regions: (1) the FE mesh of the continuum region has 138 nodes and 108 2D four-noded plane elements, shown in Figure 9.3a, and (2) the atomic region is further divided into 22 groups, as shown in Figure 9.3b. There are 528 atoms in group 1 and in group 17 and 1056 atoms in each of the other 20 groups. There are 22,176 atoms in the atomic region and, equivalently, more than 1.36 million atoms in the continuum region. One may take a close look and find that group i and its corresponding *node i* (i = 1, 2, 3,…., 17) form 17 node/group pairs. For the purpose of presentation, one may consider, and view later, the FE mesh of the entire specimen has 143 nodes (143 = 138 + 5 groups) and 120 elements [120 = 108 + 2(5 + 1)], as shown in Figure 9.3c. In this view, one may say this specimen had a crack and the crack tip is located at the centroid of group 18. In this work, the time steps are set at $\Delta t^a = 20\ \tau = 0.4838$ femtoseconds, and $\Delta t^c = 20\ \Delta t^a$. It is emphasized that from the concurrent multiscale modeling, one can obtain the positions of the centroids, the atomistic temperatures, and the virial stress tensors of the 22 groups. Therefore, later in the *Tecplot,* one may see the graphic representations of a mixture of Cauchy stress tensors and continuum temperatures at 121 nodes (121 = 138 – 17) together with virial stress tensors and atomistic temperatures at 22 groups.

Several cases have been analyzed, and the results are presented as follows.

Case 1:
The boundary conditions are specified as
Along the top edge,

$$T = T^{ref} + 100\ K$$
$$u_y = \begin{cases} 20\dfrac{t}{t^r}\,(\text{Bohr})\,,\ t \le t^r \\ 20 \quad (\text{Bohr})\,,\ t \ge t^r \end{cases} \tag{9.155}$$

Along the bottom edge,

$$T = T^{ref} - 100\ K$$
$$u_y = \begin{cases} -20\dfrac{t}{t^r}\,(\text{Bohr})\,,\ t \le t^r \\ -20 \quad (\text{Bohr})\,,\ t \ge t^r \end{cases} \tag{9.156}$$

where $t^r = 1000\ \Delta t^c = 20{,}000\ \Delta t^a$ and the reference temperature is set at $T^{ref} = 300\ K$.

In this work, $\Delta T \equiv T - T^{ref}$ is named as temperature variation; there are temperature gradient and elongation along the y-axis. It seems that this is a static problem. But in principle, there is no static problem in multiscale modeling, and besides, there is no damping built in the theory, except we put in artificial damping in the relaxation stage so that we have an idealized starting point, i.e., nearly zero atomic forces, atomistic temperature, and virial stresses, to begin with.

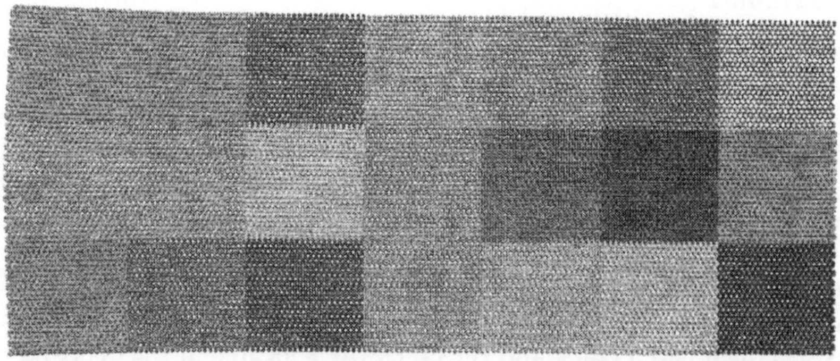

Figure 9.4 **(See color insert.)** Visual molecular dynamics (VMD) plot at $t = 2000\ \Delta t^c$.

The positions of 22,176 atoms are shown in the VMD plot (Humphrey et al., 1996) (Figure 9.4). Because, relatively speaking, the applied loading (in this case, 20 Bohr) is small, there is no crack opening in the atomic region, and therefore, only one VMD plot is shown at $t = 2000\ \Delta t^c$. The displacements (Bohr), continuum and atomistic temperature variations (Kelvin), Cauchy stress, and virial stress (Hartree/Bohr^3) are shown in the *Tecplot*, Figure 9.5. The matching between atomistic temperatures and continuum temperatures of the 17 node/group pairs is surprisingly good. The virial stress S_{22} at $t = 2000\ \Delta t^c$ for group i, $i \in [18, 19, 20, 21, 22]$ (cf. Figure 9.3b) is shown in Figure 9.6. In fact, the centroid of group 18 is the crack tip. Figure 9.6 shows the stress distribution ahead of the crack tip. It is seen that there is no stress singularity, no stress concentration either. In short, the concepts behind *linear elastic fracture mechanics* (LEFM) and MD are totally different in this respect.

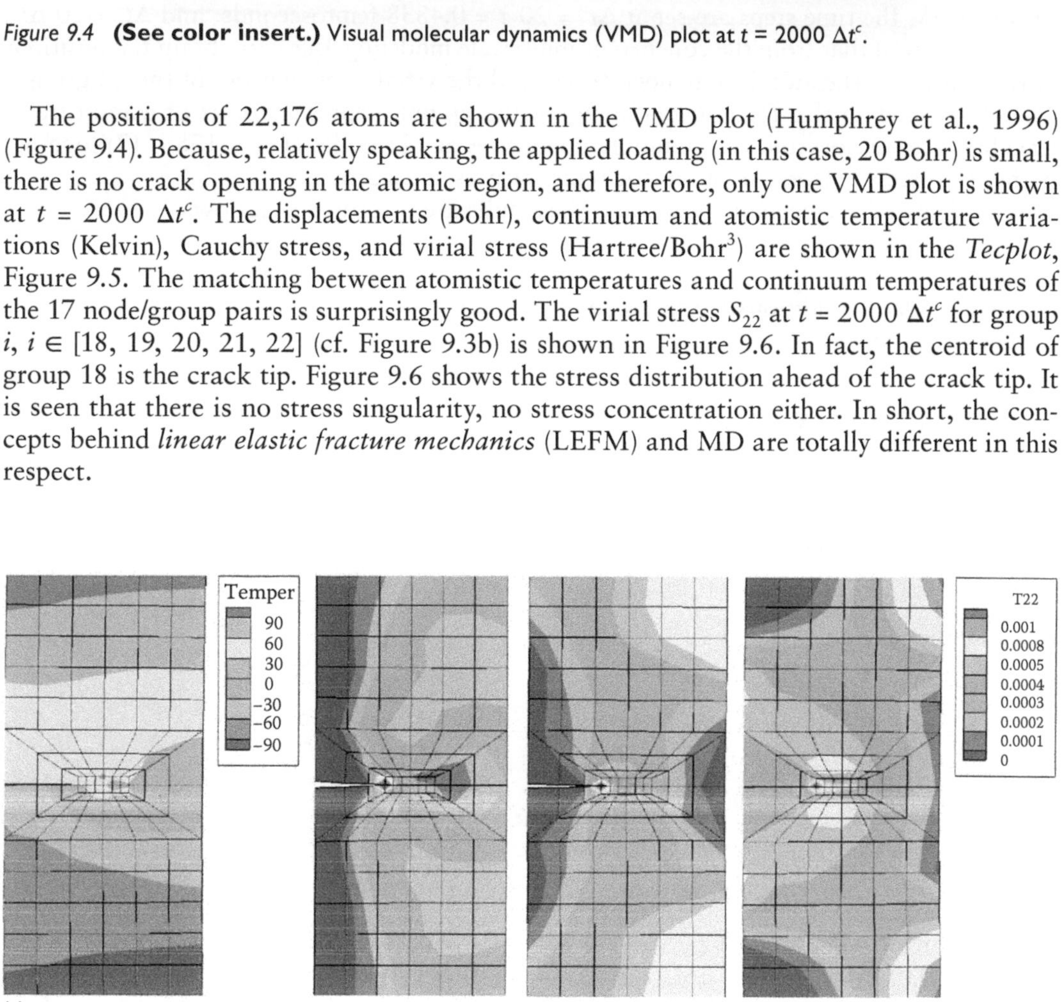

Figure 9.5 **(See color insert.)** (a) Continuum and atomistic temperature variations at $t = 2000\ \Delta t^c$; Cauchy stress σ_{22} and virial stress S_{22} (Hartree/Bohr^3) at (b) $t = 1200\ \Delta t^c$, (c) $t = 1500\ \Delta t^c$, (d) $t = 2000\ \Delta t^c$.

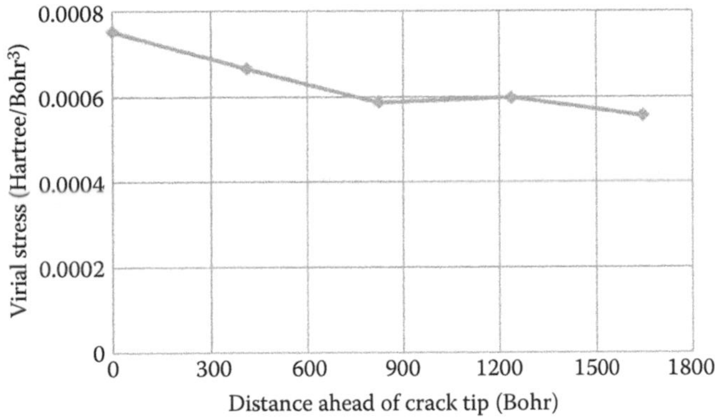

Figure 9.6 **(See color insert.)** Virial stress S_{22} in ahead of crack tip.

Case 2:
The boundary conditions are specified as follows:
Along the top edge,

$$T = T^{ref}$$
$$u_y = 50\frac{t}{t^f}\text{ Bohr} \tag{9.157}$$

Along the bottom edge,

$$T = T^{ref}$$
$$u_y = -50\frac{t}{t^f}\text{ Bohr} \tag{9.158}$$

where $t^f = 2000\ \Delta t^c = 40000\ \Delta t^a$. It means there is no temperature gradient and the elongation is monotonically increasing with respect to time, from $u_y = 0$ at $t = 0$ to $u_y = 50$ Bohr at $t = t^f$. The positions of 22,176 atoms at different time steps are shown in the VMD plot (Figure 9.7). Because the elongation is purposefully specified to be very large, one may see the crack opening, crack branching, crack propagations along multiple fronts, and crack closure.

The displacements, temperatures (continuum and atomistic), Cauchy stress, and virial stresses are shown in the *Tecplot* (Figure 9.8). At $t = 900\ \Delta t^c$, it is seen that the stress is high around at group 18, where the crack tip is located. At $t = 1150\ \Delta t^c$, clearly the location of the high stress is shifting along with the propagating crack fronts. At $t = 1900\ \Delta t^c$, crack front reaches group 9 (cf. Figure 9.7c), and it is seen that the high-stress area is shifted to the node corresponding to group 9 (cf. Figure 9.8c).

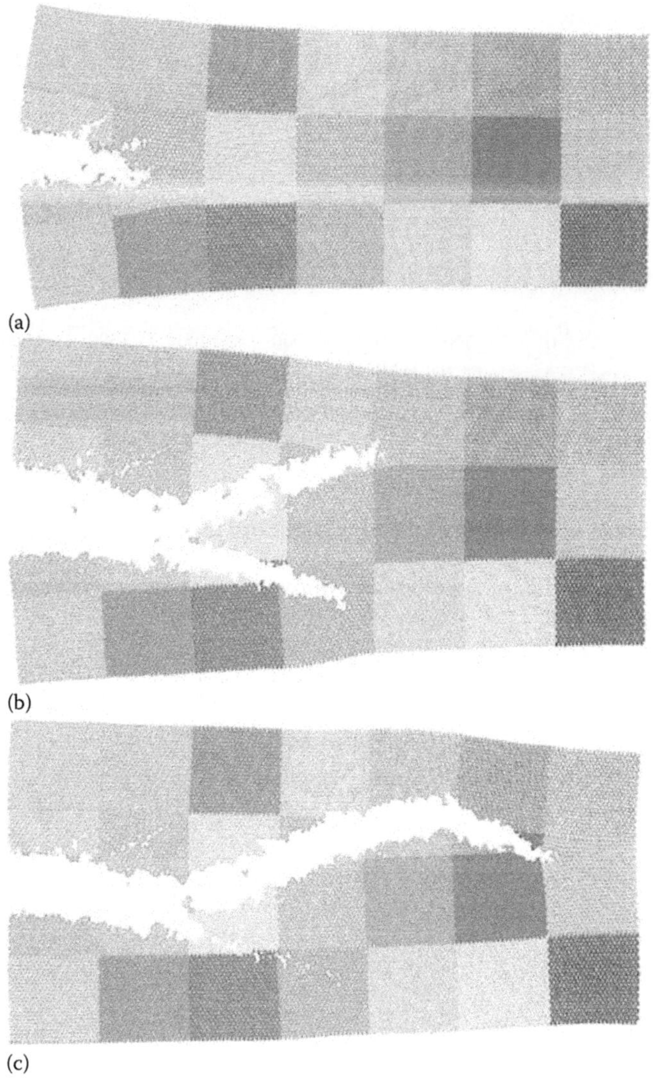

Figure 9.7 **(See color insert.)** VMD plots at various time steps: (a) $t = 900\ \Delta t^c$, (b) $t = 1150\ \Delta t^c$, and (c) $t = 1900\ \Delta t^c$.

Case 3:
The boundary conditions are specified as follows.
Along the top edge,

$$T = T^{ref}$$

$$u_y = \begin{cases} 40\dfrac{t}{t^r}\ \text{(Bohr)}, & t \le t^r \\ 40 + 10\sin\left(2\pi\dfrac{t - t^r}{\bar{t}}\right)\text{(Bohr)}, & t \ge t^r \end{cases} \tag{9.159}$$

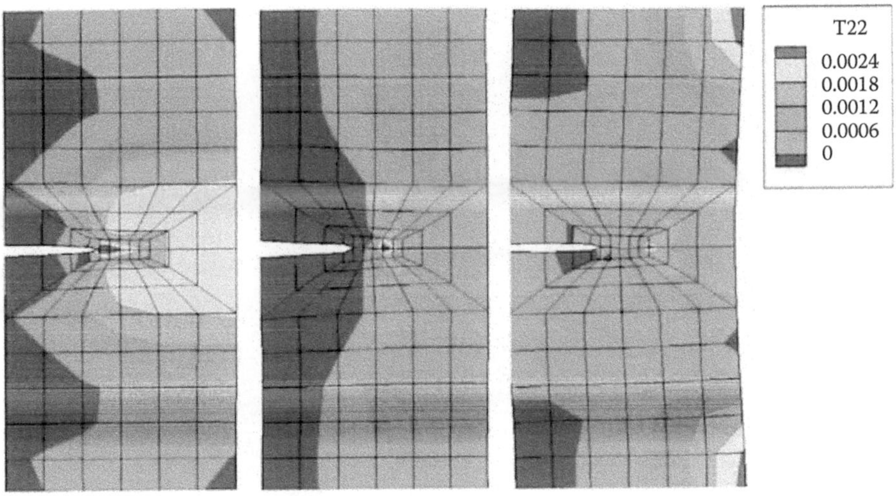

Figure 9.8 **(See color insert.)** Cauchy stress σ_{22} and virial stress S_{22} (Hartree/Bohr^3) at (a) t = 900 Δt^c, (b) t = 1150 Δt^c, and (c) t = 1900 Δt^c.

Along the bottom edge,

$$T = T^{ref}$$
$$u_y = \begin{cases} -40\dfrac{t}{t^r}\,(\text{Bohr}), & t \le t^r \\ -40 - 10\sin\left(2\pi\dfrac{t-t^r}{\bar{t}}\right)(\text{Bohr}), & t \ge t^r \end{cases} \tag{9.160}$$

where $t^r = 500\ \Delta t^c = 10000\ \Delta t^a$, $\bar{t} = 50\ \Delta t^c = 1000\Delta t^a$. It means there is no temperature gradient and the elongation, after time becomes larger than t^r, is a constant plus a sinusoidal function with the period being $\bar{t} = 1000\ \Delta t^a$. It is seen that, from Equations 9.159 and 9.160, the time average of $|u_y|$ equals to 40 *Bohr*. The purpose is to simulate the effect of cyclic loading. The positions of 22,176 atoms at different time steps are shown in the VMD plot (Figure 9.9). Because of the cyclic loading with very high frequency, one may also observe the crack opening and crack propagations along multiple fronts. It means, in critical regions, MD simulation can be and should be utilized to investigate problems in fracture mechanics and fatigue crack growth. The Cauchy stress σ_{22} in the continuum region and the virial stress S_{22} are shown in the *Tecplot* (Figure 9.10). Notice that the virial stress tensor $\boldsymbol{S}$ evaluated at current volume is equivalent to Cauchy stress tensor $\boldsymbol{\sigma}$. Similar to Case 2, one may observe that the high-stress area shifts with the propagating crack fronts. As usual, the matching between atomistic temperatures and continuum temperatures of the 17 node/group pairs is surprisingly good. This actually is a verification of the computer software in handling the interfacial conditions.

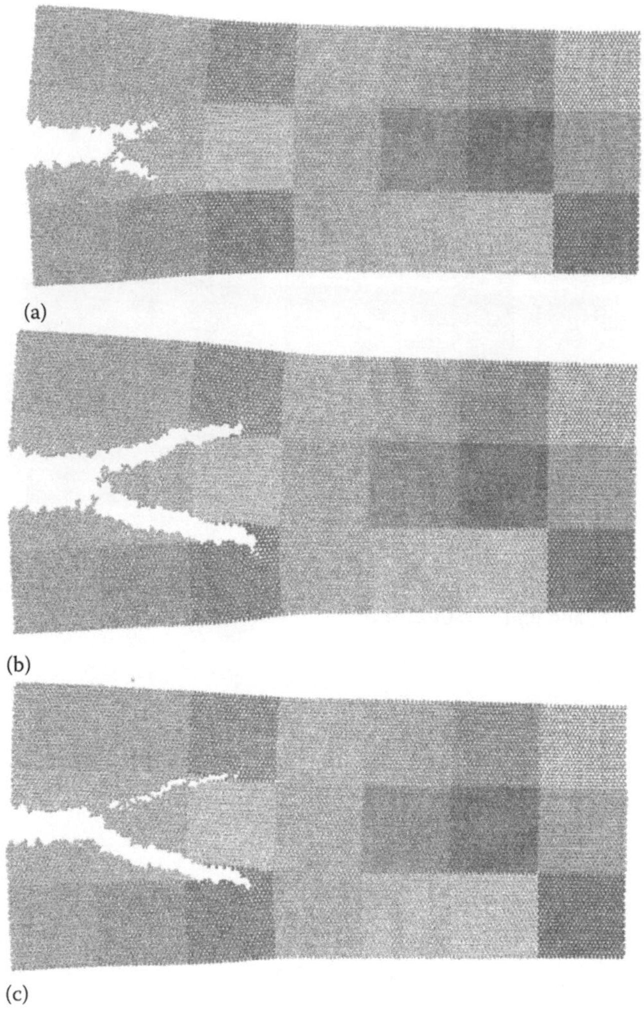

Figure 9.9 **(See color insert.)** VMD plots at various time steps: (a) $t = 1000\ \Delta t^c$, (b) $t = 1150\ \Delta t^c$, and (c) $t = 1900\ \Delta t^c$.

9.7 CLOSING REMARKS

Nowadays, due to the development of massively parallel computers, numerical simulation has emerged as a powerful tool to investigate material and structural behaviors (Oden et al., 2006). Basically, there exist two fundamental physical models, discrete and continuous, that provide foundations for all material modeling. At nanoscale, in MD simulation, the material body is viewed as a collection of atoms moving under the influence of interatomic forces. Unfortunately, due to large number of particles involved as well as the complex nature of atomistic interactions, the application of MD over a realistic range of length and time scales is not feasible. Even worse, the occurring time of concerned physical event is too far away in the future comparing with the small time step needed in MD simulations. On the other hand, at macroscale, materials are modeled by continuum physics, which is simply invalid for material systems at nanoscale because, to say the least, stress–strain relation cannot replace interatomic potential. Also, it is emphasized that the treatment of temperature is

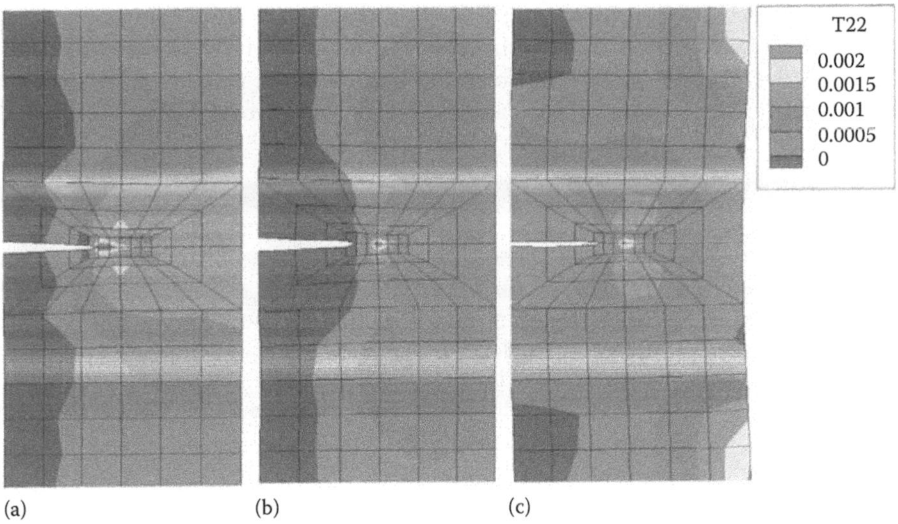

Figure 9.10 **(See color insert.)** Cauchy stress σ_{22} and virial stress S_{22} (Hartree/Bohr^3) at (a) $t = 1000\ \Delta t^c$, (b) $t = 1150\ \Delta t^c$, and (c) $t = 1900\ \Delta t^c$.

totally different from that in MD, stress singularity at the crack tip as predicted by LEFM simply does not exist in MD, and crack propagation occurs naturally in MD and does not need any criteria. From a practical viewpoint, simulation of strongly coupled multiscale systems becomes necessary. Therefore, we developed a multiple-length/time scale theory connecting atoms to GC. As indicated in Section 9.2, MD of multiphysics is adopted for the simulation of atoms in the critical region. The numerical results of sample problems show the distribution of virial stress tensor and atomistic temperature, crack opening and crack propagation with multiple crack fronts, and motion of individual atoms. Meanwhile, the displacement and temperature fields and their derivables, Cauchy stresses, and heat fluxes in the continuum region are obtained through FE analysis based on small-strain thermoelasticity. The material properties used in the FE analysis, including elastic constants, thermal expansion coefficients, thermal conductivity, and specific heat, are obtained from MD simulation—usually named as the sequential multiscale modeling. The continuum theory used in this work is thermoelasticity. Of course, it can be generalized. Then the problem is reduced to the finding of needed and specified material properties through sequential multiscale modeling.

The pivot of the concurrent multiscale modeling is hinged at the interfacial conditions, described and formulated in Section 9.6. In the interface, there are several atomic groups, and each corresponds to a node in the continuum region. For each node/group pair, we assume that (1) the sum of heat flow is vanishing and (2) the sum of interactive force is vanishing and the node is anchored at the centroid of the group (cf. Equations 9.126–9.128). The assumption, which says the sum of heat flow from the node/group pair is vanishing, is considered as logical and reasonable. The sum of interactive force being zero is also logical and reasonable. Whether one can improve or generalize the assumption "the node is anchored at the centroid of the group" is left as an open question. If one is willing to accept these two assumptions, then the challenge due to the huge differences in length and time scales between atomic region and continuum region is resolved in principle. For example, in this work, the time step for MD simulation is set at $\Delta t^a = 20\ \tau = 0.4838$ femtoseconds and the time step for FE analysis of the continuum region is set at $\Delta t^c = 20\ \Delta t^a$; i.e., the time ratio is 20. This ratio can be enlarged. The smallest length scale in the continuum region is the distance between the atomic groups. Beyond that, there is no limitation as far as FE size is concerned. It is noticed that the interfacial conditions have nothing to do with the material properties. In other words, the material in the atomic region can be different from the material in the continuum region, and therefore, this work is readily applicable for nanocomposites.

REFERENCES

Arkhipov, A., Freddolino, P. L., and Schulten, K. 2006. Stability and dynamics of virus capsids described by coarse-grained modeling, *Structure*, **14**(12), 1767–1777.

Arkhipov, A., Yin, Y., and Schulten, K. 2008. Four-scale description of membrane sculpting by BAR domains, *Biophys. J.*, **95**(6), 2806–2821.

Boresi, A. P., Chong, K. P., and Lee, J. D. 2011. *Elasticity in Engineering Mechanics*, Wiley, New York.

Chen, Y., Lee, J. D., and Eskandarin, A. 2006. *Meshless Methods in Solid Mechanics*, Springer, New York.

Eringen, A. C. 1980. *Mechanics of Continua*, R. E. Krieger, Melbourne, Australia.

Eringen, A. C. 1999. *Microcontinuum Field Theories I: Foundation and Solids*, Springer-Verlag, New York.

Gu, J., Bai, F., Li, H., and Wang, X. 2012. A generic force field for protein coarse-grained molecular dynamics simulation, *Int. J. Mol. Sci.*, **13**(11), 14451–14469.

Humphrey, W., Dalke, A., and Schulten, K. 1996. VMD—visual molecular dynamics, *J. Mol. Graphics*, **14**(1), 33–38.

Lee, J. D., Li, J., Zhang, Z., and Wang, L. 2017. Sequential and concurrent multiscale modeling of multiphysics: From atoms to continuum, in *Micromechanics and Nanomechanics of Composite Solids*, Meguid, S. A., and Weng, G. J. (eds.), Springer, New York.

Lee, J. D., and Robert, K. P. 2017. Multiscale atomistic modeling of fracture subjected to cyclic loading, *J. Micromech. Mol. Phys.*, **1**(3–4), 1640009.

Li, J., and Lee, J. D. 2014a. Stiffness-based course-grained molecular dynamics, *J. Nanomech. Micromech.*, **4**, 1–8.

Li, J., and Lee, J. D. 2014b. Reformulation of the Nose–Hoover thermostat for heat conduction simulation at nanoscale, *Acta Mech.*, **225**, 1223–1233.

Li, J., Wang X., and Lee, J. D. 2012. Multiple time scale algorithm for multiscale modeling, *Comput. Model. Eng. Sci.*, **85**, 463–480.

Li, X. 2010. A coarse-grained molecular dynamics model for crystalline solids, *Int. J. Numer. Methods Eng.*, **83**(8–9), 986–997.

Liu, I. S. 2002. *Continuum Mechanics*, Springer, New York.

Liu, I. S., and Lee, J. D. 2016. On material objectivity of intermolecular force in molecular dynamics, *Acta Mech.*, **228**(2), 731–738.

Mori, H. 1965. Transport, collective motion, and Brownian motion, *Prog. Theor. Phys.*, **33**(3), 423–455.

Oden, J. T. et al. 2006. *Simulation-Based Engineering Science*, NSF Blue Ribbon Panel Report, http://www.nsf.gov/pubs/reports/sbes_final_report.pdf

Roldan, R., Castellanos-Gomez, A., Cappelluti, E., and Guinea, F. 2015. Strain engineering in semiconducting two-dimensional crystals, *J. Phys. Condens. Matter.*, **27**, 333201.

Rudd, R. E., and Broughton, J. Q. 1998. Coarse-grained molecular dynamics and the atomic limit of finite elements, *Phys. Rev. B*, **58**(10), R5893.

Rudd, R. E., and Broughton, J. Q. 2005. Coarse-grained molecular dynamics: Nonlinear finite elements and finite temperature, *Phys. Rev. B*, **72**(14), 144104.

Shih, A. Y. et al. 2006. Coarse grained protein-lipid model with application to lipoprotein particles, *J. Phys. Chem. B*, **110**(8), 3674–3684.

Shih, A. Y. et al. 2007. Assembly of lipoprotein particles revealed by coarse-grained molecular dynamics simulations, *J. Struct. Biol.*, **157**(3), 579–592.

Subramaniyan, A. K., and Sun, C. T. 2008. Continuum interpretation of virial stress in molecular simulations, *Int. J. Solids Struct.*, **45**(14–15), 4340–4346.
Tersoff, J. 1988. New empirical approach for the structure and energy of covalent systems, *Phys. Rev. B*, **37**, 6991–7000.
Tersoff, J. 1989. Modeling solid-state chemistry: Interatomic potentials for multicomponent system, *Phys. Rev. B*, **39**, 5566–5566.
Truesdell, C., and Noll, W. 2004. *The Non-Linear Field Theories of Mechanics*. Springer-Heidelberg, Berlin.
Truesdell, C., and Toupin, R. 1960. The classical field theories, in *Principles of Classical Mechanics and Field Theory/Prinzipien der Klassischen Mechanik und Feldtheorie*, Flugge, S., Springer-Heidelberg, Berlin, pp. 226–858.
Wang, C. C. 1970a. A new representation theorem for isotropic functions. Part I: An answer to Professor GF Smith's criticism of my papers on representations for isotropic functions, *Arch. Rational Mech. Anal.*, **36**(3), 166–197.
Wang, C. C. 1970b. A new representation theorem for isotropic functions. Part II: An answer to Professor GF Smith's criticism of my papers on representations for isotropic functions. *Arch. Rational Mech. Anal.*, **36**(3), 198–223.
Wang, C. C. 1971. Corrigendum to 'representations for isotropic functions,' *Arch. Rational Mech. Anal.*, **43**, 392–395.
Yang, Z., Lee, J. D., and Eskandarian, A. 2016a. Objectivity in molecular dynamics simulations, *Int. J. Terraspace Sci. Eng.*, **8**(**1**), 79–92.
Yang, Z., Lee, J. D., Liu, I. S., and Eskandarian, A. 2016b. On non-equilibrium molecular dynamics with Euclidean objectivity, *Acta Mech.*, 1–18.
Zwanzig, R. 1973. Nonlinear generalized Langevin equations, *J. Stat. Phys.*, **9**(3), 215–220.

BIBLIOGRAPHY

Chen, J., and Lee, J. D. 2011. The Buckingham Catastrophe in multiscale modelling of fracture, *Int. J. Theor. Appl. Multiscale Mech.*, **2**(1), 3–11.
de Groot, S. R., and Suttorp, L. G. 1972. *Foundations of Electrodynamics*, North-Holland Pub. Co., Amsterdam.
Hoover, W. G. 1985. Canonical dynamics: Equilibrium phase-space distributions, *Phys. Rev. A*, **31**(3), 1695.
Liang, T., Phillpot, S. R., and Sinnott, S. B. 2009. Parametrization of a reactive many-body potential for Mo–S systems, *Phys. Rev. B*, **79**, 245110.
Liang, T., Phillpot, S. R., and Sinnott, S. B. 2012. Erratum: Parametrization of a reactive many-body potential for Mo-S systems, *Phys. Rev. B*, **85**, 199903.
Nosé, S. 1984. A unified formulation of the constant temperature molecular dynamics methods, *J. Chem. Phys.*, **81**, 511–519.
Nosé, S. 1984. A molecular dynamics method for simulations in the canonical ensemble, *Mol. Phys.*, **53**, 255–268.
Rahman, A. 1964. Correlations in the motion of atoms in liquid argon, *Phys. Rev.*, **136**(2A), A405–A411.
Rappe, A. K., Casewit, C. J., Colwell, K. S., Goddard, W. A., and Skiff, W. M. 1992. UFF, a full period table force field for molecular mechanics and molecular dynamics simulations, *J. Am. Chem. Soc.*, **114**, 10024–10035.
Stillinger, F. H., and Weber, T. A. 1985. Computer simulation of local order in condensed phases of silicon, *Phys. Rev. B*, **31**, 5262.

Sundararaman, S. [illegible] residual stress in [illegible] composites [illegible]

Tersoff, J. 1988. New empirical approach for the structure and energy of covalent systems. *Phys. Rev. B* 37, 6991–7000.

Tersoff, J. 1989. Modeling solid-state chemistry: Interatomic potentials for multicomponent systems. *Phys. Rev. B*, 39, 5566–5568.

Truesdell, C. and Noll, W. 2004. *The Non-Linear Field Theories of Mechanics*, Springer, Heidelberg, Berlin.

Truesdell, C. and [illegible] *Encyclopedia of Physics/Handbuch der Physik* [illegible] Flügge, S., Springer, Heidelberg, Berlin, pp. 226–[illegible]

Wang, C.-C. 1970. A new representation theorem for isotropic functions, Part I [illegible] Professor G. F. Smith's criticism of my papers on representations for isotropic functions [illegible] *Arch. Rational Mech. Anal.* [illegible]

Wang, C.-C. 1970. A new representation theorem for isotropic functions. Part II [illegible] functions [illegible]

[illegible]

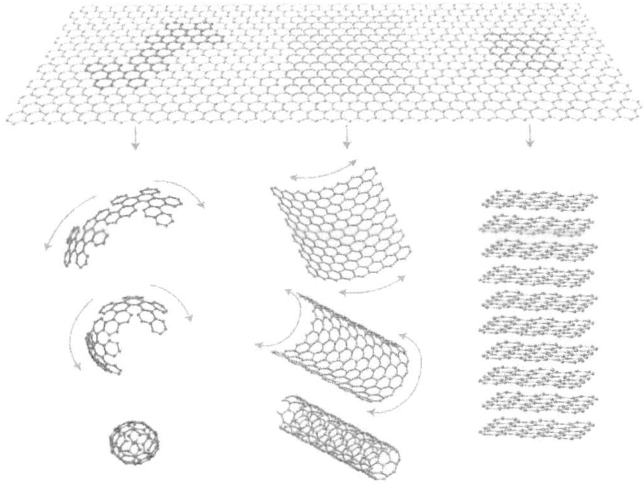

Figure 8.1 Mother of all graphitic forms. Graphene is a 2D building material for carbon materials of all other dimensionalities. It can be wrapped up into a buckyball, rolled into a nanotube or stacked into a 3D graphite.

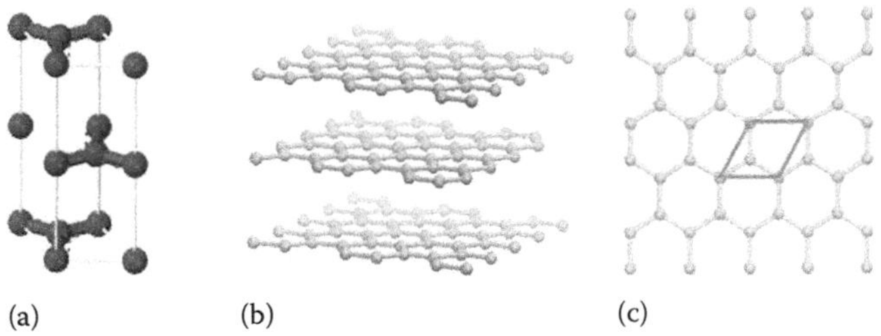

Figure 8.2 (a) Graphite unit cell; (b) few layer graphene sheet; (c) in-plane view of single layer graphene, red line marks the primitive cell.

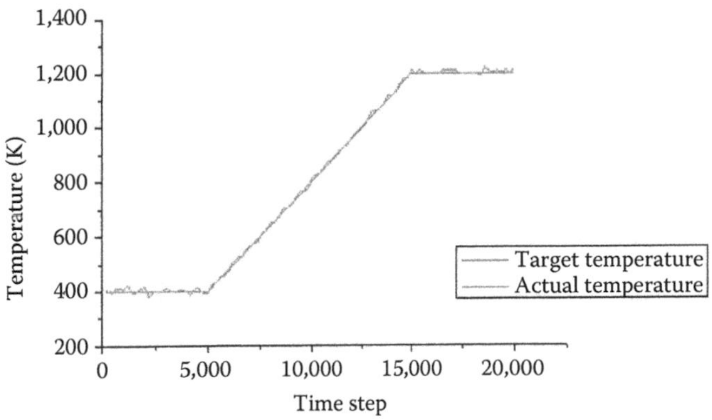

Figure 8.4 Target and actual temperature under a thermostat that raises temperature from T = 400 K to T = 1,200 K.

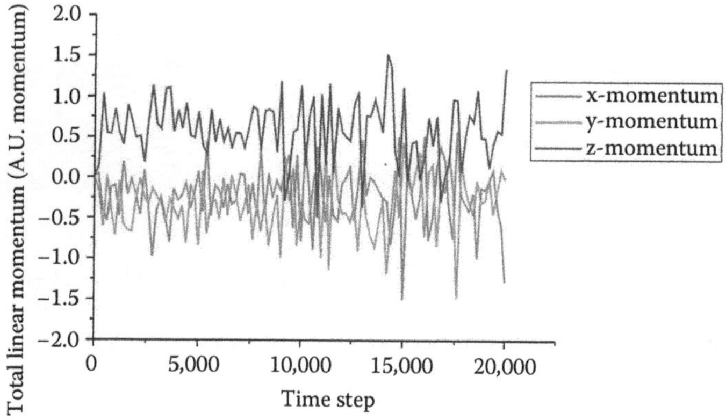

Figure 8.5 History of linear momentum of a material body under the effect of a magnetic field B = (1, 1, 1), with thermostat that raises temperature from T = 400 K to T = 1,200 K.

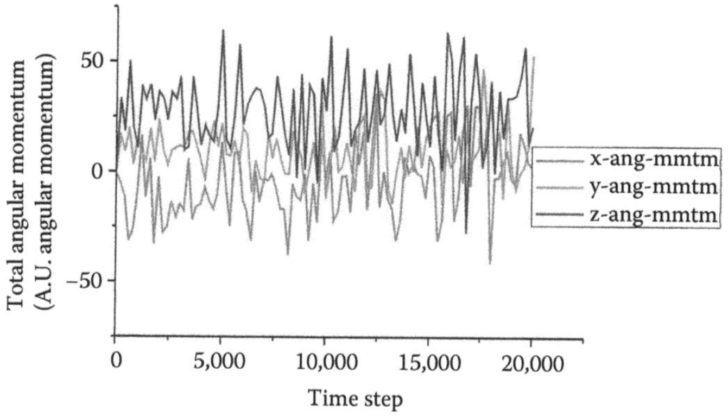

Figure 8.6 History of angular momentum of a material body under the effect of a magnetic field B = (1, 1, 1), with thermostat that raises temperature from T = 400 K to T = 1,200 K.

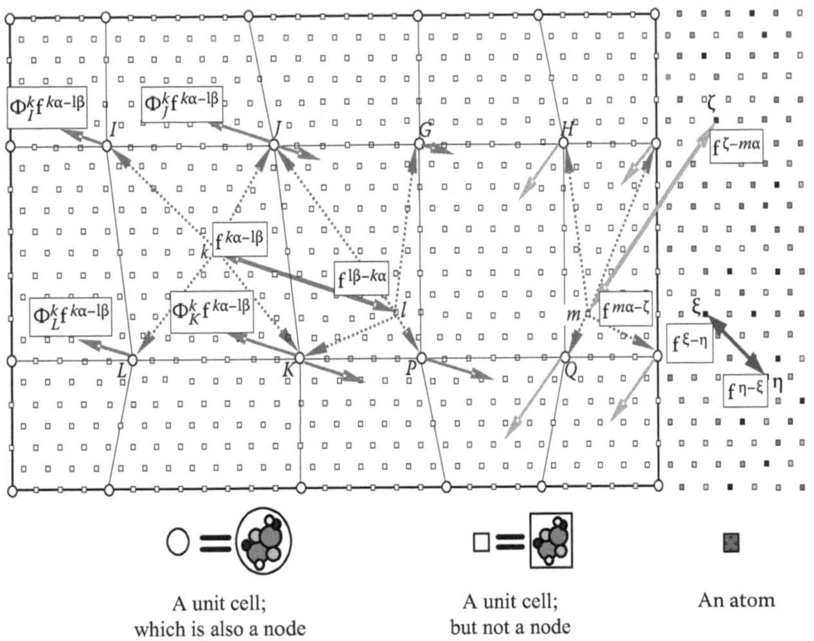

Figure 9.1 Illustrations of atom-based continuum region and atomic region. Circle = finite element node, square = unit cell, but not node, colored square = atom.

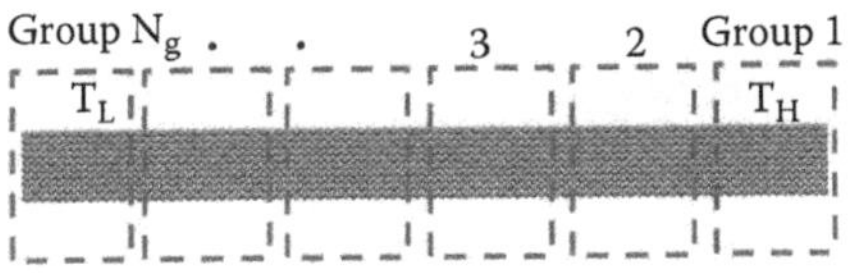

Figure 9.2 Depiction of the groups may or may not be controlled by Nosé–Hoover thermostat.

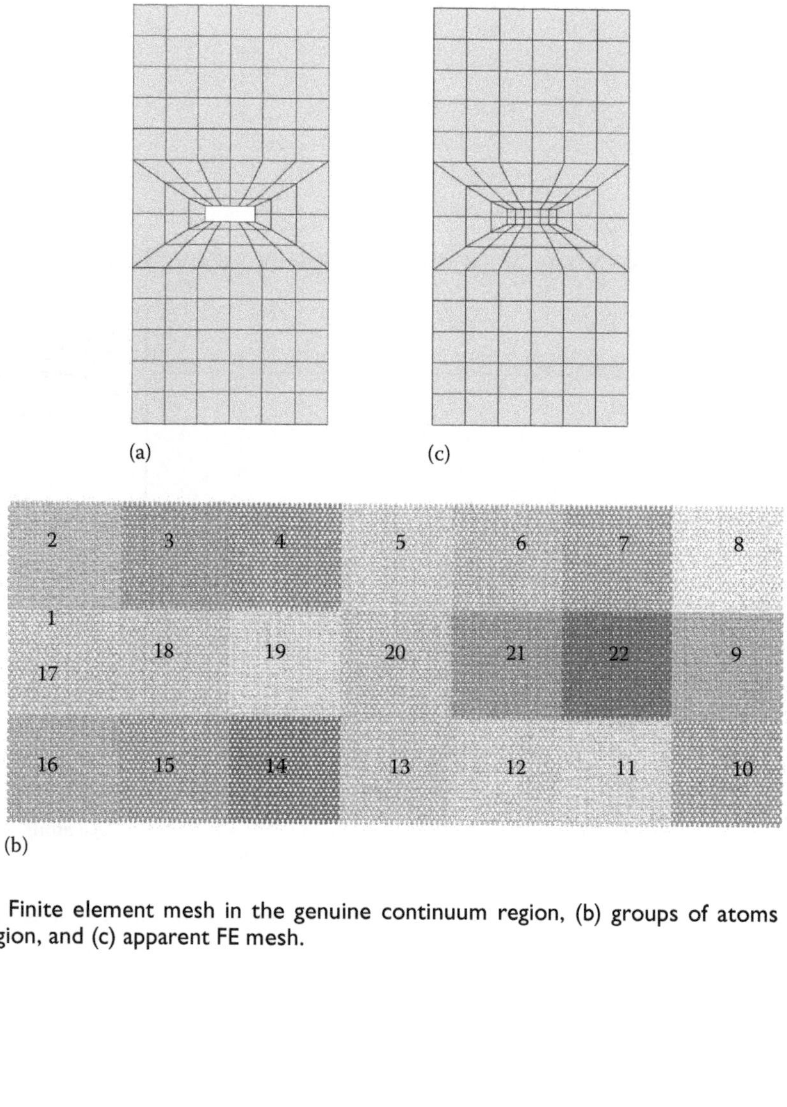

Figure 9.3 (a) Finite element mesh in the genuine continuum region, (b) groups of atoms in the atomic region, and (c) apparent FE mesh.

Figure 9.4 Visual molecular dynamics (VMD) plot at $t = 2000\ \Delta t^c$.

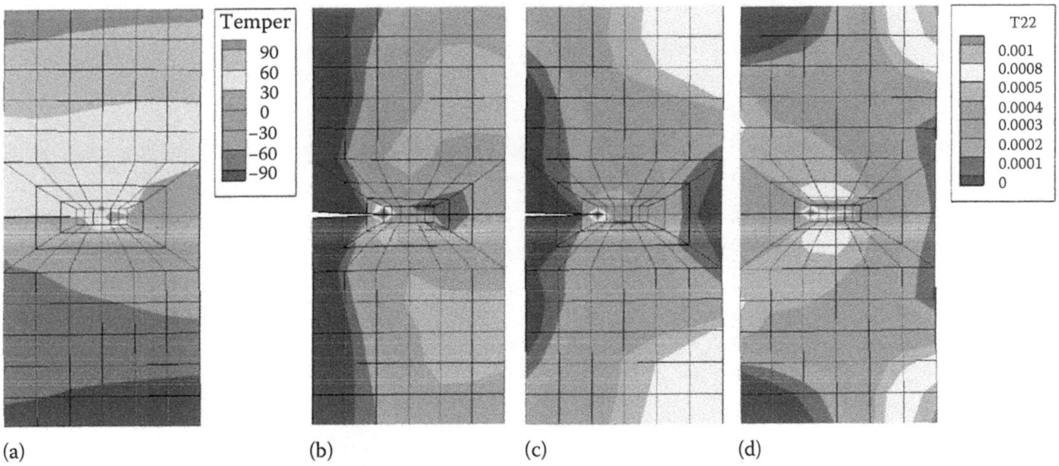

Figure 9.5 (a) Continuum and atomistic temperature variations at $t = 2000\ \Delta t^c$; Cauchy stress σ_{22} and virial stress S_{22} (Hartree/Bohr^3) at (b) $t = 1200\ \Delta t^c$, (c) $t = 1500\ \Delta t^c$, (d) $t = 2000\ \Delta t^c$.

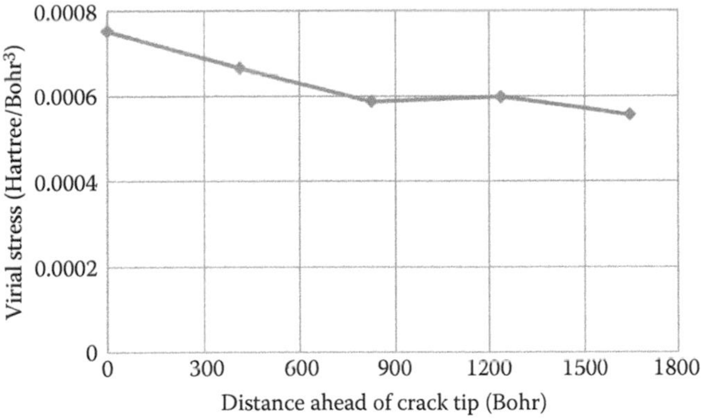

Figure 9.6 Virial stress S_{22} in ahead of crack tip.

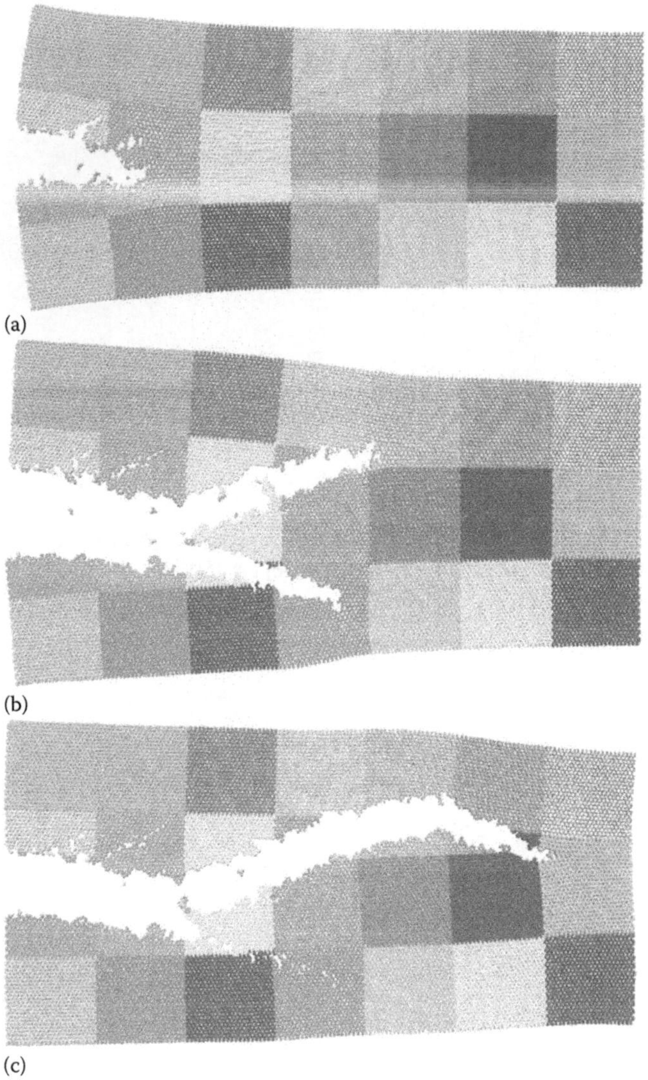

(a)

(b)

(c)

Figure 9.7 VMD plots at various time steps: (a) t = 900 Δt^c, (b) t = 1150 Δt^c, and (c) t = 1900 Δt^c.

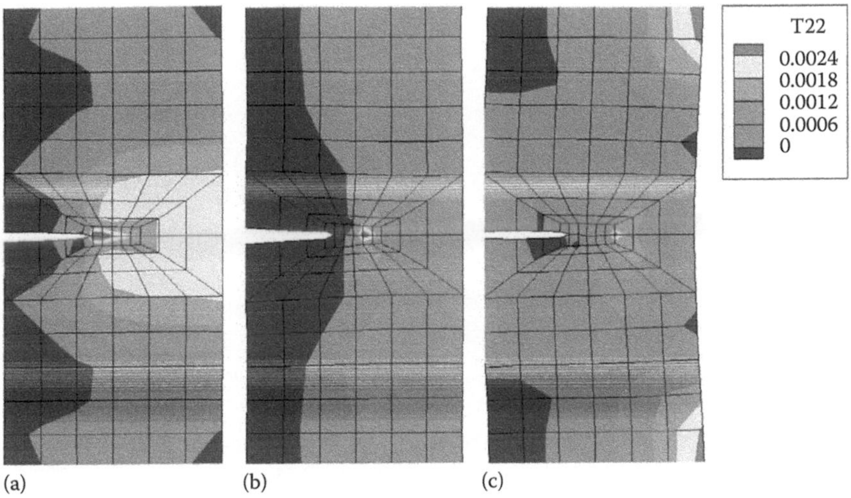

Figure 9.8 Cauchy stress σ_{22} and virial stress S_{22} (Hartree/Bohr^3) at (a) t = 900 Δt^c, (b) t = 1150 Δt^c, and (c) t = 1900 Δt^c.

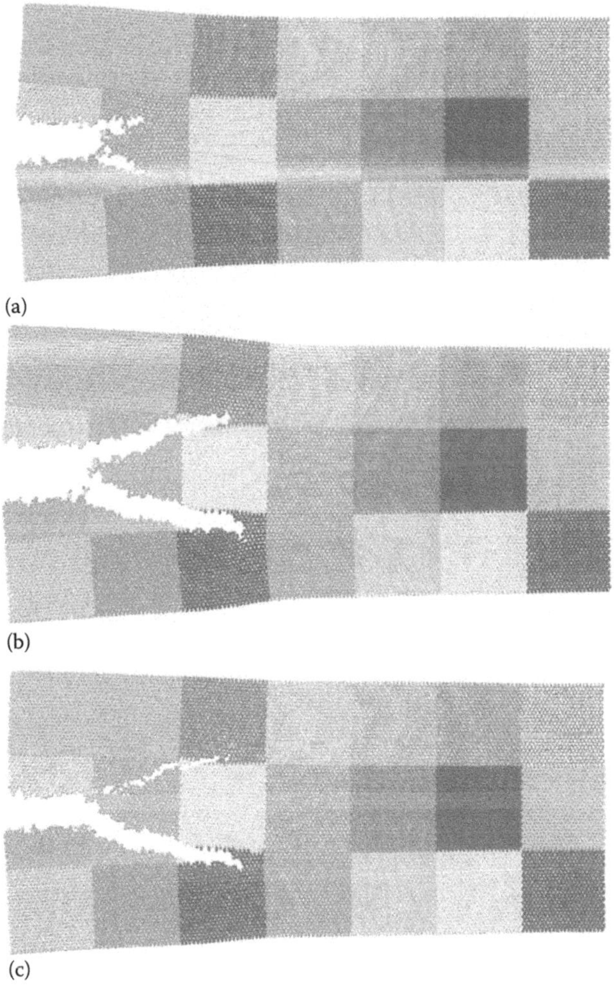

Figure 9.9 VMD plots at various time steps: (a) t = 1000 Δt^c, (b) t = 1150 Δt^c, and (c) t = 1900 Δt^c.

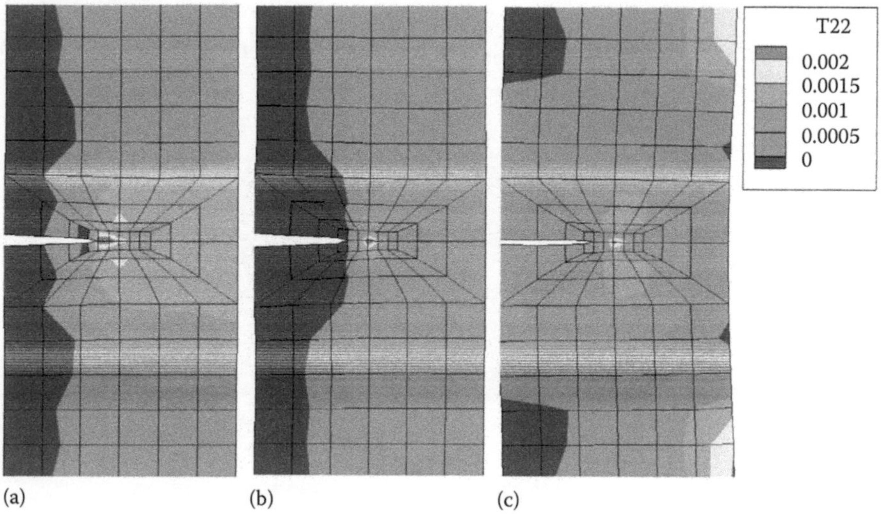

Figure 9.10 Cauchy stress σ_{22} and virial stress S_{22} (Hartree/Bohr^3) at (a) t = 1000 Δt^c, (b) t = 1150 Δt^c, and (c) t = 1900 Δt^c.

Author Index

A

B

C

D

E

F

S

T

U

V

W

X

Y

Z

Subject Index

G

H

I

J

K

L

M

R

S

T

U

V

W

Y

Milton Keynes UK
Ingram Content Group UK Ltd.
UKHW052020071024
449327UK00027B/2347

9 780367 886257